MATHEMATICS AT WORK: DECIMALS

Bertrand B. Singer

Samuel Gompers Vocational and Technical High School

Gregg Division/McGraw-Hill Book Company

New York / St. Louis / Dallas / San Francisco / Auckland / Bogotá / Düsseldorf
Johannesburg / London / Madrid / Mexico / Montreal / New Delhi / Panama
Paris / São Paulo / Singapore / Sydney / Tokyo / Toronto

Library of Congress Cataloging in Publication Data

Singer, Bertrand B
Mathematics at work, decimals.

1. Fractions, Decimal. I. Title.
QA117.S56 513'.2 76-56461
ISBN 0-07-057489-8

Mathematics at Work: Decimals

2 3 4 5 6 7 8 9 10 11 WCWC 8 7 6 5 4 3 2 1 0 9 8 7

Cover illustrations by Dennis G. Purdy

The **editors** for this book were Gerald O. Stoner and Claire Hardiman.
The **designer** was Dennis G. Purdy.
The **production supervisor** was Regina R. Malone.
It was set in Century Schoolbook by Black Dot, Inc.

CONTENTS

PREFACE

The *Mathematics at Work* series consists of three texts: *Fractions, Decimals,* and *Algebra*. Each of the texts concentrates on two main objectives:
1. To provide a source through which a student may explore careers in the industrial trades
2. To enhance this exploration via the application of the basic mathematics needed for success in the world of work

Each course is complete unto itself, with a huge variety of problems, examples, photographs, reviews, and tests covering most of the industrial trades. This wealth of diversified vocational experiences makes these texts particularly suitable for introductory or exploratory courses in vocational and technical schools. The texts are also appropriate for remedial arithmetic and remedial reading courses at other levels.

All problems are preceded by letters indicating the trade for which they are most suitable. This practice enables instructors to pick out easily those problems most closely related to the interests of their classes. These letters are:

(A) Auto mechanics
(B) Business and clerical trades
(C) Construction trades
(E) Electrical and electronic trades
(F) Farming and agricultural trades
(G) General trade areas such as welding, air conditioning, drafting, sheet metal, etc.
(M) Metal trades

New words and concepts are introduced through photographs and "on-the-job" language that is the equivalent of actual vocational experience. By expanding on any of the many subjects offered in the problems, instructors may enrich the course according to students' interests.

These courses may provide attainable goals with heightened pupil interest. The courses are noncumulative, thus increasing the pupil's chances for overall success.

Every attempt has been made to make these truly practical courses. The emphasis is placed on the functional role of mathematics as a problem-solving tool in the industrial trades. In this vein, any mathematical concept which does not have an immediate and practical application has been eliminated.

Each new concept, trade or mathematical, is developed from fundamental principles and then applied to a variety of practical situations. Each job develops either a new mathematical operation or a new trade concept and is illustrated by a series of examples that increase in both mathematical difficulty and scope of application.

This illustration method has been further enhanced by programming the last example of each job, as well as the summaries and reviews. These Self-Tests serve as a final check on students' understanding, forcing them to stop, think, and check *each* step of the problem, thus reinforcing the concepts learned in the previous illustrations. A tremendous advantage of the Self-Tests is that students have a chance to reach their own conclusions *before* they get the answer. Each Self-Test becomes a diagnostic test to pinpoint the precise error in their thinking before the students tackle the problems of the job.

The problems in each job have been carefully graded to provide a measure of success for all students. Other problems are sufficiently difficult to challenge even the gifted student. Answers have been given in the Answer Section for the odd-numbered problems to increase the self-help features of each course. Continued drill and review in both trade and mathematical concepts are obtained by extending the trade theory and application. For example, in *Decimals*, the metric system is introduced early in Job 3 so that it may be continually reviewed and applied in all the following jobs.

An answer key for the even-numbered problems and tests (for instructors only) will be available.

Diagrams and photographs are used extensively to provide an easy familiarity with the language and tools of the various trades. A programmed summary and set of review problems as well as many tests are included in each job. The review jobs attempt to attack the problems from a slightly different point of view in order to round up any loose ends in the student's mind. Modern standards have been used throughout these books.

Finally, I wish to thank Mr. Gerald O. Stoner of McGraw-Hill for his many valuable suggestions and contributions.

Bertrand B. Singer

JOB 1 Introduction to Decimals

The decimal system is an extremely fast and accurate system to use for most mathematical calculations. The hand calculator shown in Fig. 1-1 is a marvel of speed and capability that gives answers in extremely accurate decimal figures. It contains keys for converting to the metric system

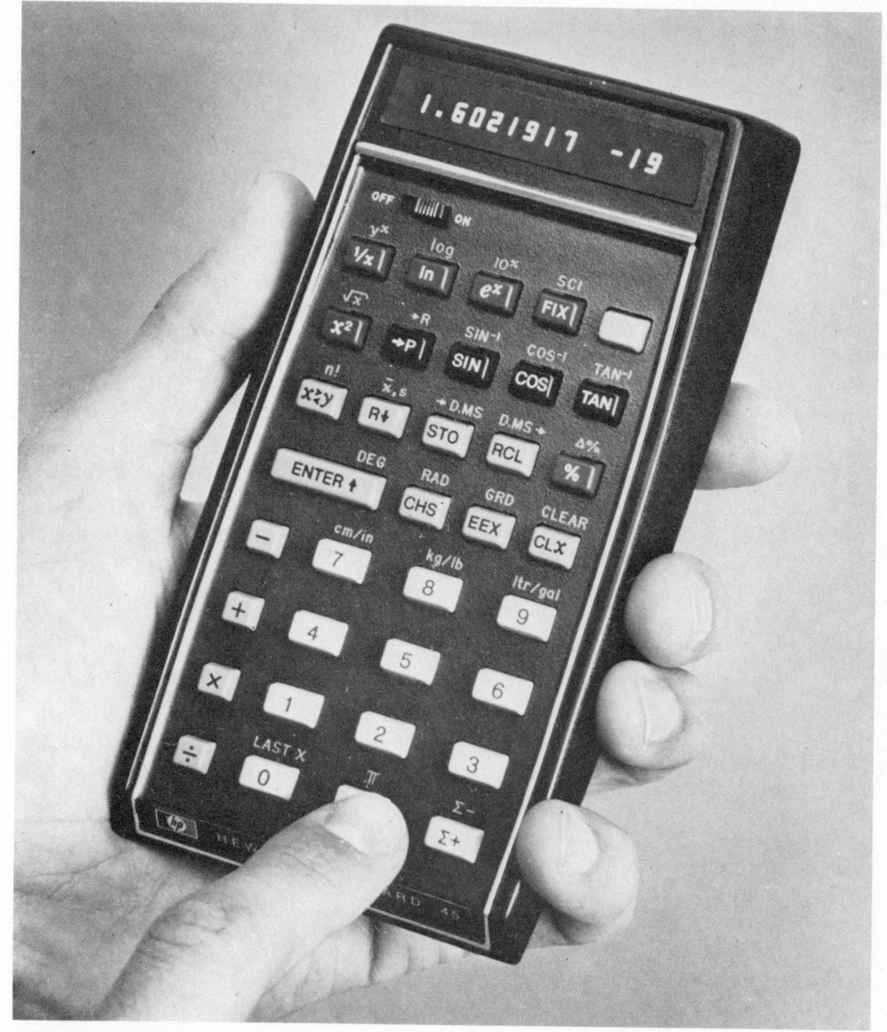

Fig. 1-1 Battery-powered scientific calculator. (Courtesy Hewlett-Packard Company)

of measurement—a decimal system which the United States is expected to adopt completely within the next few years. Also, the accuracy required by modern industry demands the use of the decimal system. A typical machine part, dimensioned partly in the decimal system, is shown in Fig. 1-2.

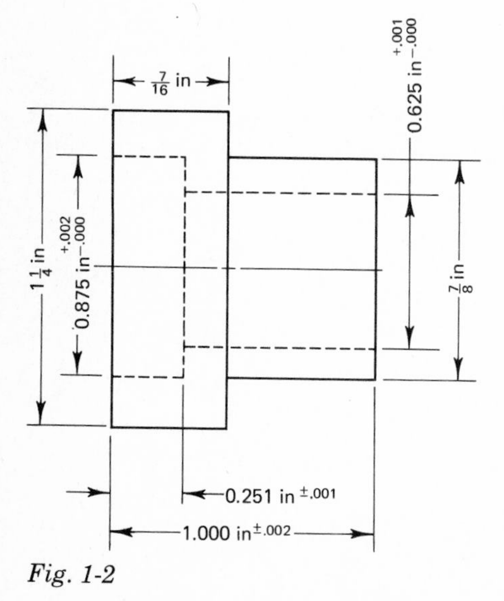

Fig. 1-2

A decimal fraction is a fraction in which the denominator is not written, but the denominator's value is indicated by the position of a decimal point (.) in the numerator. The denominator of a decimal fraction is always a number like 10, 100, 1,000, etc.

The number of digits to the right of the decimal point tells us whether the decimal is to be read as tenths, hundredths, thousandths, etc.

When a decimal has *one* digit to the right of the decimal point, the decimal is read as that many *tenths*. Thus,

$0.3 = \frac{3}{10}$ and is read as 3 *tenths*

$0.9 = \frac{9}{10}$ and is read as 9 *tenths*

When a decimal has *two* digits to the right of the decimal point, the decimal is read as that many *hundredths*. Thus,

$0.23 = \frac{23}{100}$ and is read as 23 *hundredths*

$0.47 = \frac{47}{100}$ and is read as 47 *hundredths*

Now, if we wish to write $\frac{3}{100}$ as a decimal fraction, the decimal point must be placed so that there will be *two* digits following it in order to represent *hundredths*. To make up these two digits, we must place a zero between the decimal point and the digit 3. The zero will now push the 3 into the second place, which means hundredths. Thus,

$\frac{3}{100} = 0.03$ and is read as 3 *hundredths*

Note: the zero *must* be placed between the decimal point and the digit. It is wrong to place the zero after the digit 3, as in 0.30, since this number would be read as 30 hundredths, *not* 3 hundredths. Similarly,

$\frac{7}{100} = 0.07$ and is read as 7 *hundredths*

$\frac{9}{100} = 0.09$ and is read as 9 *hundredths*

When a decimal has *three* digits to the right of the decimal point, the decimal is read as that many *thousandths*. Thus,

$0.123 = \frac{123}{1{,}000}$ and is read as 123 *thousandths*

$0.457 = \frac{457}{1{,}000}$ and is read as 457 *thousandths*

If we wish to write $\frac{43}{1{,}000}$ as a decimal fraction, the decimal point must be placed so that there will be *three* digits following it in order to represent *thousandths*. To make up these three digits, we must place a zero between the decimal point and the 43. Thus,

$\frac{43}{1{,}000} = 0.043$ and is read as 43 *thousandths*

$\frac{87}{1{,}000} = 0.087$ and is read as 87 *thousandths*

If we wish to write $\frac{9}{1{,}000}$ as a decimal fraction, the decimal point must still be placed so that there will be *three* digits following it. To make up these three digits, we must now place *two* zeros between the decimal point and the 9.

$\frac{9}{1{,}000} = 0.009$ and is read as 9 *thousandths*

$\frac{5}{1{,}000} = 0.005$ and is read as 5 *thousandths*

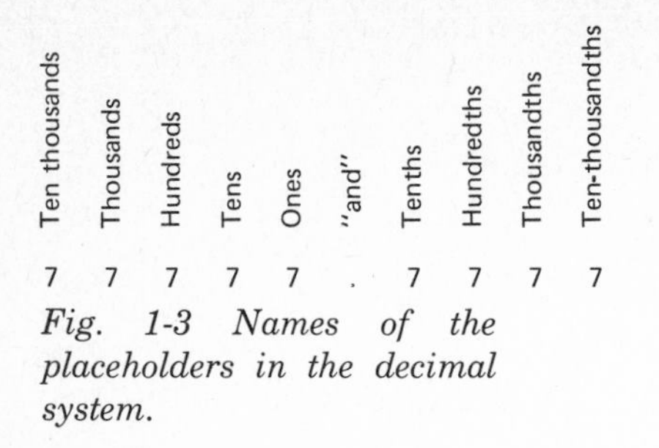

Fig. 1-3 Names of the placeholders in the decimal system.

Following this system, 4 places means ten-thousandths, 5 places means hundred-thousandths, 6 places means millionths, etc. Part of the system is shown in picture form in Fig. 1-3.

Note: zeros placed at the *end* of a decimal do *not* change the value of the decimal. They merely describe the decimal in another way. For example, 0.5 (5 tenths) = 0.50 (50 hundredths) = 0.500 (500 thousandths).

A number such as $3\frac{13}{100}$ is written as 3.13 and is read as 3 "and" 13 hundredths. In this type of number, the decimal point is read as the word "and."

Table 1-1 Some Fractions and Their Decimal Equivalents

WORDS	FRACTION	DECIMAL
Seven tenths	$\frac{7}{10}$	0.7
Forty-five hundredths	$\frac{45}{100}$	0.45
Sixty hundredths	$\frac{60}{100}$	0.60
Four hundredths	$\frac{4}{100}$	0.04
Two thousandths	$\frac{2}{1,000}$	0.002
Sixteen thousandths	$\frac{16}{1,000}$	0.016
One hundred forty-nine thousandths	$\frac{149}{1,000}$	0.149
Eight-hundred thousandths	$\frac{800}{1,000}$	0.800
Nine ten-thousandths	$\frac{9}{10,000}$	0.0009
Four and nine hundredths	$4\frac{9}{100}$	4.09

COMPARING THE VALUE OF DECIMALS

When we compare the value of various decimal fractions, we must first be certain that they have the same denominators. This means that the decimals must be written so that they have the same number of decimal places.

Example 1-1 Which is larger, 0.3 or 0.25?

Solution Since 0.3 has only one decimal place and 0.25 has two decimal places, we must change 0.3 into a two-place decimal by adding a zero. This does *not* change the value. It merely expresses it differently (0.3 or 3 tenths is the same as 0.30 or 30 hundredths). Now we can see that 0.30 (30 hundredths) is larger than 0.25 (25 hundredths).

Problems

Write the following fractions as decimal fractions:

1. $\frac{7}{10}$
2. $\frac{29}{100}$
3. $\frac{114}{1,000}$
4. $\frac{3}{10}$
5. $\frac{6}{1,000}$
6. $\frac{9}{1,000}$
7. $\frac{18}{1,000}$
8. $\frac{3}{1,000}$
9. $\frac{11}{1,000}$
10. $\frac{4}{10}$
11. $\frac{13}{1,000}$
12. $\frac{74}{100}$
13. $\frac{45}{1,000}$
14. $\frac{316}{1,000}$
15. $\frac{545}{10,000}$
16. $\frac{23}{100}$
17. $\frac{62}{10,000}$
18. $\frac{7}{100}$
19. $\frac{5}{10,000}$
20. $\frac{500}{1,000}$

Arrange each of the following groups of decimals in order, starting with the largest:

21. 0.007, 0.16, 0.4
22. 0.2, 0.107, 0.28
23. 0.496, 0.8, 0.02
24. 0.8, 0.06, 0.040
25. 0.5, 0.051, 0.18
26. 0.90, 0.018, 0.06
27. 0.1228, 0.236, 0.4
28. 0.006, 0.05, 0.3
29. 0.19, 0.004, 0.08
30. 0.060, 0.40, 0.0080
31. 0.02, 0.004, 0.13
32. 0.0026, 0.092, 0.025

Fig. 1-4 Checking the ring side clearance of the piston-ring groove. (Courtesy Chevrolet Division of the General Motors Corporation)

33. (*A*) An auto mechanic checked a piston for ring-groove wear by first cleaning the groove and then inserting a new ring. He then checked the clearance between the ring and the side of the groove with a feeler gage, as shown in Fig. 1-4, recording a clearance of 0.0031 in. If the manufacturer's specifications for the ring side clearance on this 1969 Camaro permit any value between 0.0012 in and 0.0032 in, should this piston be discarded? Why?
34. (*A*) The state of charge in an automobile battery may be checked by measuring the specific gravity of the electrolyte with a hydrometer. Using the table below, find the approximate state of charge if the hydrometer indicates a specific gravity of 1.250.

APPROXIMATE GRAVITY	STATE OF CHARGE
1.260–1.290	Fully charged
1.230–1.260	About three-fourths charged
1.200–1.230	About half charged
1.170–1.200	About one-fourth charged
1.140–1.170	Almost run down
1.110–1.140	Discharged

CHANGING MIXED NUMBERS TO DECIMALS

When a mixed number is read as a decimal, the word "and" appears as a decimal point. For example, $2\frac{7}{10}$ is read as two *and* seven tenths and is written as 2.7. A whole number may be written as a decimal if a decimal point is placed at the right end of the number. For example, the number 4 means 4.0 or 4.00 or 4.000.

Problems

Change the following mixed numbers to decimals:

1. $2\frac{3}{10}$
2. $18\frac{5}{100}$
3. $3\frac{144}{1,000}$
4. $1\frac{17}{100}$
5. $2\frac{25}{1,000}$
6. $7\frac{35}{100}$
7. $2\frac{20}{1,000}$
8. $3\frac{9}{10}$
9. $1\frac{2}{1,000}$
10. $62\frac{90}{100}$
11. $9\frac{145}{1,000}$
12. $3\frac{27}{1,000}$

JOB  Changing Fractions to Decimals

We shall discover that many of the answers to our problems will be fractions like $\frac{1}{8}$ ampere (A), $\frac{7}{40}$ in, or $\frac{3}{13}$ ohm (Ω). These will be perfectly correct mathematical answers, but they will be completely worthless to a practical mechanic or electrician. Micrometers and electrical measuring instruments express values as decimals, not as ordinary fractions. Also, the manufacturers of electrical and mechanical components give the value of the parts in terms of decimals.

Suppose we worked out a problem and found that the current in an electric circuit should be $\frac{1}{8}$ A. Then, using an ammeter, we tested the circuit and found that 0.125 A flowed. Is our circuit correct? How would we know? How can we compare $\frac{1}{8}$ and 0.125?

The easiest way is to change the fraction $\frac{1}{8}$ into its equivalent decimal and then compare the decimals.

RULE 1 **To change a fraction into a decimal, divide the numerator by the denominator.**

Example 2-1 Change $\frac{1}{8}$ into an equivalent decimal.

Solution $\frac{1}{8}$ means $1 \div 8$. To write this as a long-division example, place the numerator inside the long-division sign and the denominator outside the sign.

$$8\overline{)1}$$

We cannot divide 8 into 1, but remember that every whole number may be written with a decimal point at the *end* of the number. We can add as many zeros as we want without changing the value. Our problem now looks like this:

$$8\overline{)1.000}$$

1. Put the decimal point in the answer directly above its position in the number 1.000.

$$\begin{array}{r} . \\ 8\overline{)1.000} \end{array}$$

2. Try to divide the 8 into the first digit. The 8 does not divide into 1. Then try to divide the 8 into the first two digits. The 8 divides into 10 once. Place this number 1 in the answer directly above the *last* digit of the number into which the 8 was divided.

$$\begin{array}{r} 0.1 \\ 8\overline{)1.000} \end{array}$$

3. Multiply this 1 by the divisor 8 and place the answer (8) as shown below. Draw a line and subtract.

$$\begin{array}{r} 0.1 \\ 8\overline{)1.000} \\ \underline{8} \\ 2 \end{array}$$

4. Bring down the next digit (0) and divide this new number (20) by the 8. The 8 will divide into 20 two times.

$$\begin{array}{r} 0.1 \\ 8\overline{)1.000} \\ \underline{8\downarrow} \\ 20 \end{array}$$

5. Place this 2 in the answer directly above the last digit brought down.

$$\begin{array}{r} 0.12 \\ 8\overline{)1.000} \\ \underline{8\downarrow} \\ 20 \end{array}$$

6. Multiply the 2 by the divisor 8, and repeat steps 3 to 5. The answer comes out as 0.125. This means that $\frac{1}{8}$ is equal to 0.125.

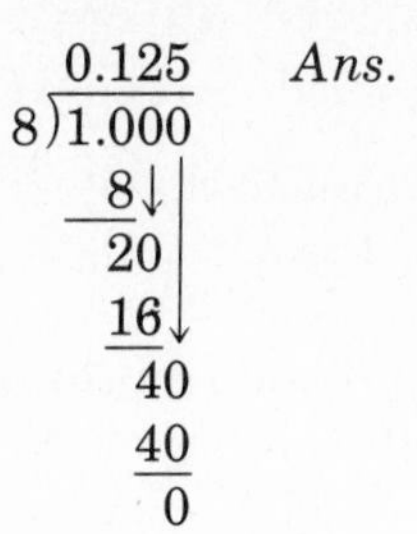

Example 2-2 Change $\frac{7}{40}$ into an equivalent decimal.

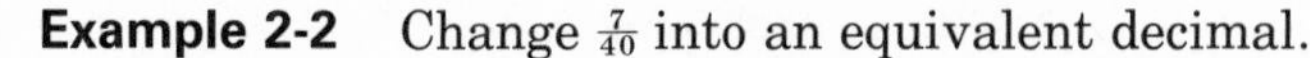

Solution

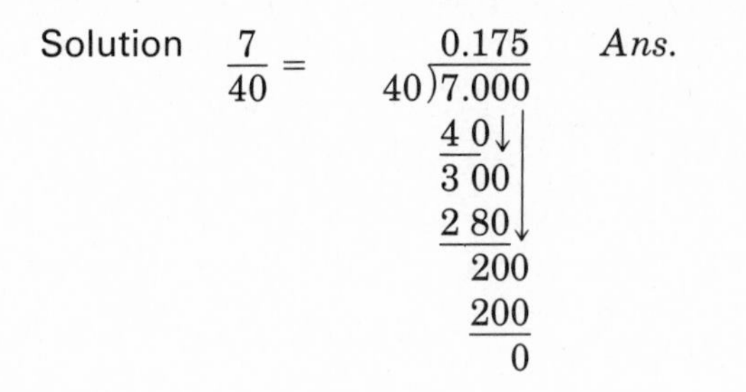

Example 2-3 Change $\frac{6}{13}$ into an equivalent decimal.

Solution The answer does not come out even.

```
 6          0.461
 --  =   13)6.000
 13         5 2
              80
              78
               20
               13
                7
```

We see that there is a remainder. If the remainder is more than half the divisor, we drop it and add an extra unit to the last place of the answer. Since 7 is more than half of 13, $0.461\frac{7}{13}$ becomes

```
  0.461
+     1
  0.462   Ans.
```

If any remainder is less than half of the divisor, drop it completely and leave the answer unchanged. For example,

$0.236\frac{5}{12} = 0.236$ (since 5 is less than half of 12)

$0.483\frac{1}{4} = 0.483$ (since 1 is less than half of 4)

An important question may have occurred to you by now. "If it doesn't come out even, how long should I continue to divide?" The answer to this depends on the use to which the answer is to be put. Some jobs require five or six decimal places, while others need only one place or even none at all. For example, the capacitor in the tuning circuit of a radio receiver should be worked out to an answer like 0.00025 microfarad (μF). The pitch diameter of a screw thread requires a measurement like 0.498 in, as shown in Fig. 2-1. A bias resistor of 203.4 Ω is just as well written as 203 Ω or even 200 Ω, because precision is not needed in this application.

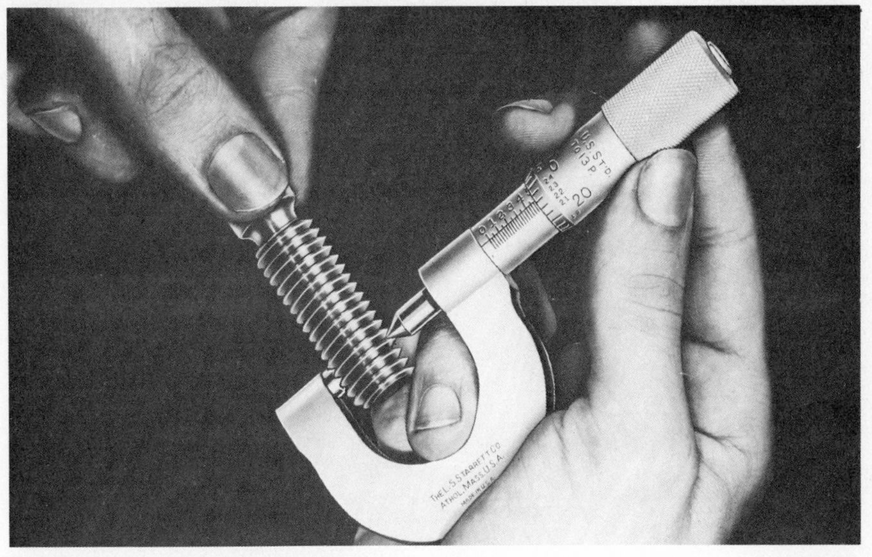

Fig. 2-1 A screw thread micrometer measures the pitch diameter in thousandths of an inch. (Courtesy The L. S. Starrett Company)

DEGREE OF ACCURACY

A very general rule for the number of decimal places required in an answer is given in Table 2-1.

Table 2-1 Decimal Places Required in Answers

ANSWER	NUMBER OF DECIMAL PLACES	EXAMPLES
Less than 1	3	0.132, 0.008
1–10	2	3.48, 6.07
10–100	1	28.3, 52.9
100 up	None	425, 659

Example 2-4 If a section of land equals 640 acres, what part of a section, expressed as a decimal, is represented by 96 acres?

Solution 96 acres represents $\frac{96}{640}$ of a section.

$$\frac{96}{640} = 640\overline{)96.00}^{\;.15}$$

```
        .15
640 )96.00
     64 0↓
     32 00
     32 00
         0
```

Therefore, 96 acres represents 0.15 of a section. *Ans.*

SELF-HELP FEATURES

The summaries and the various steps in some of the illustrative problems that follow are incomplete. The correct answers will be found at the right side of the page. To achieve the maximum benefit from this type of instruction, please follow these instructions carefully.

1. Place the response shield over the answers at the right of the page.

2. Read the statement on the left side of the page, noting the blank space where you are to respond.

3. Write the correct response in the blank space.

4. Slide the response shield down to uncover the correct answer.

5. If you have responded correctly, continue to the next line.

6. If your response is incorrect, *stop!* Review the preceding material to discover the reason for your error.

7. When you are satisfied that you understand, draw a line through the incorrect answer and write the correct answer above it.

8. Repeat steps 2 to 5 above.

The following example is a self-test type. Please use your response shield.

Self-Test 2-5 Find the weight per horsepower of an automobile which develops 345 horsepower (hp) and weighs 4,416 pounds (lb).

Solution To find the weight per horsepower, divide the ______ by the ______. This division may be expressed as the fraction $\frac{?}{345}$

Writing the fraction as a long-division example, we get $?\overline{)\ ?}$

Dividing,

$$\begin{array}{r} 12.8 \text{ lb/?} \quad \textit{Ans.} \\ 345\overline{)4416.0} \\ \underline{345\downarrow} \\ 966 \\ \underline{690\downarrow} \\ 276\ 0 \\ \underline{276\ 0} \\ 0 \end{array}$$

weight
horsepower

4,416

345; 4,416

hp

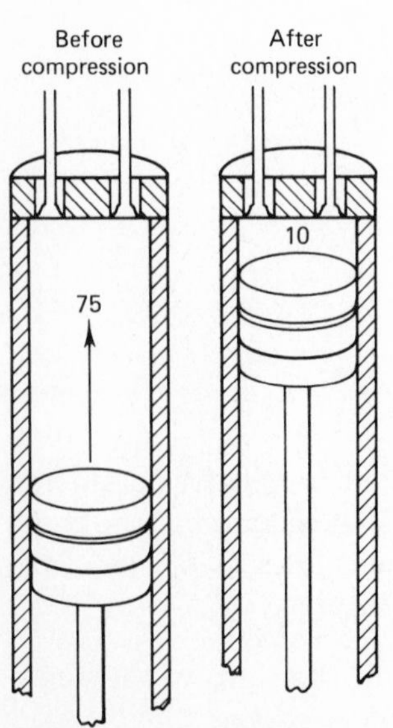

Fig. 2-2 Find the compression ratio of the cylinder.

Problems

Change the following fractions to equivalent decimals:

1. $\frac{1}{4}$	6. $\frac{3}{10}$	11. $\frac{4}{9}$	16. $\frac{9}{16}$
2. $\frac{3}{8}$	7. $\frac{3}{20}$	12. $\frac{13}{15}$	17. $\frac{1}{50}$
3. $\frac{5}{8}$	8. $\frac{2}{7}$	13. $\frac{3}{32}$	18. $\frac{1}{200}$
4. $\frac{1}{3}$	9. $\frac{7}{8}$	14. $\frac{2}{15}$	19. $\frac{5}{26}$
5. $\frac{2}{5}$	10. $\frac{3}{16}$	15. $\frac{25}{40}$	20. $\frac{9}{64}$

21. (*F*) A farmer planted a total of 40 acres, 16 acres in corn and 24 acres in wheat. The part of the total acreage planted in corn is $\frac{16}{40}$. Express this fraction as a decimal.
22. (*F*) Out of a tract of 200 acres, 90 acres of timberland were clear cut. Express this part of the total ($\frac{90}{200}$) as a decimal.
23. (*A*) The compression ratio of an engine cylinder is the comparison, by division, of the volume occupied by a fuel mixture before compression to the volume occupied after compression. Express the compression ratio of the cylinder shown in Fig. 2-2 as a decimal.
24. (*F*) A feed mixture is made of 36 lb of corn silage and 60 lb of alfalfa. Express the ratio of silage to alfalfa ($\frac{36}{60}$) as a decimal.

25. (*E*) The current in the parallel electric circuit shown in Fig. 2-3 divides in the ratio of $\frac{12}{30}$. Express the ratio as a decimal.

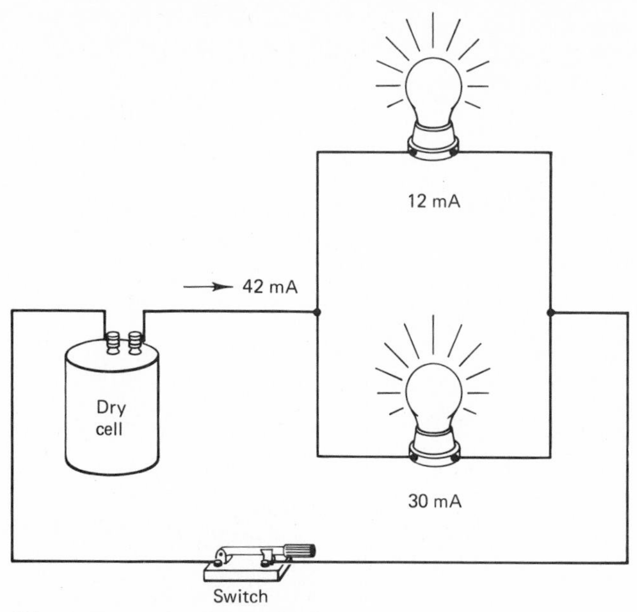

Fig. 2-3 A simple parallel circuit.

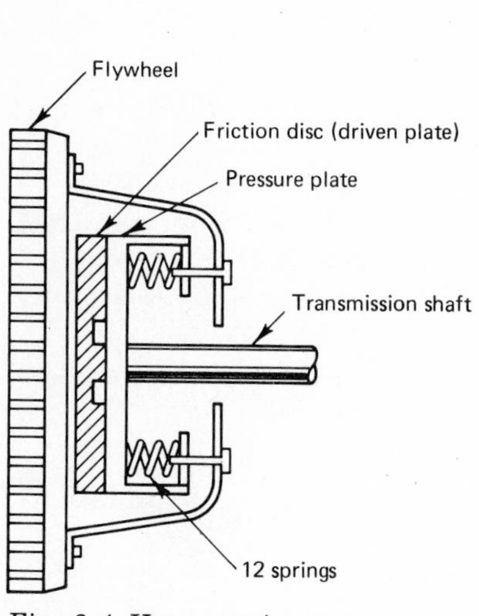

Fig. 2-4 Heavy spring pressure forces the pressure plate against the driven plate.

26. (*E*) A 40-watt (W) lamp uses about $\frac{3}{8}$ A. Express the current used as a decimal.
27. (*A*) The points on the distributor of an 8-cylinder engine are in contact 240° out of a total of 360°. What decimal part of the total does this represent?
28. (*A*) The force exerted on the clutch pressure plate shown in Fig. 2-4 is 900 lb. If the pressure plate has 12 springs, find the force exerted by each spring.
29. (*E*) The ratio of 1 kilowatt (kW) of power to 1 hp is $\frac{1,000}{750}$. Express the ratio as a decimal.
30. (*M, A*) The screw thread in a micrometer (Fig. 2-5) advances $\frac{1}{40}$ in in one complete turn. Express this distance as a decimal.

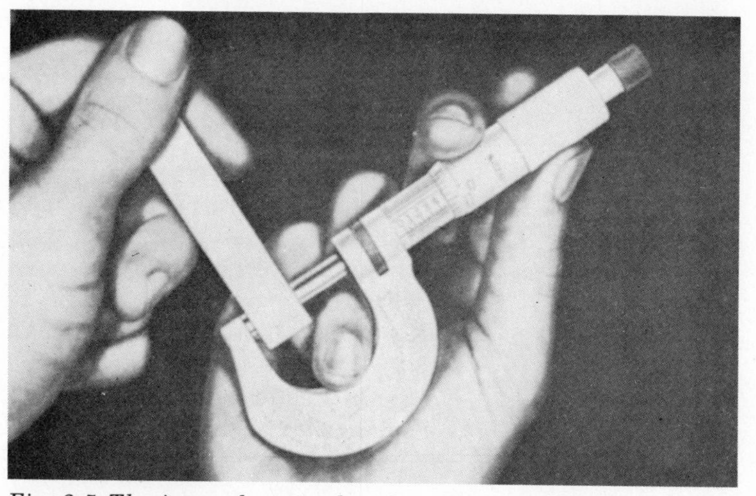

Fig. 2-5 The internal screw thread in the micrometer moves the spindle $\frac{1}{40}$ in in one complete turn. (Courtesy The L. S. Starrett Company)

31. (A) The steering gear reduction on an automobile allows the wheels to turn 5° when the steering wheel is turned through 80°. Express this reduction ($\frac{5}{80}$) as a decimal.
32. (C) In Fig. 2-6, the *pitch* of a roof or rafter is defined as the ratio of the *rise* to the *span* of the roof. Express the pitch $\frac{8}{24}$ as a decimal.

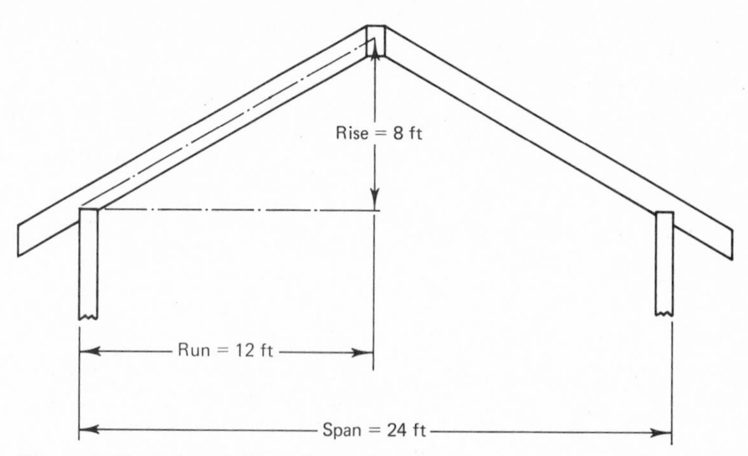

Fig. 2-6 Find the pitch of the roof.

33. (F, C) A certain brand of wire fencing weighs 2 lb/ft. How many feet of fencing are contained in a roll that weighs 43 lb?
34. (C) If a can of paint will cover 250 sq ft of wall surface, how many cans are needed to cover 1,700 sq ft?
35. (G) The lift truck shown in Fig. 2-7 can handle $\frac{3}{25}$ of the complete load at a time. Express this fraction as a decimal.

Fig. 2-7 A lift truck at work in a mill yard.
(Courtesy Clark Equipment Company)

36. (C) A highway crew poured 634 ft of concrete in 5 days. This length is $\frac{634}{5,280}$ of a mile. Express this ratio as a decimal.

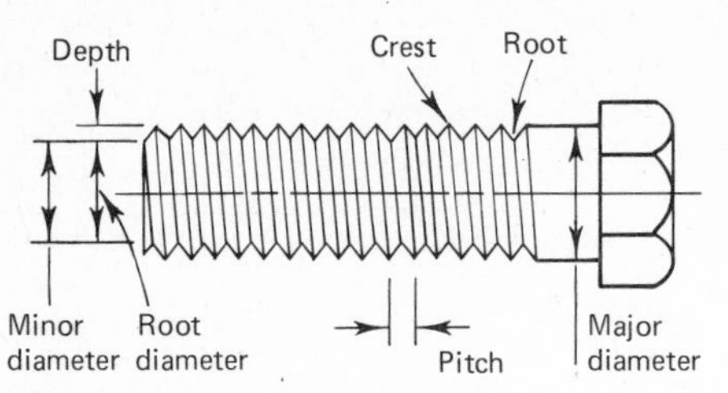

Fig. 2-8 Important parts of a screw thread.

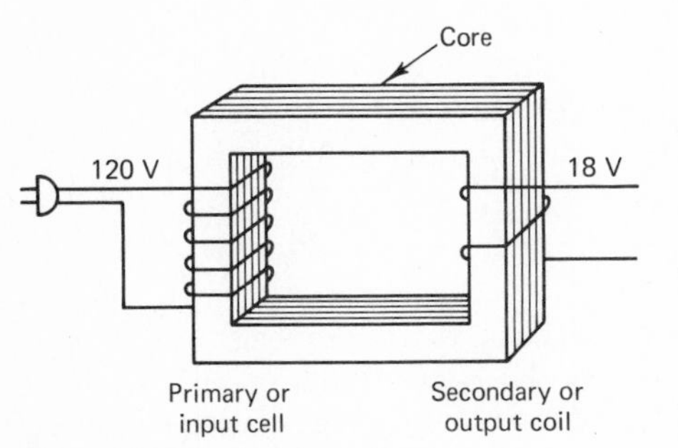

Fig. 2-9 Basic transformer construction.

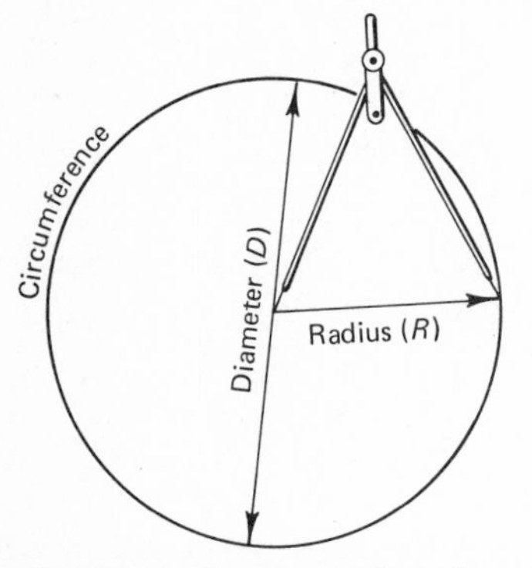

Fig. 2-10 In a circle, the diameter is equal to twice the radius.

37. (*A*) The rear-axle ratio of an automobile is $\frac{34}{8}$. Express the ratio as a decimal.
38. (*C*) A carpenter who could ordinarily lay 1,400 shingles per day could only lay 868 shingles because of bad weather. Express as a decimal the part of her normal work that she completed.
39. (*C*) A mason plastered 81 square yards (sq yd) of wall out of a total of 450 sq yd to be plastered. What decimal part of the total area did he complete?
40. (*F, B*) A fertilizer manufacturer has 45,200 lb of phosphoric acid on hand. If he uses it at the rate of 800 lb/day, how many days will it last?
41. (*A*) A marine diesel engine has a piston displacement of 780 cu in and delivers 273 hp at 2,600 revolutions per minute (rpm). How many horsepower per cubic inch does this represent?
42. (*M, A*) As shown in Fig. 2-8, the pitch of a thread is the distance between corresponding points on two adjacent threads. If the pitch of a thread is $\frac{2}{7}$ in, express the pitch as a decimal.
43. (*E*) A bell transformer reduced the primary voltage of 120 volts (V) to the 18 V delivered by the secondary, as shown in Fig. 2-9. Express the voltage ratio ($\frac{120}{18}$) as a decimal.
44. (*G*) In any circle (Fig. 2-10), the ratio of the circumference to the diameter is approximately $\frac{22}{7}$. Give the decimal equivalent of this fraction to two decimal places.
45. (*G*) The $\frac{3}{16}$-in-*OD* (outside diameter) copper tubing used in refrigerators weighs 65 lb/1,000 ft. Find the number of feet in 1 lb of this tubing.
46. (*F*) On a feed lot a steer consumed 496 lb of feed in 32 days. Find the consumption per day.
47. (*G*) A baseball player got 68 hits in 225 times at bat. Find his batting average to three decimal places.
48. (*A*) A 245-hp engine lost 38 hp in the automatic transmission. Express this loss ($\frac{38}{245}$) as a decimal.
49. (*G*) The energy efficiency ratio of a window air conditioner is found by dividing its Btu rating by its wattage rating. Find the efficiency ratio of a 6,500-Btu air conditioner that uses 900 W of energy. Would you buy this appliance if the minimum acceptable ratio were 6?
50. (*A*) A certain rotary engine developed 180 hp at 5,000 rpm, while a standard V8 developed 195 hp at 4,800 rpm. Find the horsepower per rpm for each engine to three decimal places.

USING THE DECIMAL EQUIVALENT CHART

There are some fractions that are used very often. These are the fractions which represent the parts of an inch on a ruler, like $\frac{1}{16}$, $\frac{3}{8}$, $\frac{5}{32}$, $\frac{9}{64}$, etc. Since they are so widely used, a table of decimal equivalents has been prepared. In order to find the decimal equivalent of a fraction of this type, refer to Table 2-2 on page 13.

Problems

Use the table to find the decimal equivalent of each of the following fractions.

1. $\frac{5}{8}$
2. $\frac{7}{32}$
3. $\frac{9}{16}$
4. $\frac{41}{64}$
5. $2\frac{1}{4}$
6. $3\frac{11}{32}$
7. $1\frac{9}{64}$
8. $4\frac{3}{4}$

Use the table to find the fraction nearest in value to the following decimals.

9. 0.37	11. 0.449	13. 0.035	15. 0.41
10. 0.785	12. 0.88	14. 0.525	16. 0.615

Write each of the following decimals as a mixed number or fraction.

17. 0.75	19. 0.141	21. 2.8125	23. 1.844
18. 0.625	20. 0.5625	22. 3.641	24. 0.9375

25. (*G*) A blueprint gives a dimension of 4.323 in. Express this measurement to the nearest sixty-fourth of an inch.
26. (*A*) The original diameter of the cylinders in an automobile engine was $4\frac{1}{4}$ in. After 25,000 mi, cylinder *A* measured $4\frac{5}{16}$ in and cylinder *B* measured 4.258 in. Which cylinder showed the greatest wear?
27. (*A, M*) Will a $\frac{3}{8}$-in "C" wrench fit a nut measuring 0.355 in across the flats?
28. (*E*) A 100-W lamp requires 0.91 A, and a table radio requires $\frac{7}{8}$ A. Which item uses more current?
29. (*G*) Player *A* got 19 hits out of 64 at-bats, while player *B* got 20 hits out of 60 at-bats. Which player has the better batting average?
30. (*A*) An adjustment screw on a carburetor must be at least 0.56 in long. Will a $\frac{17}{32}$-in-long screw be acceptable?
31. (*G, A, M*) In Fig. 2-11, is the 0.31-in-diameter hole in the washer large enough to accept a $\frac{5}{16}$-in-diameter bolt?
32. (*A*) The valve tappet clearance on an automobile engine should be 0.016 in. Could a mechanic without a feeler gage safely use a $\frac{1}{64}$-in thick shim instead?
33. (*M, G*) Change each of the measurements shown on Fig. 2-12 to decimal values.

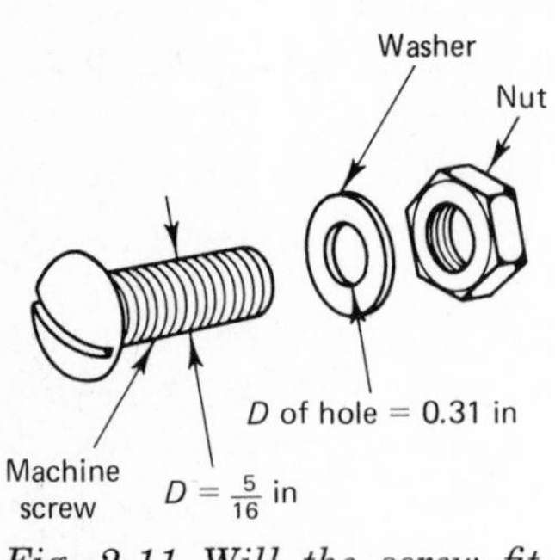

Fig. 2-11 Will the screw fit through the washer?

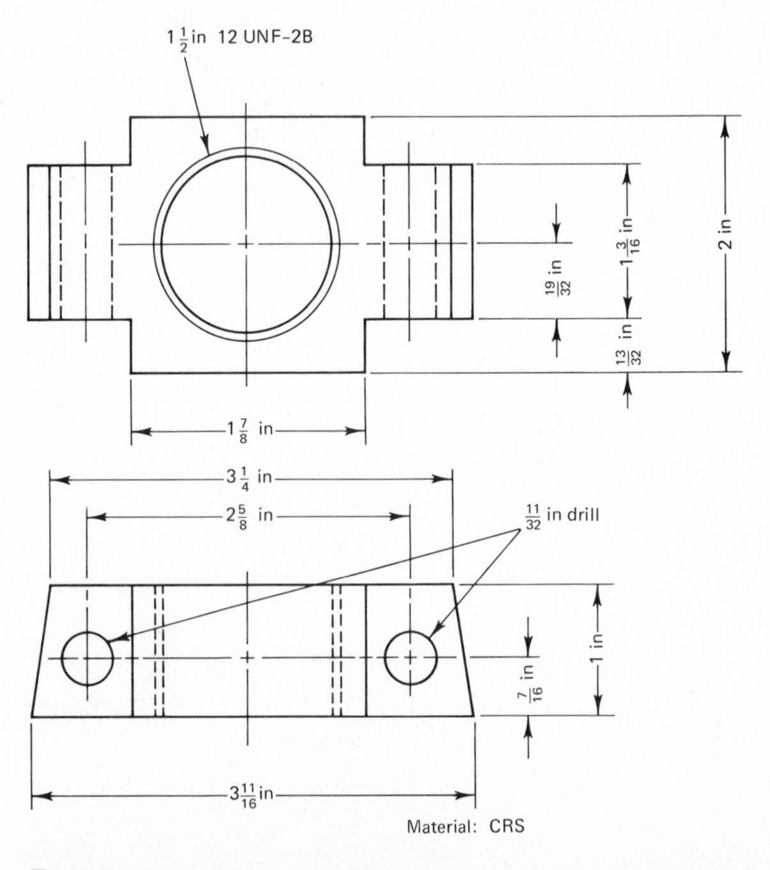

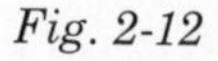
Fig. 2-12

34. (*M, A, G*) The distance between any two threads on an 11-pitch thread is $\frac{1}{11}$ in. Express this measurement to the nearest sixty-fourth of an inch.

ROUNDING OFF NUMBERS

As you can see from Table 2-2, some fractions have been worked out to five or six decimal places. Such accuracy of measurement is hardly ever required. We usually expect to work with measurements expressed to the nearest thousandth (three decimal places), to the nearest hundredth (two decimal places), or to the nearest tenth (one decimal place). When an answer works out to more decimal places than is needed, we must *round off* the decimal to the accuracy required by the job.

Table 2-2 Table of Decimal Equivalents

FRACTIONS			DECIMAL	FRACTIONS			DECIMAL
		1/64	0.015625			33/64	0.515625
	1/32	2/64	0.03125		17/32	34/64	0.53125
		3/64	0.046875			35/64	0.546875
1/16	2/32	4/64	0.0625	9/16	18/32	36/34	0.5625
		5/64	0.078125			37/64	0.578125
	3/32	6/64	0.09375		19/32	38/64	0.59375
		7/64	0.109375			39/64	0.609375
1/8	4/32	8/64	0.125	5/8	20/32	40/64	0.625
		9/64	0.140625			41/64	0.640625
	5/32	10/64	0.15625		21/32	42/64	0.65625
		11/64	0.171875			43/64	0.671875
3/16	6/32	12/64	0.1875	11/16	22/32	44/64	0.6875
		13/64	0.203125			45/64	0.703125
	7/32	14/64	0.21875		23/32	46/64	0.71875
		15/64	0.234375			47/64	0.734375
1/4	8/32	16/64	0.25	3/4	24/32	48/64	0.75
		17/64	0.265625			49/64	0.765625
	9/32	18/64	0.28125		25/32	50/64	0.78125
		19/64	0.296875			51/64	0.796875
5/16	10/32	20/64	0.3125	13/16	26/32	52/64	0.8125
		21/64	0.328125			53/64	0.828125
	11/32	22/64	0.34375		27/32	54/64	0.84375
		23/64	0.359375			55/64	0.859375
3/8	12/32	24/64	0.375	7/8	28/32	56/64	0.875
		25/64	0.390625			57/64	0.890625
	13/32	26/64	0.40625		29/32	58/64	0.90625
		27/64	0.421875			59/64	0.921875
7/16	14/32	28/64	0.4375	15/16	30/32	60/64	0.9375
		29/64	0.453125			61/64	0.953125
	15/32	30/64	0.46875		31/32	62/64	0.96875
		31/64	0.484375			63/64	0.984375
1/2	16/32	32/64	0.5	1	32/32	64/64	1.0

RULE 2 **To round off a decimal:**

1. **Break the decimal at the required number of decimal places.**
2. **If the digit immediately to the right of the break is less than 5, drop all digits to the right of the break.**
3. **If the digit immediately to the right of the break is more than 5, drop all digits to the right of the break and add 1 to the digit immediately to the left of the break.**

Example 2-6 Express the fraction $\frac{13}{64}$ as a decimal rounded off to the nearest thousandth.

Solution From Table 2-2, the value of $\frac{13}{64} = 0.203125$.

1. Since the nearest thousandth means 3 decimal places, we break the decimal as shown below.

   0.203 ⸾ 125

2. The digit immediately to the right of the break is 1, which is less than 5. Drop all digits to the right of the break.

3. The answer is 0.203 (correct to the nearest thousandth).

Example 2-7 Express the fraction $\frac{23}{32}$ as a decimal rounded off to the nearest hundredth.

Solution From Table 2-2, the value of $\frac{23}{32} = 0.71875$.

1. Since the nearest hundredth means two decimal places, we break the decimal as shown below.

   0.71 ⸾ 875

2. The digit immediately to the right of the break is 8, which is more than 5. Drop all digits to the right of the break and add 1 unit to the digit to the left of the break.

$$\begin{array}{r} 0.71 \\ +\quad 1 \\ \hline 0.72 \end{array}$$ *Ans.* (correct to the nearest hundredth)

Example 2-8 Express the fraction $\frac{3}{7}$ as a decimal correct to the nearest hundredth.

Solution Since the nearest hundredth means two decimal places, we divide the 3 by 7 until we get one more decimal place than is needed.

$$\begin{array}{r} 0.428 \\ 7\overline{)3.000} \\ \underline{2\,8} \\ 20 \\ \underline{14} \\ 60 \\ \underline{56} \\ 4 \end{array}$$ (Stop dividing)

1. Since the nearest hundredth means two decimal places, we break the decimal as shown below.

   0.42 ⸾ 8

2. The digit immediately to the right of the break is 8, which is more than 5. Drop the digit 8 and add 1 unit to the digit to the left of the break.

$$\begin{array}{r} 0.42 \\ +\quad 1 \\ \hline 0.43 \end{array}$$ *Ans.* (correct to the nearest hundredth)

Self-Test 2-9 Express the fraction $\frac{4}{7}$ as a decimal correct to the nearest thousandth.

Solution Since the nearest thousandth means (how many?) decimal places, we divide the ______ by the ______ until we get one more decimal place than is needed. | 3 | 4, 7

$$\begin{array}{r} 0.5714 \\ 7\overline{)4.0000} \\ \underline{3\,5} \\ 50 \\ \underline{49} \\ 10 \\ \underline{7} \\ 30 \\ \underline{28} \\ 2 \end{array}$$ (stop dividing)

1. Since the nearest thousandth means 3 decimal places, we break the decimal as shown below.

   0 . 5 7 1 4 (indicate break) | 0.571 ⦚ 4

2. The digit immediately to the right of the break is ______, which is (more)(less) than 5. Drop the digit 4 and add (1)(0) to the digit to the left of the break. | 4, less | 0

3. The answer is ______. | 0.571

Some difficulty has been encountered in rounding off decimals when the digit after the break is a 5. To clarify the situation, The American National Standards Institute established the following rules.

RULE 3 **If the digit immediately to the right of the break is a 5 followed by zeros, then the digit immediately to the left of the break is**

a. Left *unchanged* if it is *even*.

b. *Increased* by 1 if it is *odd*.

Example 2-10 Round off 18.650 to the nearest tenth.

Solution

1. Since the nearest tenth means 1 decimal place, we break the decimal as shown below.

   18.6 ⦚ 50

2. The digit immediately to the right of the break is a 5, followed by a zero. Since the digit immediately to the left of the break (6) is even, it is left unchanged.

3. The answer is 18.6.

Example 2-11 Round off 27.735 to the nearest hundredth.

Solution

1. Since the nearest hundredth means 2 decimal places, we break the decimal as shown below.

   27.73 ⦚ 5

2. The digit immediately to the right of the break is a 5, followed by unwritten zeros. Since the digit immediately to the left of the break (3) is odd, it is increased by 1.

3. The answer is 27.74.

RULE 4 **If the digit immediately to the right of the break is a 5 followed by a nonzero number, add 1 to the digit immediately to the left of the break.**

Example 2-12 Round off 0.7654 to the nearest hundredth.

Solution

1. Since the nearest hundredth means 2 decimal places, we break the decimal as shown below.

   0.76 ⦚ 54

2. The digit immediately to the right of the break is a 5 followed by a 4. Therefore, add 1 to the digit immediately to the left of the break.

3. The answer is 0.77.

Problems

Round off each of the following as directed:

1. 0.7852 to the nearest thousandth
2. 8.076 to the nearest hundredth
3. 45.672 to the nearest tenth
4. 0.08732 to the nearest thousandth
5. 18.016 to the nearest tenth
6. 9.933 to the nearest hundredth
7. 4.0645 to the nearest thousandth
8. 0.08753 to the nearest thousandth
9. 26.735 to the nearest hundredth
10. 6.12357 to the nearest hundredth

Change each of the following fractions to a decimal, rounded off as directed:

11. $\frac{5}{16}$ to the nearest thousandth
12. $\frac{19}{32}$ to the nearest hundredth
13. $\frac{3}{8}$ to the nearest hundredth
14. $\frac{31}{64}$ to the nearest thousandth
15. $\frac{21}{32}$ to the nearest hundredth
16. $\frac{3}{4}$ to the nearest tenth
17. $\frac{7}{8}$ to the nearest hundredth
18. $\frac{31}{32}$ to the nearest hundredth
19. $\frac{27}{64}$ to the nearest thousandth
20. $\frac{25}{32}$ to the nearest ten-thousandth
21. $\frac{4}{9}$ to the nearest thousandth
22. $\frac{5}{7}$ to the nearest thousandth
23. $\frac{2}{3}$ to the nearest hundredth
24. $\frac{7}{11}$ to the nearest hundredth

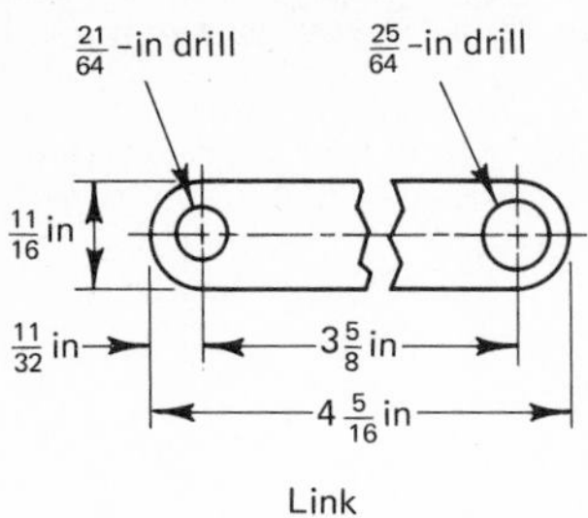

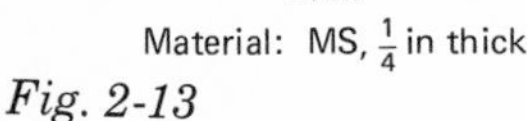

Fig. 2-13

25. $\frac{12}{21}$ to the nearest thousandth
26. (*M, G*) Change each of the dimensions shown in Fig. 2-13 to decimals, rounded off to the nearest thousandth.
27. (*G*) The *point* is a unit of measurement used in the printing industry. It equals approximately $\frac{1}{72}$ in. Change this to a decimal correct to the nearest ten-thousandth of an inch.
28. (*M, G*) The plate in Fig. 2-14 has four holes. (a) Which hole, *A* or *B*, has its center closer to the right edge? (b) Which hole, *C* or *D*, has its center closer to the left edge?

(See Answer Key for TEST 1—Introduction to Decimals)

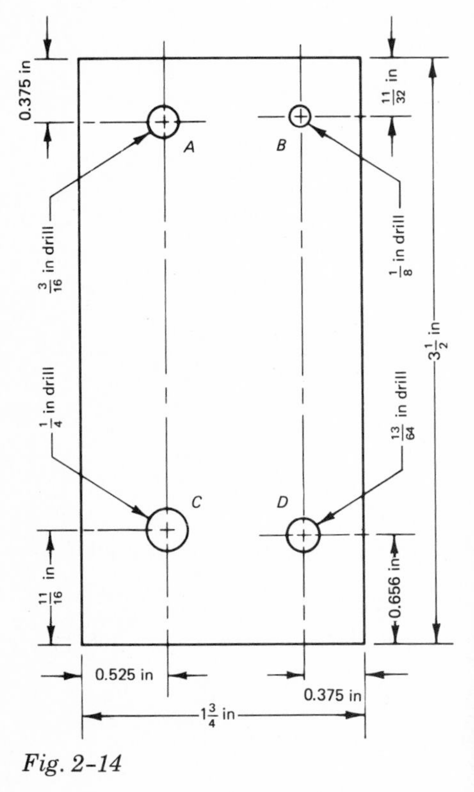

Fig. 2-14

JOB  3 Introduction to the Metric System of Measurement

The metric system is a decimal system of measurement that is the practical standard throughout most of the world. It is expected that this system, which has been formally adopted by the United States, will come into general use within the next few years. Many industries in the United States have already converted to metric measurements. Weather forecasters on television often give the temperature in both degrees Fahrenheit and degrees Celsius. Signs giving distances in both miles and kilometers are being erected all over the country.

Fig. 3-1 Highway signs often give the distances in both miles and kilometers. (Courtesy The Automobile Club of New York)

The basic unit of length in the metric system is the meter (m), which is equal to about 39.37 in.

If we divide a meter into 1,000 parts, each part is called a *millimeter* (mm), since the prefix "milli" means one one-thousandth of a quantity.

If we divide a meter into 100 parts, each part is called a *centimeter* (cm), since the prefix "centi" means one one-hundredth of a quantity.

If we divide a meter into 10 parts, each part is called a *decimeter* (dm), since the prefix "deci" means one-tenth of a quantity.

The relationship among these three quantities is shown in Fig. 3-2.

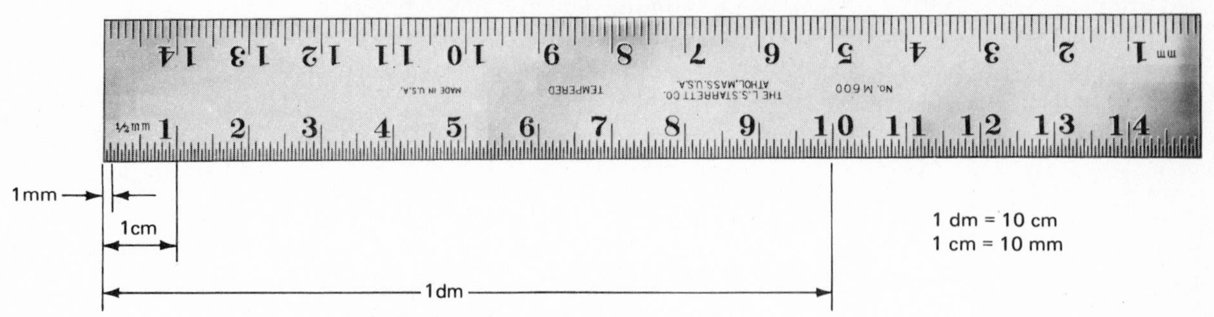

Fig. 3-2 A metric scale. (Courtesy The L. S. Starrett Company)

For measurements larger than a meter,

1 dekameter (dam) = 10 meters, since the prefix "deka" means 10 times as large.

1 hektometer (hm) = 100 meters, since the prefix "hekto" means 100 times as large.

1 kilometer (km) = 1,000 meters, since the prefix "kilo" means 1,000 times as large.

The set of lengths in the metric system from 1,000 to $\frac{1}{1,000}$ is given in Table 3-1.

Table 3-1 Lengths in the Metric System

1 cm = 10 mm	or	1 mm = $\frac{1}{10}$ cm = 0.1 cm
1 dm = 10 cm	or	1 cm = $\frac{1}{10}$ dm = 0.1 dm
1 m = 10 dm	or	1 dm = $\frac{1}{10}$ m = 0.1 m
1 m = 100 cm	or	1 cm = $\frac{1}{100}$ m = 0.01 m
1 m = 1,000 mm	or	1 mm = $\frac{1}{1,000}$ m = 0.001 m
1 dam = 10 m	or	1 m = $\frac{1}{10}$ dam = 0.1 dam
1 hm = 100 m	or	1 m = $\frac{1}{100}$ hm = 0.01 hm
1 km = 1,000 m	or	1 m = $\frac{1}{1,000}$ km = 0.001 km

Table 3-1 is described in picture form in Fig. 3-3.

Before we try to change any of these measurements from one unit to another, let us look at a simple method for multiplying and dividing by numbers like 10, 100, 1,000, etc.

MULTIPLYING BY 10, 100, 1,000, ETC.

RULE 1 **To multiply values by numbers like 100, 1,000, 1,000,000, etc., move the decimal point one place to the right for every zero in the multiplier.**

Example 3-1 1.59 × 10 = 1⊙5.9 = 15.9 (move point one place to the right)

4.76 × 100 = 4⊙76. = 476. or 476 (move point two places to the right)

A decimal point is ordinarily not written at the end of a whole number. However, it may be written if desired. In the following example, the point must be written in together with two zeros so that we can move the decimal point past the required number of places.

34 × 100 = 34⊙00. = 3,400. or 3,400

If the decimal point must be moved past more places than are available, zeros are added at the end of the number to make up the required number of places. This is shown in the following example.

0.0259 × 1,000,000 = ⊙025900. = 25,900

Self-Test 3-2 Multiply: 25.2 × 100

Solution To multiply, move the decimal point to the (left)(right). There are ______ zeros in the number 100. The point will be moved ______ places. Zeros (are)(are not) needed to account for the correct number of places. The answer is 25.2 × 100 = ______.

right
2
2, are
2,520

Problems

1. 0.0072 × 1,000
2. 45.76 × 100
3. 3.09 × 1,000
4. 0.0045 × 100
5. 37 × 100
6. 0.08 × 1,000
7. 0.0006 × 1,000
8. 0.00056 × 1,000,000
9. 27 × 100
10. 15 × 1,000
11. 0.00078 × 100
12. 7.8 × 1,000,000
13. 15.4 × 1,000
14. 4.5 × 10
15. 0.005 × 100
16. 6 × 10,000
17. 0.008 × 10,000
18. 0.009 × 100,000
19. 2.34 × 1,000
20. 0.0008 × 1,000,000

DIVIDING BY 10, 100, 1,000, ETC.

RULE 2 **To divide values by numbers like 10, 100, 1,000, etc., move the decimal point one place to the left for every zero in the divisor.**

Example 3-3 17.4 ÷ 10 = 1.7⊙4 = 1.74 (move point one place left)

45 ÷ 100 = .45⊙ = 0.45 (move point two places left)

6.5 ÷ 1,000 = .006⊙5 = 0.0065 (move point three places left)

Self-Test 3-4 Divide 8.5 by 1,000.

Solution To divide, move the decimal point to the (left)(right). There are ______ zeros in the number 1,000. The point will be moved ______ places. Zeros (are)(are not) needed to account for the correct number of places. The answer is 8.5 ÷ 1,000 = ______.

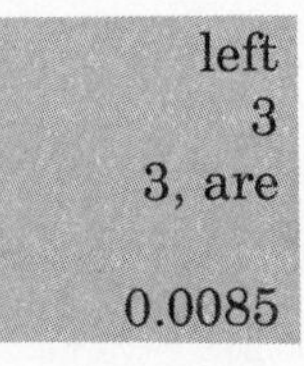

Problems

1. 6,500 ÷ 100
2. 7,500 ÷ 1,000
3. 880,000 ÷ 1,000
4. 32 ÷ 100

5. $6 \div 100$
6. $17.8 \div 10$
7. $835 \div 1{,}000$
8. $550 \div 1{,}000{,}000$
9. $653.8 \div 1{,}000$
10. $100{,}000{,}000 \div 1{,}000{,}000$
11. $0.45 \div 1{,}000$
12. $0.08 \div 10$
13. $8.5 \div 1{,}000$
14. $7 \div 1{,}000{,}000$
15. $\frac{28.6}{1{,}000}$
16. $\frac{15.6}{100}$
17. $\frac{398}{10{,}000}$
18. $\frac{0.08}{10}$
19. $\frac{2}{1{,}000}$
20. $\frac{600}{1{,}000}$

CHANGING UNITS OF MEASUREMENT

RULE 3 1. **To change large units into small units, *multiply*.**
2. **To change small units into large units, *divide*.**

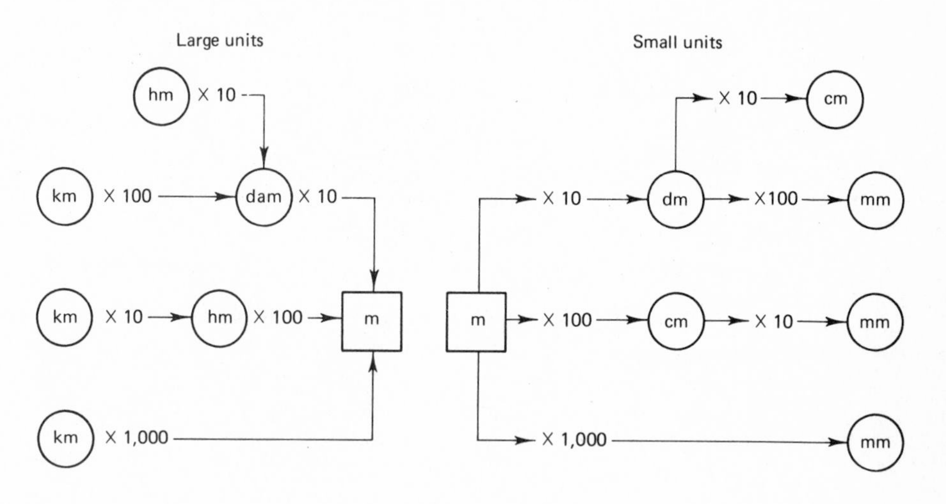

Fig. 3-3 Rules for changing units in the metric system.

Example 3-5 Change the following units of measurement.

a. Change 8 cm to millimeters.
There are 10 mm in each centimeter, and since we are changing large units (centimeters) to small units (millimeters), we *multiply* by 10.

$8 \text{ cm} = 8 \times 10 = 80 \text{ mm}$ *Ans.*

b. Change 42 mm to centimeters.
There are 10 mm in each centimeter, and since we are changing small units (millimeters) to large units (centimeters), we *divide* by 10.

$42 \text{ mm} = 42 \div 10 = 4.2 \text{ cm}$ *Ans.*

c. Change 2 m to decimeters.
There are 10 dm in each meter, and since we are changing large units (meters) to small units (decimeters), we *multiply* by 10.

$2 \text{ m} = 2 \times 10 = 20 \text{ dm}$ *Ans.*

d. Change 300 cm to meters.
There are 100 cm in each meter, and since we are changing small units (centimeters) to large units (meters), we *divide* by 100.

300 cm = 300 ÷ 100 = 3 m *Ans.*

e. Change 5 m to millimeters.
There are 1,000 mm in each meter, and since we are changing large units (meters) to small units (millimeters), we *multiply* by 1,000.

5 m = 5 × 1,000 = 5,000 mm *Ans.*

f. Change 125 m to dekameters.
There are 10 m in each dekameter, and since we are changing small units (meters) to large units (dekameters), we *divide* by 10.

125 m = 125 ÷ 10 = 12.5 dam *Ans.*

g. Change 8.4 hm to meters.
There are 100 m in each hektometer, and since we are changing large units (hektometers) to small units (meters), we *multiply* by 100.

8.4 hm = 8.4 × 100 = 840 m *Ans.*

h. Change 6,450 m to kilometers.
There are 1,000 m in each kilometer, and since we are changing small units (meters) to large units (kilometers), we *divide* by 1,000.

6,450 m = 6,450 ÷ 1,000 = 6.45 km *Ans.*

Self-Test 3-6 Change the following units of measurement.

a. Change 85 mm to centimeters.
Since there are ______ mm in each centimeter,
85 mm (×)(÷) by 10 = ______ cm *Ans.*

10
÷, 8.5

b. Change 2 m to centimeters.
Since there are ______ cm in each meter,
2 m (×)(÷) by 100 = ______ cm *Ans.*

100
×, 200

c. Change 4.5 km to meters.
Since there are ______ m in each kilometer,
4.5 km (×)(÷) by 1,000 = ______ m *Ans.*

1,000
×, 4,500

Self-Test 3-7 State the lengths of the distances indicated in Fig. 3-4.

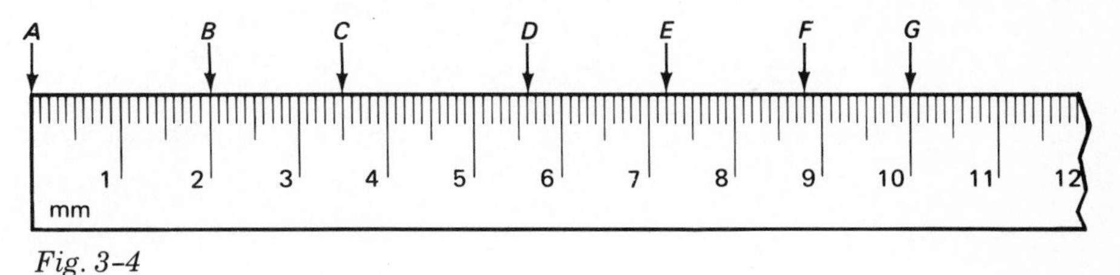

Fig. 3-4

Solution The lengths are:

AB = 20 mm or ______ cm
AC = ______ mm or 3.5 cm
AD = ______ mm or ______ cm
AE = ______ mm or ______ cm
AF = ______ mm or ______ cm
AG = ______ mm or ______ cm

2
35
56, 5.6
72, 7.2
88, 8.8
100, 10

Problems

1. Using the ruler shown in Fig. 3-5, find the indicated lengths in millimeters and centimeters.

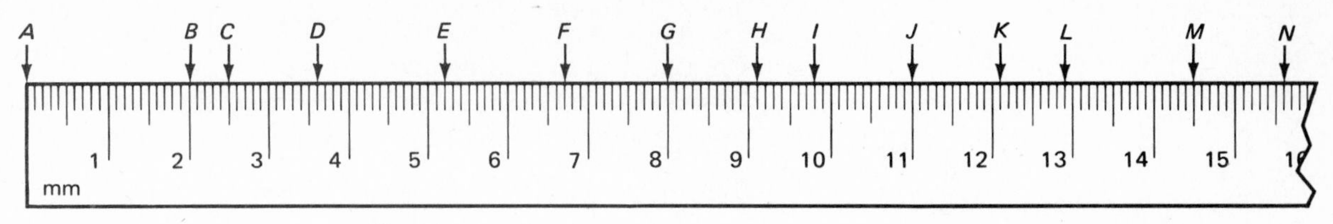

Fig. 3-5

	mm	cm		mm	cm		mm	cm
AB =			*AH* =			*BC* =		
AC =			*AI* =			*CD* =		
AD =			*AJ* =			*GH* =		
AE =			*AK* =			*JK* =		
AF =			*AL* =			*KL* =		
AG =			*AM* =			*MN* =		

Change the following measurements:

2. 35 cm to millimeters
3. 48 mm to centimeters
4. 2.4 dm to centimeters
5. 80 cm to decimeters
6. 36 m to decimeters
7. 3 dm to meters
8. 5 m to centimeters
9. 325 cm to meters
10. 2.45 m to millimeters
11. 6,500 mm to meters
12. 125 mm to centimeters
13. 6.8 cm to millimeters
14. 4.2 hm to dekameters
15. 36 dam to hektometers
16. 75 km to hektometers
17. 8 hm to kilometers
18. 3.9 hm to meters
19. 100 m to hektometers
20. 20 km to meters
21. 3,498 m to kilometers
22. 1.5 dam to decimeters
23. (*G*) Arrange in order of size, starting with the largest:

    0.15 m, 20 cm, 180 mm

24. (*A*) The cylinder bore (diameter) of the 1971 Volkswagen is 85.6 mm and the stroke is 69 mm. Express both measurements in centimeters.
25. (*A*) In a vacuum test on an engine operating at 3,000 ft, the vacuum gage read 34.56 cm. Express this measurement in millimeters.
26. (*A*) The 1974, 8-350 Camaro engine uses a spark plug with a gap of 0.089 cm. Change this measurement to millimeters.

Remove the metric ruler from the back cover of this book to make the following measurements. Indicate the measurements in both millimeters and centimeters.

27. (G, C)

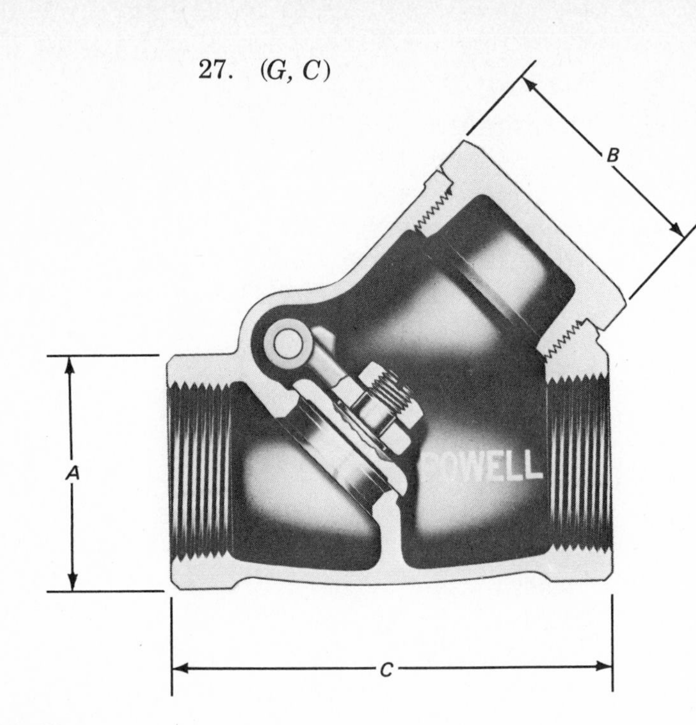

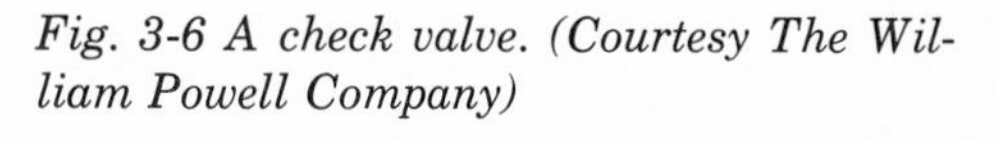

Fig. 3-6 A check valve. (Courtesy The William Powell Company)

28. (M, G)

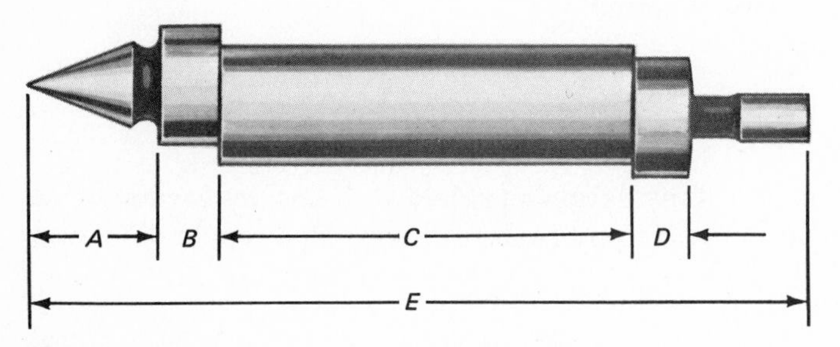

Fig. 3-7 An edge finder. (Courtesy The L.S. Starrett Company)

29. (M, A, G)

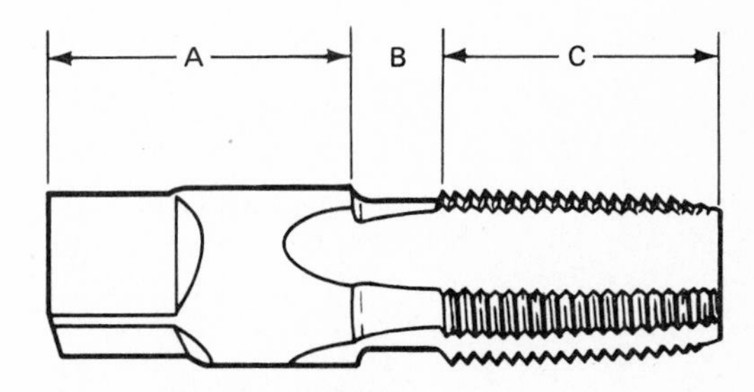

Fig. 3-8 A taper pipe tap.

30. (G, C)

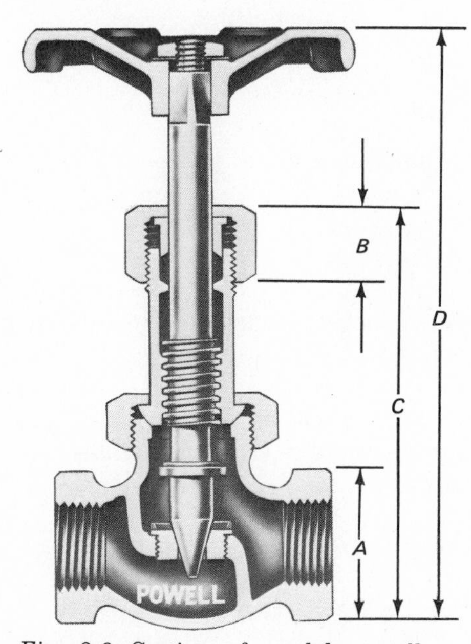

Fig. 3-9 Section of a globe needle valve. (Courtesy The William Powell Company)

31. (M, G)

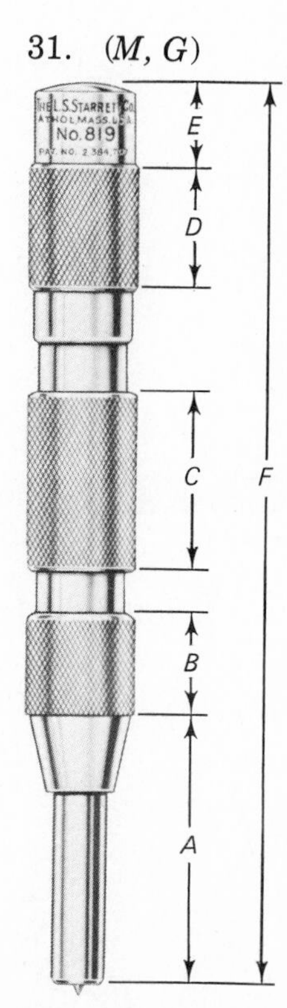

Fig. 3-10 An automatic centering punch. (Courtesy The L.S. Starrett Company)

JOB 4 Adding Decimals

When adding decimals, be sure to write the numbers so that the decimal points are in a straight vertical line. Empty spaces should be filled in with zeros to help keep the digits in a straight column.

Example 4-1 Add 2.52 + 0.007 + 13.03 + 0.7 + 26

Solution Note that the whole number 26 should be written as 26.000.

```
 2.52            2.520
 0.007           0.007
13.03     or    13.030
 0.7             0.700
26.             26.000
------          ------
                42.257     Ans.
  |
  |<----Lined-up decimal points
```

Example 4-2 Find the total thickness of insulation on shielded radio wire covered with resin (0.55 mm), lacquered cotton braid (0.49 mm), and copper shielding (0.225 cm).

Solution **1.** Change 0.225 cm to millimeters.

$$0.225 \text{ cm} = 0.225 \times 10 = 2.25 \text{ mm}$$

2. The total thickness is the sum of the individual thicknesses.

```
0.55 mm
0.49 mm
2.25 mm
-------
3.29 mm     Ans.
 |
 |<-------- Lined-up decimal points
```

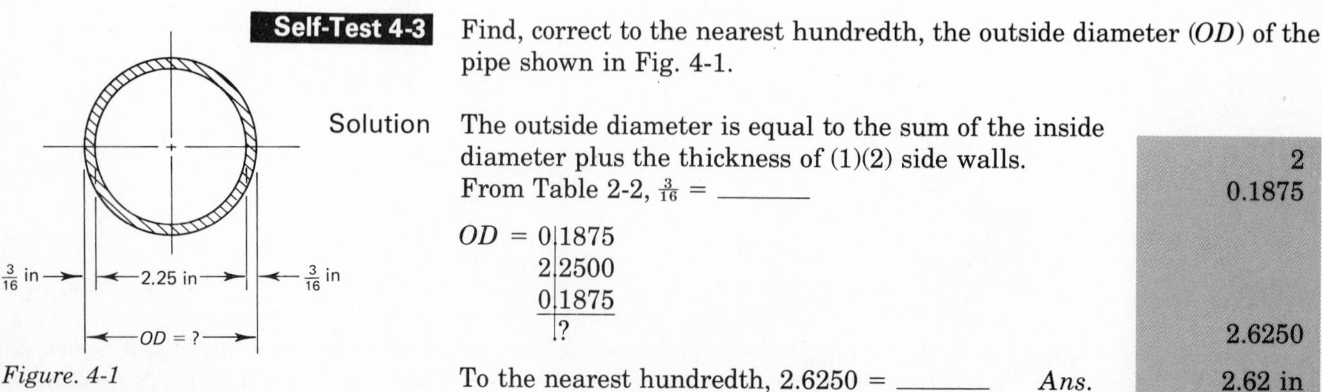

Figure. 4-1

Self-Test 4-3 Find, correct to the nearest hundredth, the outside diameter (*OD*) of the pipe shown in Fig. 4-1.

Solution The outside diameter is equal to the sum of the inside diameter plus the thickness of (1)(2) side walls.
From Table 2-2, $\frac{3}{16}$ = ______

```
OD = 0.1875
     2.2500
     0.1875
     ------
      ?
```

To the nearest hundredth, 2.6250 = ______ *Ans.*

2
0.1875
2.6250
2.62 in

Problems

Add the following decimals:

1. 3.28 + 9.5 + 0.634 + 0.078
2. 56.09 + 13 + 4.876 + 49.007
3. 13 + 3.072 + 0.7 + 6.06
4. 54 + 0.033 + 0.713 + 8.05
5. 0.087 + 6.18 + 4 + 1.7

Express the sum of the following decimals to the nearest hundredth:

6. 6.089 + 1.25 + 3.4 + 0.123
7. 0.625 + 3.75 + 1.1375 + 0.075
8. 1.50 + 2.875 + 1.5625 + 0.72
9. 2.25 + 1.075 + 0.875 + 1.094
10. 12 + 2.565 + 1.07 + 2.625

Express the sum of the following decimals to the nearest tenth:

11. 34.08 + 3.125 + 15 + 6.25
12. 2.125 + 3.4 + 2.65 + 0.55
13. 3.75 + 12.2 + 2.375 + 7
14. 0.065 + 0.52 + 0.25 + 1.5
15. 4.56 + 3.05 + 5.75 + 12.4

16. (*E*) Find the total current in amperes drawn by the following appliances by adding the currents: electric iron (4.12 A), electric clock (0.02 A), light bulb (0.91 A), and radio (0.5 A).
17. (*F*) Four steers dressed out at the following weights: 684.5 lb, 719.4 lb, 727.8 lb, and 698.3 lb. Find their total dressed weight.
18. (*M*) A tapered pin has a small-end diameter of 1.625 cm. If the large-end diameter is 0.6875 cm larger, find the diameter at the large end.
19. (*A, E, B*) What is the cost to rewind a motor if the following materials and labor were used: top sticks at $0.13, #17 wire at $1.75, #26 wire at $0.88, armature lacquer at $0.56, and labor at $11.25?
20. (*M*) Find the total length of the T-slot bolt shown in Fig. 4-2.

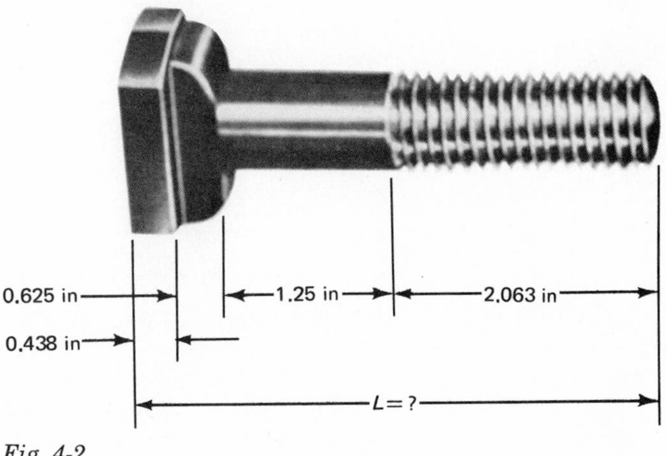

Fig. 4-2

21. (*E*) If #14 copper wire has a diameter of 0.0641 in and #10 wire is 0.0378 in larger, find the diameter of #10 copper wire.
22. (*A, M*) An engine cylinder whose diameter is $4\frac{1}{2}$ in. (Fig. 4-3) is to be rebored to 0.062 in. oversize. Find the diameter after reboring.

DRILL & WIRE GAUGE.
THE L.S.STARRETT CO.
ATHOL, MASS.,U. S. A.
DECIMAL EQUIVALENTS

1 .228 28 .140 29 .136 60 .040
2 .221 27 .144 30 .128 59 .041
3 .213 26 .147 31 .120 58 .042
4 .209 25 .149 32 .116 57 .043
5 .205 24 .152 33 .113 56 .046
6 .204 23 .154 34 .111 55 .052
7 .201 22 .157 35 .110 54 .055
8 .199 21 .159 36 .106 53 .059
9 .196 20 .161 37 .104 52 .063
10 .193 19 .166 38 .101 51 .067
11 .191 18 .169 39 .099 50 .070
12 .189 17 .173 40 .098 49 .073
16 .177 41 .096 48 .076
15 .180 42 .093 47 .078
14 .182 43 .089 46 .081
13 .185 44 .086 45 .082
No.186

Fig. 4-4 A drill and wire gage. (Courtesy The L.S. Starrett Company)

Fig. 4-3 Scraping a cylinder block before reboring. (Courtesy Fel-Pro Inc.)

23. (*F, G*) The lengths of the sides of a field are 419.3 ft, 582.7 ft, 385.4 ft, and 716.9 ft. Find the total number of feet of fencing required to enclose the field.
24. (*G, M*) Using the Drill & Wire Gauge shown in Fig. 4-4, add the diameters of the following sets of drills.
    a. #9, #22, #38, and #47
    b. #1, #7, #13, and #40
25. (*G, M*) In the cut plate shown in Fig. 4-5, find (a) dimension *A*, (b) dimension *B*, and (c) the total length *L*.

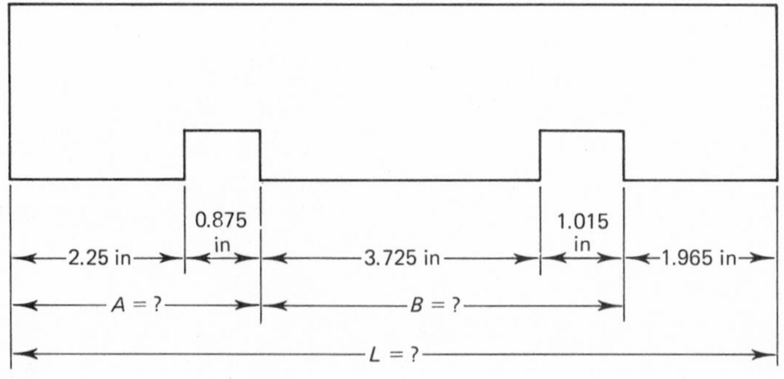

Fig. 4-5

26. (*M, A*) The following thicknesses of Jo blocks (Job 7) were wrung together: 0.350 in, 0.1007 in, 1.0000 in, and 0.147 in. Find the total thickness of the combination.
27. (*B, A*) An auto parts dealer had sales for the week as follows: Monday, $524.20; Tuesday, $756.75; Wednesday, $743.69; Thursday, $845.25; and Friday, $926.45. Find the total sales for the week.
28. (*A, M*) The diameter of a motor shaft bearing is 0.12 mm larger than the shaft of the motor. What is the diameter of the bearing if the shaft has a diameter of 2.25 cm?
29. (*E*) Find the total resistance of the leads of an electrical installation if the resistances are 0.054 Ω, 1.004 Ω, and 1.5 Ω, respectively.

30. (*E, B*) What is the cost to repair three electrical outlets if each outlet requires one loom box at $0.94, one toggle switch at $0.42, one hickey at $0.13, and 1 hr of labor at $8.75?
31. (*A, M*) What is the smallest combination of thickness gages shown in Fig. 4-6 that will add up to a thickness of 0.1095 in?

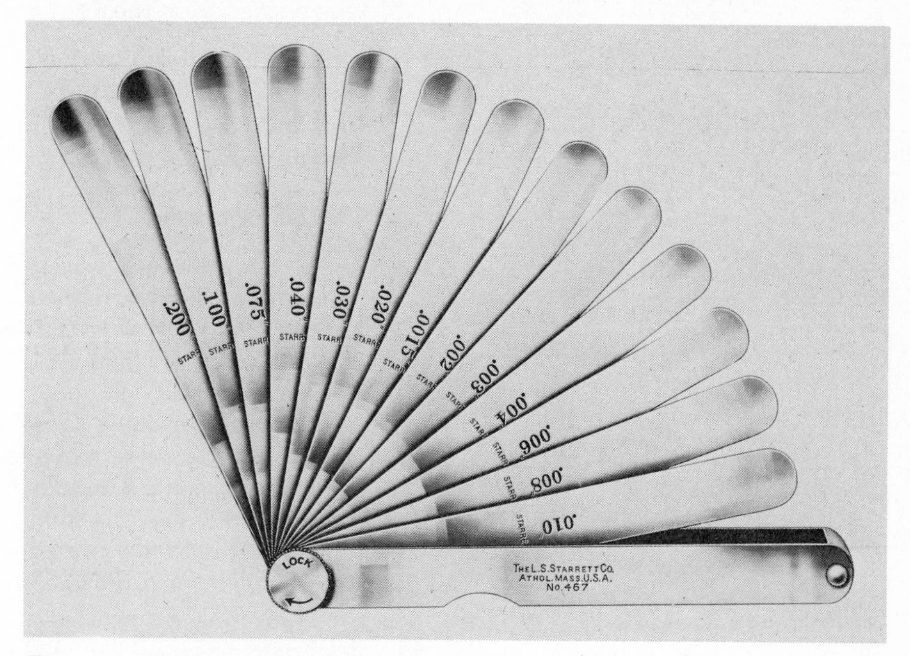

Fig. 4-6 Thickness gages. (Courtesy The L.S. Starrett Company)

32. (*M, A*) In the 3-pitch gear shown in Fig. 4-7, the addendum is 0.3333 in and the dedendum is 0.3856 in. Find the whole depth of the tooth.

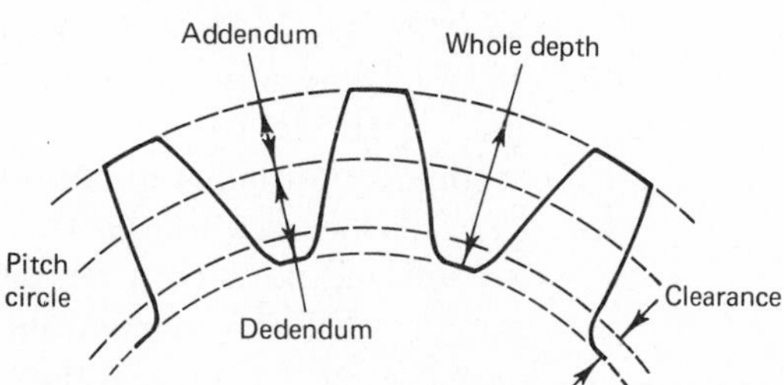

Fig. 4-7 Part of a spur gear.

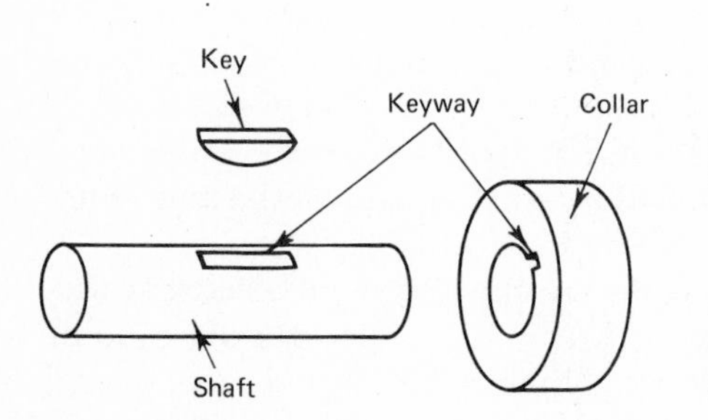

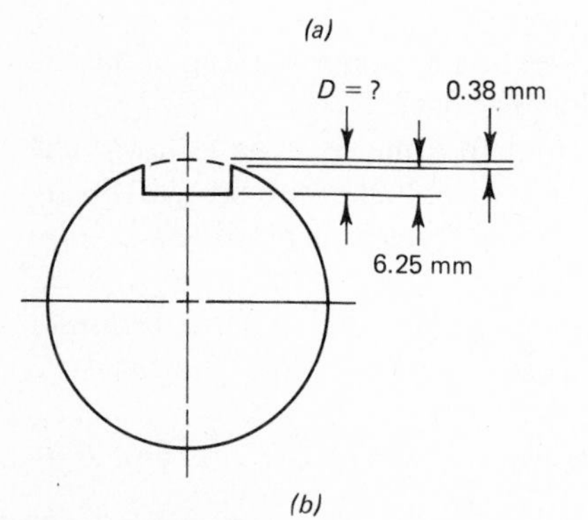

Fig. 4-8 Keys are used to lock collars to shafts.

33. (*M*) Keys are used to lock pulleys, collars, and gears to shafts so that they will turn together. The grooves that are cut into the shaft and collar are called *keyways* (Fig 4-8*a*). Find the depth of the key seat shown in Fig. 4-8*b*.
34. (*F, A*) Three tractors used fuel at the following rates:
Gasoline tractor—6.714 gal/hr
Propane tractor—8.528 gal/hr
Diesel tractor—5.219 gal/hr
Find the total fuel consumption per hour.
35. (*M, A*) A bushing is to be force-fitted into a gear hub. The hole in the gear is bored to a diameter of 1.5 in. If the diameter of the bushing should be 0.001 in larger, find the diameter of the bushing.
36. (*M*) The two feed screws of a bench lathe should measure 10.48 cm and 14.89 cm long, respectively. Allowing a total of 14 mm for cutting and facing, how long a piece of metal is needed to make the two parts?

37. (*M*) A shaper (Fig. 10-4) took three cuts over a piece of work of 0.108 in, 0.092 in, and 0.043 in, respectively. Find the total thickness of metal removed.
38. (*C, E*) To determine the size of entrance wire required for a house, it is necessary to find the total wattage required. Find the total wattage needed from the data below.

Lighting: 1,400 sq ft @ 0.003 kW/sq ft	=	4.2 kW
Central air conditioner	=	9.2 kW
Electric stove	=	13.7 kW
Washer-dryer	=	4.6 kW
Two appliance circuits	=	3.0 kW
Two $\frac{1}{6}$-hp workshop motors	=	0.5 kW

39. (*B, A*) Find the total cost to overhaul an automobile engine at the following rates: Bearings at $14.85, Rings at $30.90, Clutch set at $24.65, water pump at $29.50, carburetor kit at $4.95, tune-up kit at $3.87, 8 spark plugs at $1.12 each, and labor at $125.00.
40. (*M, A*) What size hole must a washer have in order to be 0.012 in larger than a machine screw measuring $\frac{1}{4}$ in in diameter?
41. (*E*) The emitter current of a transistor is always equal to the sum of the collector current and the base current. Find the emitter current if the collector current is 0.03 A and the base current is 0.0015 A.
42. (*M*) A shaft is 1.5625 in in diameter. What size drill should be used to bore a hole for it if 0.0156 in is allowed for clearance on all sides? (See Fig. 4-9.)

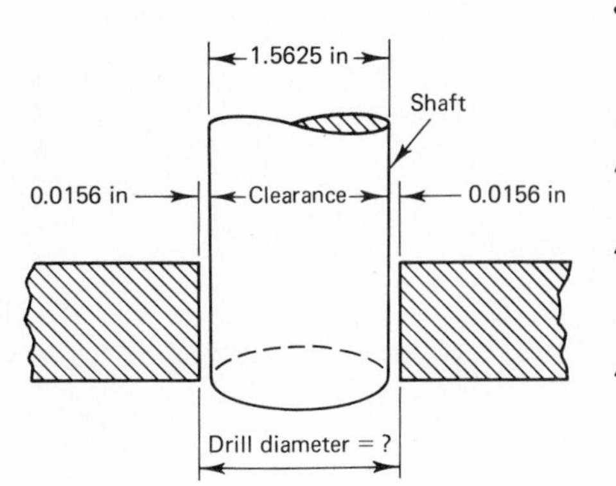

Fig. 4-9

43. (*C*) A pipe has an inside diameter of 0.6875 in. Its walls are 0.0625 in thick. (a) Find the outside diameter. (b) What size drill should be used to make a hole to carry the pipe if 0.0625 in is allowed for clearance on all sides?
44. (*A*) The valve tappet clearance on an automobile engine should be 0.019 in. What is the smallest combination of the following gages that should be used to check the clearance: 0.007 in, 0.004 in, 0.001 in, 0.002 in, 0.009 in, and 0.012 in?
45. (*A, M*) An automobile engine has a cylinder bore of $3\frac{3}{8}$ in. If not more than 0.010 in wear is permissible before the cylinder should be rebored, what is the maximum permissible diameter of the cylinder?
46. (*A*) The specification for the valve stem clearance on each side for a certain engine is between 0.001 in and 0.0015 in. If the valve stem shown in Fig. 4-10 measures 0.348 in, what should be the maximum diameter of the valve guide?

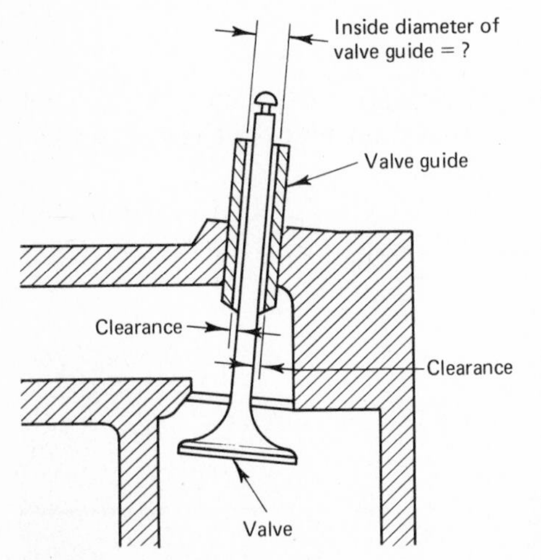

Fig. 4-10 The valve stem is held in position by the valve guide.

47. (*E, B, C*) The following materials were charged to an electrical wiring job: conduit, $4.25; #8 wire, $1.75; BX cable, $18.50; conduit fittings, $3.85; outlet boxes, $6.58; switches, $5.72; and tape, solder, and pipe clips, $4.25. What was the total amount charged for materials?
48. (*G, E*) A carbon brush 0.9687 in thick has a copper plating of 0.0156 in on each side. What is the total thickness?
49. (*E*) The insulation to be used in a slot in a motor is as follows: fish paper, 0.0156 in, Tufflex, 0.0313 in, varnished cambric, 0.0156 in, and top stick, 0.125 in. What is the total thickness of all the insulation?
50. (*E*) A 0.00035-μF tuning capacitor is in parallel with a trimmer capacitor of 0.000075 μF, as shown in Fig. 4-11. Find the total capacity by adding the capacitances.
51. (*G, M*) In Fig. 4-12, find dimension E if $A = 0.1875$ in and $B = 0.5625$ in.
52. (*G, M*) In Fig. 4-12, find dimension F if $A = 0.2655$ in, $B = 0.785$ in, and $C = 1.345$ in.

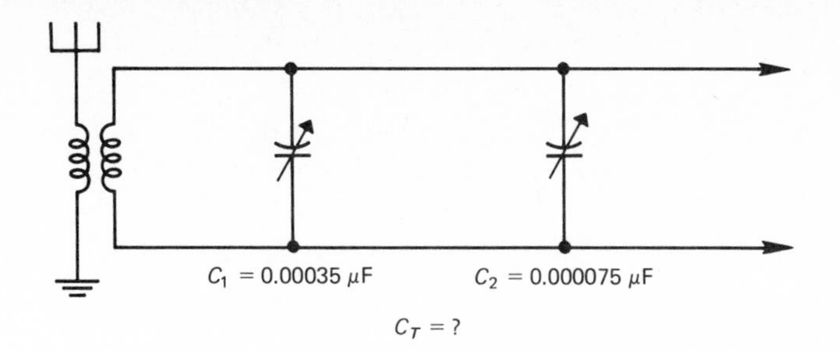

Fig. 4-11 A simple tuning circuit.

53. (*G, M*) In Fig. 4-12, find dimension G if $A = 0.375$ in, $B = 0.500$ in, $C = 2.075$ in, and $D = 0.945$ in.

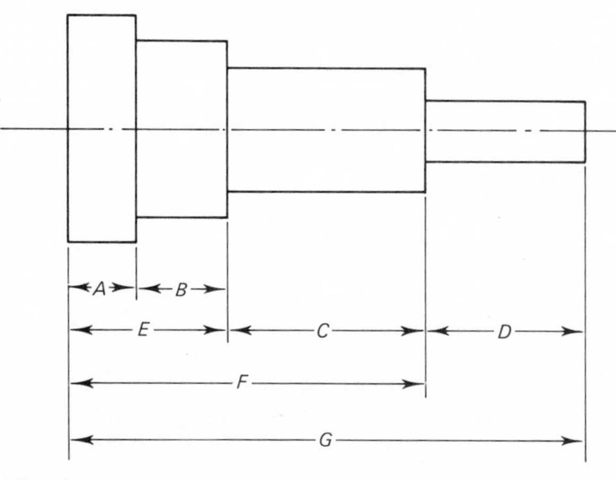

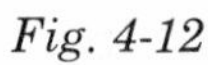

Fig. 4-12

54. (*M, G*) Number 8 (B & S) gage sheet steel is 0.1285 in thick, and Number 25 (B & S) gage sheet copper is 0.0179 in thick. Find the total thickness of 2 sheets of steel and 3 sheets of copper.
55. (*M*) A counterbore 7.5 cm in diameter was enlarged in the boring operation shown in Fig. 4-13. If the boring bit took a cut 2 mm deep, find the finished diameter of the counterbore.

Fig. 4-13 A boring operation on a lather. (Courtesy LeBlond Inc.)

JOB 5 Subtracting Decimals

When subtracting decimals, write the numbers in a column as for addition, lining up the decimal points in a straight vertical column.

Example 5-1 Subtract 2.36 from 4.79.

Solution The number after the word *from* is written on top. The number after the word *subtract* is written underneath.

$$\begin{array}{r} 4.79 \\ -\ 2.36 \\ \hline 2.43 \end{array} \quad \textit{Ans.}$$

Example 5-2 Subtract 1.04 from 3.

Solution The number 3 is written as 3.00 to locate the decimal point correctly.

$$\begin{array}{r} 3.00 \\ -\ 1.04 \\ \hline 1.96 \end{array} \quad \textit{Ans.}$$

Example 5-3 A $\frac{15}{32}$-in-diameter hole in a shaft was enlarged to 0.516 in diameter by the reaming operation shown in Fig. 5-1. Find the difference in the diameters.

Fig. 5-1 A reaming operation on a lathe. (Courtesy LeBlond Inc.)

Solution From Table 2-2, $\frac{15}{32}$ = 0.46875, and rounding off makes this 0.469.

$$\begin{array}{lr} \text{New diameter} & = 0.516 \\ -\ \text{Old diameter} & = \underline{0.469} \\ \text{Difference} & = 0.047 \text{ in} \end{array} \quad \textit{Ans.}$$

Self-Test 5-4 The 1973 Chevy II 8-307 engine has a cylinder bore of $3\frac{7}{8}$ in. The piston is to have 0.0025 in clearance on all sides. Find the diameter of the piston.

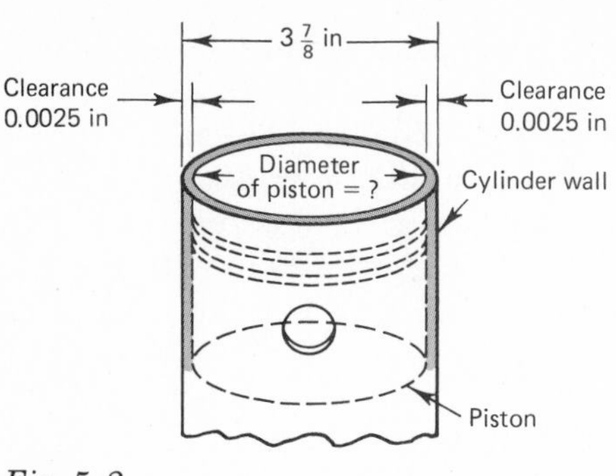

Fig. 5-2

Solution In order for the piston to move up and down easily in the cylinder, there must always be (no)(some) space between the piston and the cylinder wall. This space is called the piston _______. The clearance in this engine equals _______ in. The clearance appears on (one)(both) sides of the piston. Since the piston is (larger)(smaller) than the cylinder, the piston diameter = cylinder diameter (+)(−) both clearances.

Written as a decimal, $3\frac{7}{8}$ = ?
Two clearances = 2 × 0.0025 = ?
Subtracting, the piston diameter = ? *Ans.*

some
clearance
0.0025
both
smaller
−
3.875
0.005
3.870 in

Problems

1. 0.26 − 0.03
2. 1.36 − 0.18
3. 0.4 − 0.06
4. 0.05 − 0.004
5. 18.92 − 11.36
6. $\frac{5}{8}$ − 0.002
7. 0.627 − 0.31
8. 0.827 − 0.31
9. 3 − 0.08
10. 0.5 − 0.02
11. 6 − 0.1
12. 0.83 − $\frac{1}{2}$
13. 2.89 − 0.5
14. 12.6 − 7
15. 0.316 − 0.054
16. 14 − 8.06
17. $5\frac{1}{4}$ − 2.63
18. 3.125 − $\frac{1}{8}$

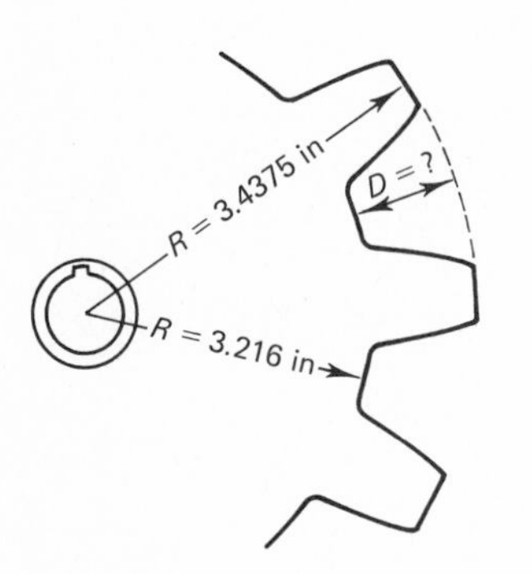

Fig. 5-3

19. (*G*) Subtract $\frac{1}{4}$ from 0.765.
20. (*G*) Find the difference between 110 and 54.9.
21. (*G*) Find the difference between (a) 0.316 and 0.012, (b) 3.006 and 1.9, (c) 0.5 and 0.11, (d) 7.07 and 1.32, and (e) 2 and 0.02.
22. (*G*) Subtract (a) 0.008 from 0.80, and (b) 0.216 from 2.16.
23. (*G*) From 2.004 subtract 1.09.
24. (*E*) What is the difference in the diameters of #1 copper wire (0.2893 in) and #7 copper wire (0.1447 in)?
25. (*B, E*) The electric meter readings on successive months were 70.08 and 76.49. Find the difference.
26. (*M*) A tapered pin has a small-end diameter of 2.125 cm and a large-end diameter of 22.7 mm. Find the difference in the diameters.
27. (*M*) Find the depth of the gear tooth shown in Fig. 5-3.
28. (*M*) A locating pin which should be 0.875 cm in diameter measures 0.918 cm in diameter. How much is it oversize?
29. (*M, A*) A hole bored 2.1875 in was reamed out to 2.1925 in. How much was its diameter enlarged?
30. (*E, C*) According to the American Wire Gage table, #8 wire has a diameter of 128.5 mil and #16 wire has a diameter of 50.82 mil. Find the difference in the diameters.
31. (*C, E, B*) The operator of the PAY loader shown in Fig. 5-4 earns \$9.87/hr. If an electrician earns \$8.95/hr, find the difference in their hourly rates of pay.

Fig. 5-4 The Hough PAY loader. (Courtesy International Harvester Company)

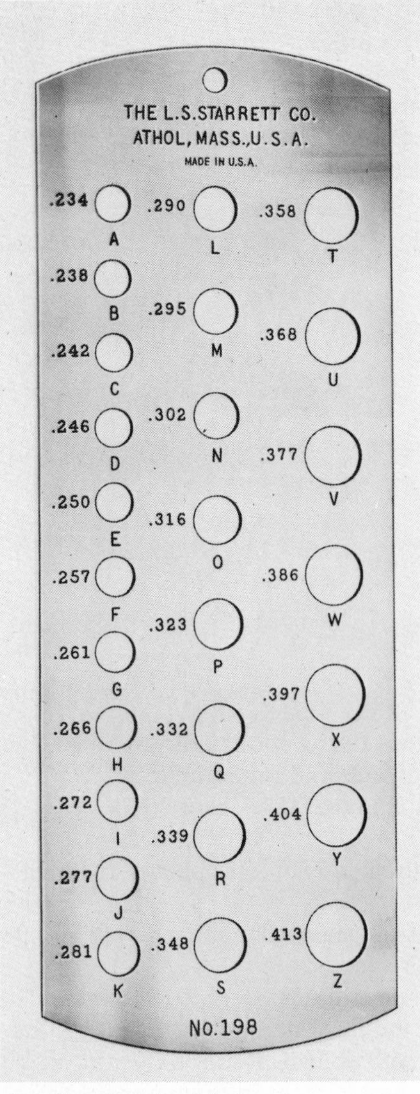

Fig. 5-5 Standard letter-size drill gage. (Courtesy The L. S. Starrett Company)

32. (*C, B*) At present, 1-in × 2-in California redwood sells for \$0.16 per running foot, while 1-in × 2-in white spruce sells for \$0.31 per running foot. Find the difference in the cost of 100 ft.
33. (*F*) A 245.8-acre farm has 219.9 acres under cultivation. How many acres are not being cultivated?
34. (*F*) In 100 lb of a fertilizer mixture commonly used on corn land are 38.5 lb of cottonseed meal, 51.4 lb of phosphoric acid, and the rest kainite (a mixture of magnesium sulphate and potassium chloride). How many pounds of kainite are used in every 100 lb of fertilizer?
35. (*C, F*) Normal 2-in lumber is actually 1.625 in thick. What thickness is lost in the dressing operation?
36. (*E*) What would be the voltage at the outlet from a 117-V line that has a voltage drop of 5.85 V from entrance panel to outlet?
37. (*A*) The original thickness of the disk brake on a car was 0.750 in. Find the permissible wear if it should be resurfaced at 0.675 in.
38. (*M, A*) Using the standard letter-size drill gage shown in Fig. 5-5, find the difference in the diameters of the following drills.
    a. E and A
    b. G and D
    c. Q and F
    d. N and F
    e. W and P
    f. Z and J
    g. X and T
    h. Y and R
    i. S and M
39. (*E*) The resistance of the windings of an electromagnet at room temperature is 28.69 Ω. If its resistance at 175 degrees Fahrenheit is 36.98 Ω, find the increase in the resistance.
40. (*G, M, A*) Using Fig. 5-6, find dimensions *A* and *B*.

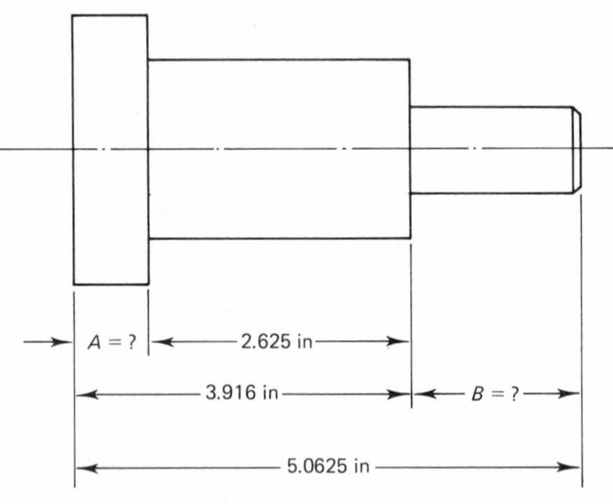

Fig. 5-6

41. (*G*) Using Table 2-2 on page 13, find, correct to the nearest thousandth, the difference between $\frac{51}{64}$ and $\frac{11}{16}$.
42. (*M*) A 10-in-long reamer is 1.2575 in in diameter at the small end and 1.5625 in at the large end. Find the total increase in diameter. Find the increase in diameter for only 1 in of the reamer's length.
43. (*A, E*) When a hydrometer is used to check a fully charged lead-acid battery, the reading should be 1.270. What is the difference between this reading and an actual reading of only 1.157?
44. (*A, M*) A cubic inch of aluminum weighs 0.0924 lb, while a cubic inch of steel weighs 0.282 lb. Find the difference in the weight of 1 cu in of each of these metals.

45. (B, A) A service station opened for business in the morning with the dials on the gas pump reading 1,457.3 gal. Seven hours later they read 2,183.2 gal. How many gallons of gas were sold?
46. (E) A 6GM6 tube used as an IF amplifier in a television receiver draws current from the 250-V tap of the power supply. If the voltage loss in the plate load is 118.6 V, what voltage is available at the plate of the tube?
47. (E) The intermediate frequency at the output of a converter stage is found by obtaining the difference between the oscillator frequency and the radio frequency. If the radio frequency is 1.1 megahertz (MHz) and the oscillator frequency is 1.555 MHz, what will be the intermediate frequency?
48. (A, M) The outside diameter of the roller bearing shown in Fig. 5-7 is $3\frac{1}{2}$ in. If the inside diameter is 1.766 in, find the thickness of the bearing.

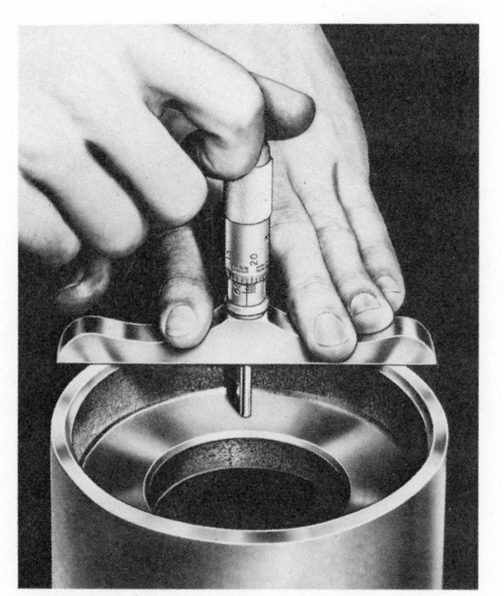

Fig. 5-8 Measuring the depth of a shoulder with the Starrett No. 445C-3RL Micrometer Depth Gage.

Fig. 5-7 A roller bearing. (Courtesy The Torrington Company)

49. (E) A circuit in a television receiver requires a 0.005-μF capacitor. How much extra capacitance connected in parallel with a 0.00035-μF capacitor will produce a total capacitance of 0.005 μF?
50. (C, G) The actual outside diameter of standard 5-in pipe is 5.563 in. If the wall thickness is 0.258 in, find the actual inside diameter.
51. (M) In Fig. 5-8, the depth of the shoulder must equal 0.625 in. How much material must be machined off the top surface if a depth micrometer measures the depth as (a) 0.652 in, (b) 0.708 in, (c) 0.623 in, (d) 0.678 in, and (e) 0.641 in?
52. (A) The cylinder in an automobile engine may wear in two different ways. (1) The cylinder may be worn to a slight oval shape due to the side thrust of the piston as it moves down the cylinder on the power stroke. This is called the *out-of-round measurement* and is made at the top of the cylinder. (2) The start of the power stroke at the top of the cylinder creates the greatest wear. This wear decreases as the piston descends, causing the cylinder to be worn into a slight taper. Thus, in Fig. 5-9,

$$\text{Out-of-round} = A - B$$
$$\text{Maximum taper} = A - C \text{ or } B - D, \text{ whichever is greater}$$

In Fig. 5-9, find (a) the out-of-round measurement, and (b) the maximum taper.
53. (G) Two rods measure 47.8 mm and 1.82 dm in length. Find the difference in their lengths.

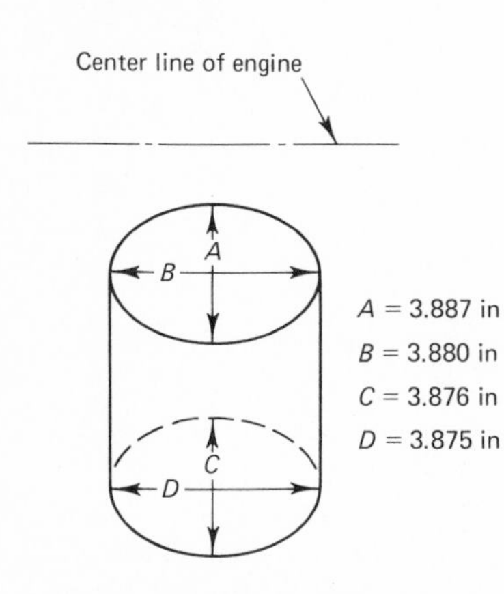

A and *C* measurements taken at right angles to the center line of the engine

B and *D* measurements taken parallel to the center line of the engine

Fig. 5-9 Measurements to determine out-of-roundness and taper in a cylinder.

54. (*A*) Piston rings are used to (1) seal the clearance between the piston and the cylinder walls and (2) prevent oil from getting up into the combustion chambers. There are two important measurements that must be taken on these rings, as shown in Fig. 5-10.

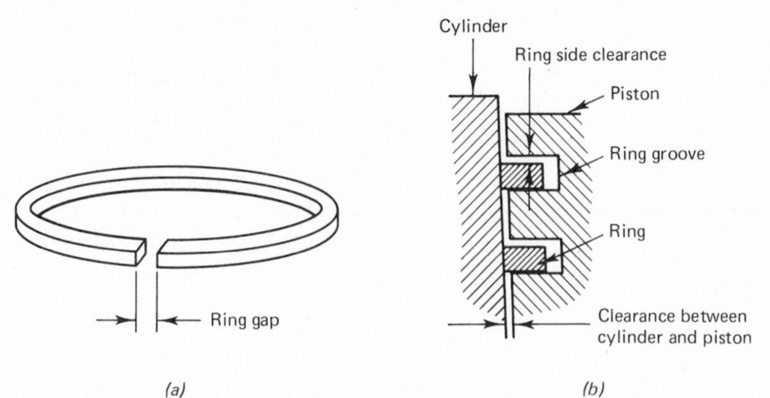

Fig. 5-10 (a) Ring gap. (b) Ring side clearance on piston rings.

a. Ring gap (Fig. 5-10*a*). If the gap is greater than the limit specified by the manufacturer, it must be discarded. If it is less than specifications, the ends should be carefully filed to 0.005 in more than the minimum measurement.
b. Ring side clearance (Fig. 5-10*b*). If the side clearance is greater than the specified limit, it means that the piston-ring groove is badly worn, and the piston must be replaced.

The ring gap data for the 1973 Cadillac Eldorado is shown in the table below. All measurements are in inches.

ENGINE	TOP COMPRESSION RING	BOTTOM COMPRESSION RING	OIL CONTROL RING
8-500	0.013–0.025	0.013–0.025	0.015–0.055

For each of the following feeler gage measurements, should the ring (a) be accepted, (b) be discarded, or (c) be filed? (d) If it should be filed, by how much?

Top compression ring: 1. 0.027 in
2. 0.018 in
3. 0.011 in

Bottom compression ring: 1. 0.022 in
2. 0.009 in
3. 0.030 in

Oil control ring: 1. 0.060 in
2. 0.011 in
3. 0.014 in

55. (*A*) The specifications for the ring side clearance on the Chevrolet Chevelle 6-250 engine are given below in inches.

YEAR	TOP COMPRESSION	BOTTOM COMPRESSION
1967	0.0020–0.0035	0.0020–0.0040
1968	0.0012–0.0027	0.0012–0.0032

If all clearances measure 0.0030 in, by how much is each clearance above or below the maximum specified value? Should any pistons be discarded? Why?

JOB 6 Dimensions and Tolerances

Mass production has been the secret of economic success in the United States. It is usually cheaper to make thousands of shafts for a machine at one time rather than making them one at a time. The factory that makes these shafts may be located in New York, but the thousands of bushings into which these shafts must fit may be manufactured in Ohio. Yet, any one of the shafts must fit with any one of the bushings. This interchangeability of parts is obtained by setting certain maximum variations from the ideal measurement.

Obviously, no mechanic can make a thousand shafts all of which are exactly 2.000 in in diameter. A small variation, *or tolerance,* is always permitted. This tolerance is the amount by which the final measurement may vary from the exact or basic dimension given. Tolerances are expressed as plus (+) or minus (−) measurements. A plus tolerance means that the actual measurement may be *more* than the basic measurement. This measurement is known as the *upper limit.* A minus tolerance means that the actual measurement may be *less* than the basic measurement. This measurement is known as the *lower limit.* The total tolerance is the sum of the plus and minus tolerance.

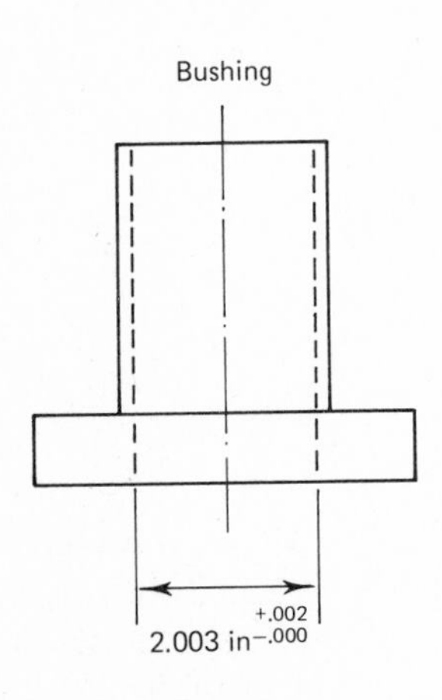

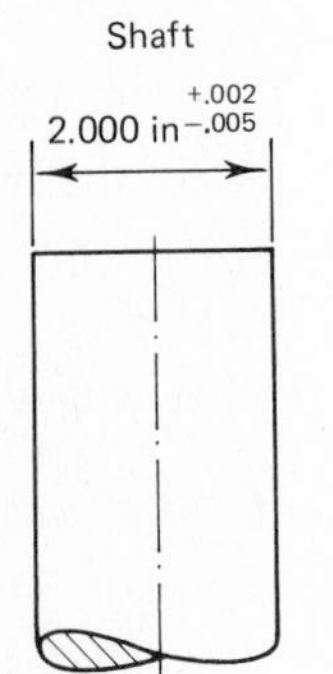

Fig. 6-1 Tolerances on mating parts.

For example, in Fig. 6-1, the basic measurement of the shaft is 2.000 in with a plus tolerance of 0.002 in and a minus tolerance of 0.005 in. This means that the largest acceptable measurement is:

2.000	(Basic dimension)
+0.002	(Plus tolerance)
2.002 in	(Maximum dimension)

The smallest acceptable measurement is:

2.000	(Basic dimension)
−0.005	(Minus tolerance)
1.995 in	(Minimum dimension)

Thus, the dimensions of the shaft may vary between a lower limit of **1.995** in and an upper limit of **2.002** in. Any measurement between these two limits will be acceptable.

The total tolerance = 0.002 + 0.005 = 0.007 in

For the bushing into which this shaft will fit:

2.003	(Basic dimension)
+0.002	(Plus tolerance)
2.005 in	(Maximum dimension)

That is, the largest acceptable measurement is 2.005 inches. The smallest acceptable measurement is:

2.003	(Basic dimension)
−0.000	(Tolerance)
2.003 in	(Minimum dimension)

Thus, the dimensions of the bushing may vary between a lower limit of **2.003** in and an upper limit of **2.005** in.

The total tolerance = **0.002** + **0.000** = **0.002** in

Note that the largest shaft (**2.002** in) will still fit into the smallest hole (**2.003** in).

Example 6-1 An inspector is checking a plug for a measurement indicated as $1.224^{\pm.005}$. Which of the following measurements is acceptable: (a) 1.215, (b) 1.230, or (c) 1.227?

Solution Upper limit / Lower limit

Upper limit		Lower limit	
1.224	1.224	(Basic dimension)	
+ .005	(Plus tolerance)	− .005	(Minus tolerance)
1.229	(Upper limit)	1.219	(Lower limit)

Therefore, any measurement which falls between 1.219 and 1.229 will be acceptable. Any other measurement will be unacceptable.

a. 1.215 is too small and unacceptable.

b. 1.230 is too large and unacceptable.

c. 1.227 falls between the acceptable limits and passes inspection.

Example 6-2 Find the upper and lower limits of the ends of the taper shown in Fig. 6-2.

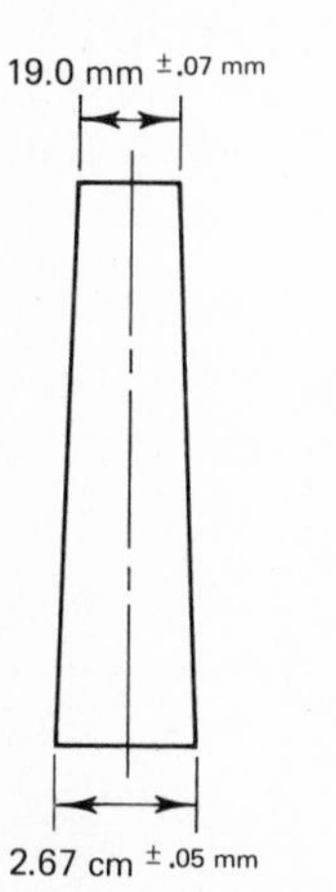

Fig.6-2 A round piece of work that gradually decreases in diameter is said to be "tapered."

Solution

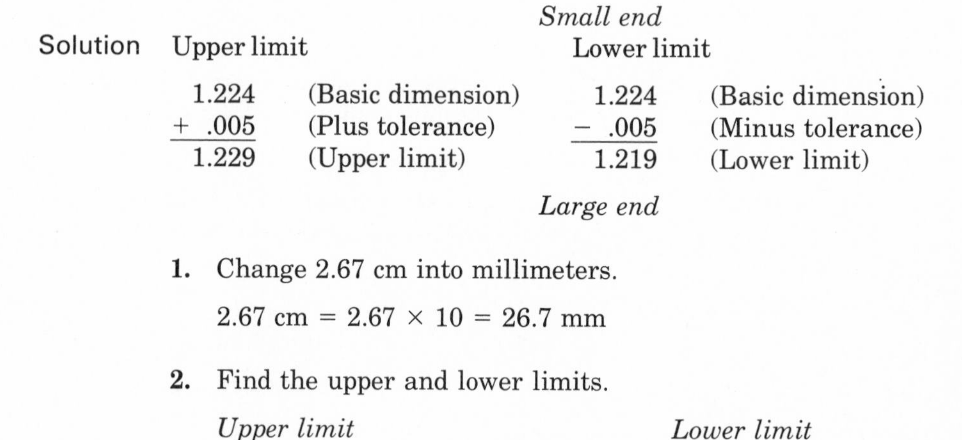

Small end

Upper limit		Lower limit	
1.224	(Basic dimension)	1.224	(Basic dimension)
+ .005	(Plus tolerance)	− .005	(Minus tolerance)
1.229	(Upper limit)	1.219	(Lower limit)

Large end

1. Change 2.67 cm into millimeters.

   2.67 cm = 2.67 × 10 = 26.7 mm

2. Find the upper and lower limits.

Upper limit		*Lower limit*	
26.7 mm	(Basic dimension)	26.7 mm	(Basic dimension)
+ .05 mm	(Plus tolerance)	− .05 mm	(Minus tolerance)
26.75 mm	(Upper limit)	26.65 mm	(Lower limit)

Example 6-3 A hole in a bushing is dimensioned as $0.528^{\pm.002}$. Which of the following pins will be certain to fit the hole *EVERY* time? (a) $0.525^{+.003}_{-.002}$ (b) $0.525^{+.000}_{-.002}$

Solution In order for a pin to fit into a hole *every* time, the *largest* pin measurement must *always* be *smaller* than the *smallest* hole.

The smallest hole = 0.528
−0.002
0.526

a. The largest pin = 0.525
+0.003
0.528

Therefore, since the largest pin (0.528) is *larger* than the smallest hole (0.526), the pin will *not* fit every time.

b. The smallest hole = 0.526
The largest pin = 0.525

$$\begin{array}{r} 0.525 \\ +0.000 \\ \hline 0.525 \end{array}$$

Therefore, since the largest pin (0.525) is *smaller* than the smallest hole (0.526), the pin will *always* fit every time.

Problems

1. (*G, M, A*) Find the upper and lower limits of a dimension indicated as $3.125^{+.003}_{-.002}$.
2. (*G, M, A*) What is the total tolerance on a dimension written as $0.8790^{+.0004}_{-.0002}$?
3. (*G, M, A*) Find the total tolerance on a piece if the limiting dimensions are 0.8788 and 0.8794 in.
4. (*G, M*) An inspector checks a part and "mikes" it at 1.345 in. If the drawing calls for a dimension $1.350^{\pm.004}$ in, will the part be accepted? Why?
5. (*G, M, A*) In the mating parts shown in Fig. 6-3, find the upper and lower limits for the dimensions *A, B, C, D, E, F, G,* and *H*.

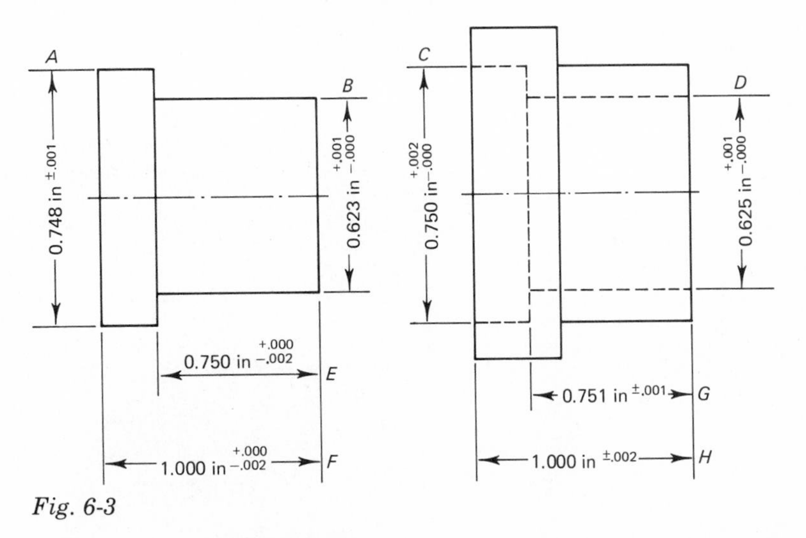

Fig. 6-3

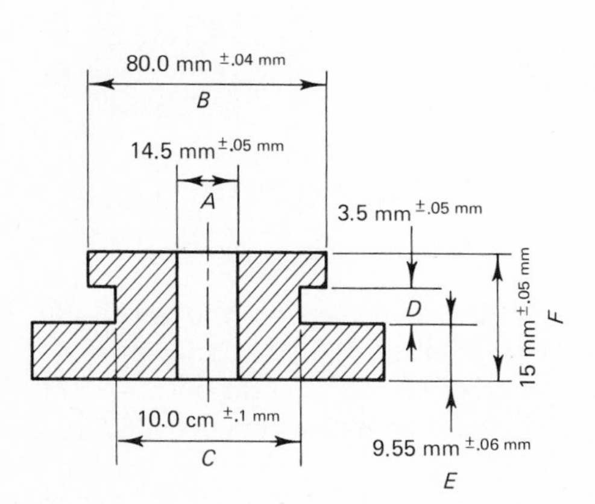

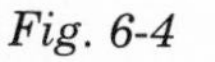
Fig. 6-4

6. (*G, M, A*) In Fig. 6-4, find the upper and lower limits for the dimensions *A, B, C, D, E,* and *F*.
7. (*M*) A shaft measuring 1.415 in in diameter is to be turned to $1.375^{+.000}_{-.004}$ in. What is the deepest cut that can be taken?
8. (*M*) A hole in a bushing is dimensioned as $0.625^{\pm.003}$ in. Which of the following pins will be certain to fit the hole *every* time: (a) $0.622^{+.005}_{-.000}$ in, (b) $0.630^{+.002}_{-.005}$ in?
9. (*G, M*) The diameter of a shaft is $0.875^{+.000}_{-.001}$ in. The diameter of the bearing in which the shaft is to fit is marked $0.876^{+.001}_{-.000}$ in. (a) What are the permissible dimensions for each part to give the greatest looseness of fit? (b) What are the permissible dimensions for each part to give the greatest tightness of fit?

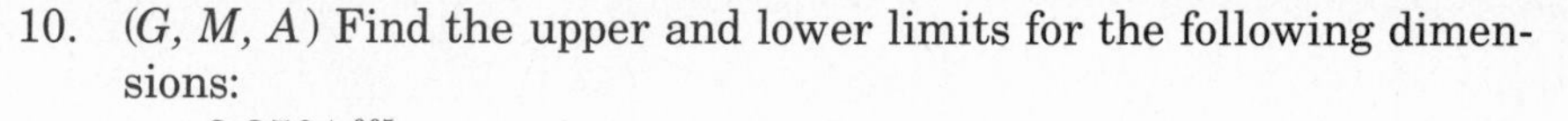

10. (*G, M, A*) Find the upper and lower limits for the following dimensions:
    a. $2.250^{\pm.005}$
    b. $1.250^{+.002}_{-.000}$
    c. $0.565^{+.003}_{-.001}$
    d. $1.500^{+.002}_{-.001}$
    e. $1.0000^{\pm.0002}$
    f. $1.475^{+.004}_{-.000}$
    g. $1.375^{+.003}_{-.005}$
    h. $1.2300^{+.0025}_{-.0000}$
    i. $1.0750^{+.0001}_{-.0004}$
    j. $0.7150^{+.0015}_{-.0000}$

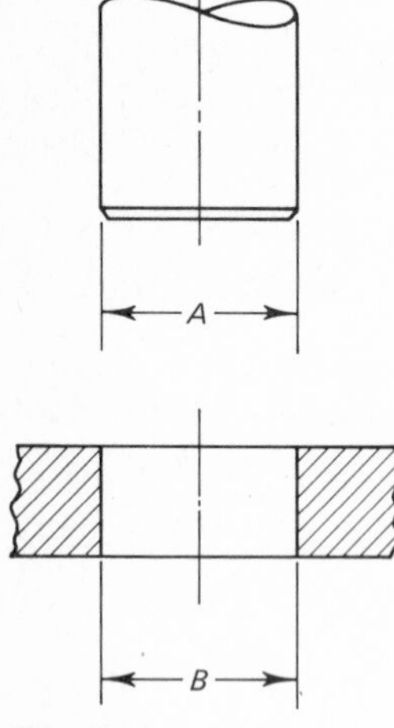

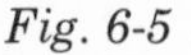
Fig. 6-5

11. (*G, M*) In Fig. 6-5, a pin with a diameter *A* is to fit into a hole with a diameter *B*. For each of the following, will the parts fit *every* time?

PROBLEM	a.	b.	c.	d.	e.
Pin diameter (*A*)	$1.248^{+.000}_{-.005}$	$0.995^{+.002}_{-.001}$	$0.498^{+.000}_{-.001}$	$0.372^{\pm.001}$	$1.074^{+.002}_{-.001}$
Hole diameter (*B*)	$1.250^{+.005}_{-.000}$	$1.000^{\pm.002}$	$0.502^{\pm.003}$	$0.375^{+.002}_{-.001}$	$1.075^{\pm.003}$

12. (*A, M*) The crankshaft journal shown in Fig. 6-6 is "miked" at 1.876 in. If the dimension of the part is $1.875^{+.002}_{-.003}$ in, is the part acceptable? Why?

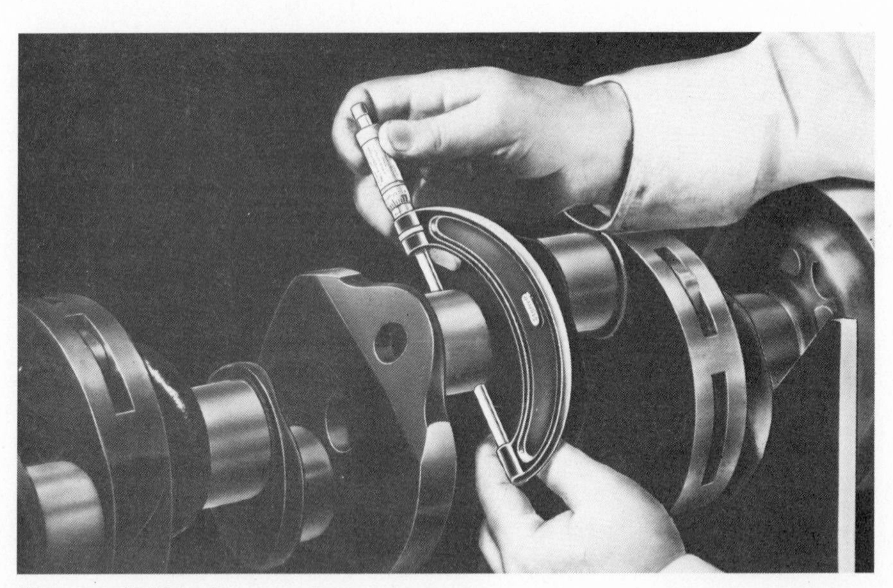

Fig. 6-6 "Miking" a crankshaft journal. (Courtesy The L. S. Starrett Company)

13. (*A*) The recommended piston-to-bore clearance on the 1973-430 Corvette engine is 0.0058 to 0.0080 in. Find the total tolerance.
14. (*A*) The dimension for the intake valve stem diameter on the 1971 Comet engine should be $0.3395^{\pm.0015}$ in. Is a measurement of 0.3408 in acceptable?
15. (*A*) A diesel cylinder block was rebored to fit a new $4.625^{+.000}_{-.001}$ liner. If the block was bored to 4.626 in, find the minimum clearance on each side.

JOB 7 Gage Blocks

Gage blocks are used when extreme accuracy is required, both for direct measurements and to check the accuracy of micrometers and other precision measuring tools. A gage block is a rectangular block of special alloy steel. The blocks are carefully hardened and their opposite surfaces ground and polished to an extremely high degree. The surfaces will actually stick to each other when one block is slid over another.

The blocks are accurate to at least 8 millionths of an inch. Because they may be *wrung* together with no lost space between them, they may be put together in many combinations to produce almost any desired dimension. For example, when the blocks shown in Fig. 7-1*a* have been wrung together, the distance between the parallel faces is equal to exactly the sum of the blocks, as shown in Fig. 7-1*b*.

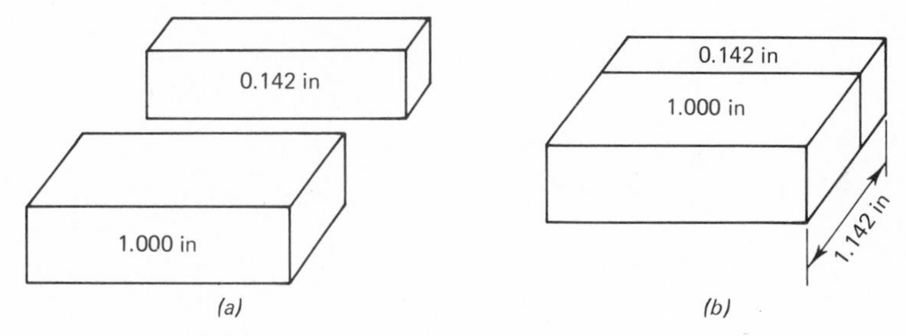

Fig. 7-1 Gage blocks are accurate to 8 millionths of an inch.

The complete set of blocks is shown in Fig. 7-2. However, the most commonly used set contains 81 blocks. Practically any 4-place decimal measurement may be produced by wringing together the appropriate blocks.

Fig. 7-2 A complete set of gage blocks. (Courtesy the L. S. Starrett Company)

The sizes available in an 81-block set are listed in Table 7-1.

Table 7-1 Sizes Available in an 81-Block Gage Set (in)

0.1001	0.1002	0.1003	0.1004	0.1005	0.1006	0.1007	0.1008	0.1009	
0.101	0.102	0.103	0.104	0.105	0.106	0.107	0.108	0.109	0.110
0.111	0.112	0.113	0.114	0.115	0.116	0.117	0.118	0.119	0.120
0.121	0.122	0.123	0.124	0.125	0.126	0.127	0.128	0.129	0.130
0.131	0.132	0.133	0.134	0.135	0.136	0.137	0.138	0.139	0.140
0.141	0.142	0.143	0.144	0.145	0.146	0.147	0.148	0.149	
0.050	0.100	0.150	0.200	0.250	0.300	0.350	0.400	0.450	0.500
0.550	0.600	0.650	0.700	0.750	0.800	0.850	0.900	0.950	
		1.000	2.000	3.000	4.000				

Three basic principles should be followed when setting up a dimension using these blocks:

1. Always use as few blocks as possible.

2. The first block selected should have a digit in the last place which is the same as the digit in the last place of the desired dimension.

3. Continue selecting blocks, eliminating the digits of the dimension as you work from right to left.

Example 7-1 Set up the dimension 2.7253 in.

Solution

	Dimension	Blocks used
1. Write the dimension.	2.7253	
2. Eliminate the 3 in the fourth place by using a block which has a 3 in the fourth place.	0.1003	0.1003
3. Subtract.	2.6250	
4. We can eliminate both the 5 and the 2 by using the block.	0.125	0.125
5. Subtract.	2.500	
6. Eliminate the 0.5.	0.500	0.500
7. Subtract.	2.000	
8. Eliminate the 2.000.	2.000	2.000
9. Subtract.	0.000	
10. Check by adding the blocks used.		2.7253

Example 7-2 Set up the dimension 1.3827 in.

Solution

	Dimension	Blocks used
1. Write the dimension.	1.3827	
2. Eliminate the 7 in the last place.	0.1007	0.1007
3. Subtract.	1.2820	
4. Eliminate the 2 in the last place.	0.142	0.142
5. Subtract.	1.140	
6. Eliminate the .14.	0.140	0.140
7. Subtract.	1.000	
8. Eliminate the 1.000.	1.000	1.000
9. Subtract.	0.000	
10. Check by adding the blocks used.		1.3827

Problems

Set up blocks for the following dimensions:

1.	0.9752	7.	0.8843	13.	1.2560
2.	0.3006	8.	0.2933	14.	1.1357
3.	2.3479	9.	2.6781	15.	2.7654
4.	2.0973	10.	3.9753	16.	2.2005
5.	1.0543	11.	2.0214	17.	2.0044
6.	0.5128	12.	0.6666	18.	0.9734

JOB 8 Review of Jobs 1 to 7

1. Decimal fractions are fractions whose (numerators)(denominators) are numbers like 10, 100, or 1,000.

   denominators

2. The denominator is shown by the number of digits to the (left)(right) of the decimal point. Each digit represents a(n) ______ in the denominator of the fraction.

   right
   zero

3. 0.4 is read as 4 ______.
   0.009 is read as 9 ______.
   Fifty-three thousandths is written as ______.

   tenths
   thousandths
   0.053

4. Decimals can be compared only when they have the ______ number of decimal places.

   same

5. The word "and" in a mixed number is written as a(n) ______ in the decimal system.

   point

6. Fractions may be changed to decimals by dividing the ______ by the ______.

   numerator, denominator

7. To round off a decimal:
   a. Break the decimal at the required number of places.
   b. If the digit to the right of the break is less than 5, (keep)(drop) all digits to the right of the break.

      drop

   c. If the digit to the right of the break is more than ______, (drop)(keep) all digits to the right of the break and (add)(subtract) 1 to (from) the digit immediately to the left of the break.

      5, drop
      add

   d. If the digit to the right of the break is a 5 followed by zeros, the digit to the left of the break will
      1. Remain unchanged if it is ______.
      2. Be increased by 1 if it is ______.

      even
      odd

8. Metric units.

a. A centimeter is (larger)(smaller) than a kilometer. — smaller

b. A decimeter is (larger)(smaller) than a dekameter. — smaller

c. 1 cm = _______ mm. — 10

d. 1 dm = _______ mm. — 100

e. 1 dam = _______ m. — 10

f. According to the chart shown in Fig. 3-3, when we change from a large measurement to a small measurement, we should (multiply) (divide). — multiply

g. When we change from a small measurement to a large measurement, we should (multiply) (divide). — divide

h. To change meters to millimeters, we should _______ by _______. — multiply, 1,000

i. To change millimeters to centimeters, we should _______ by _______. — divide, 10

j. To change centimeters to meters, we should _______ by _______. — divide, 100

9. To multiply by numbers like 10, or 100, we should move the decimal point 1 place to the (left) (right) for every _______ in the multiplier. — right, zero

10. To divide by numbers like 10, or 100, we should move the decimal point 1 place to the (left)(right) for every zero in the multiplier. — left

11. When adding or subtracting decimals, both the digits and the _______ _______ should be lined up in straight vertical columns. — decimal points

12. Tolerance:
The upper limit of a dimension is obtained by (adding)(subtracting) the tolerance (to)(from) the basic dimension. — adding, to
The lower limit of a dimension is obtained by (adding)(subtracting) the tolerance (to)(from) the basic dimension. — subtracting, from
The total tolerance on the dimension $0.455^{+.002}_{-.003}$ is equal to _______. — 0.005

13. Gage blocks.
Set up the dimension 0.3848.

Use the block	?	0.1008
Subtract.	0.2840	
Use the block	?	0.134
Subtract.	0.1500	
Use the block	?	0.150
Subtract	0.0000	

Example 8-1 A surface plate 7.25 cm thick is to be refinished. A cut 1.5 mm deep and a second cut 0.65 mm deep are taken over the plate. Find the finished thickness of the plate.

Solution **1.** The total thickness removed is equal to the sum of the two cuts.

$$\begin{array}{r} 1.5\ \text{mm} \\ +0.65\ \text{mm} \\ \hline 2.15\ \text{mm} \end{array}$$

2. The finished thickness is equal to the original thickness minus the thickness removed.

7.25 cm =	72.50 mm	(original thickness)
	− 2.15 mm	(thickness removed)
Finished thickness =	70.35 mm	*Ans.*

Example 8-2 A $1\frac{7}{16}$-in-diameter shaft is to be turned down in a lathe. How deep a cut is required to bring the diameter down to 1.374 in?

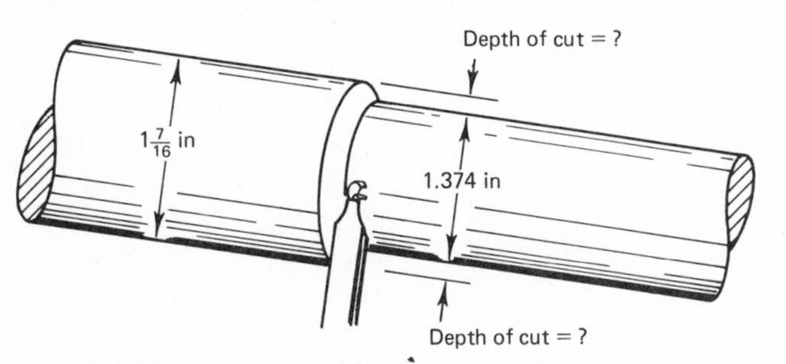

Fig. 8-1 Turning down the diameter of a steel shaft.

Solution **1.** The amount of metal to be removed is the difference between the original diameter ($1\frac{7}{16}$) and the final diameter (1.374). From Table 2-2, $1\frac{7}{16}$ = 1.4375 = 1.438.

Original diameter =	1.438 in
− Final diameter =	1.374 in
Metal removed =	0.064 in

2. But, since a cut on one side of the shaft will remove an equal amount on the other side, the depth of the cut will need only be $\frac{1}{2} \times 0.064$ = 0.032 in. *Ans.*

Self-Test 8-3 An engine cylinder is $3\frac{5}{8}$ in in diameter. The piston is to have 0.0035 in clearance on all sides. Find the diameter of the piston.

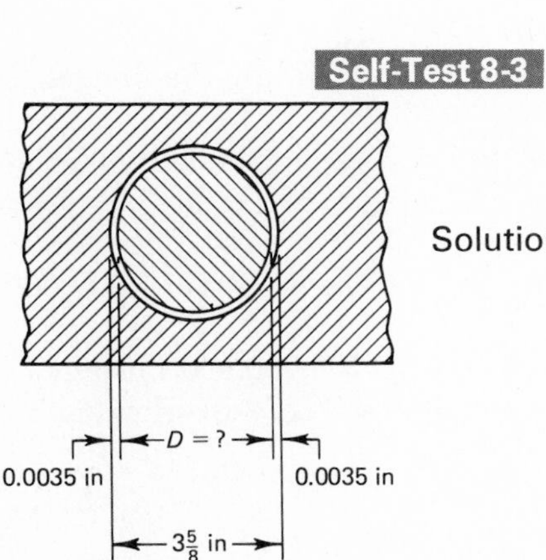

Fig. 8-2 Cross section of a piston.

Solution **1.** Looking at a cross section of the piston, the clearance of 0.0035 appears on (one)(both) sides of the piston. Therefore, the total clearance equals _______ × 0.0035 or _______ in.

both

2, 0.007

2. Diameter of piston = cylinder diameter (+)(−) total clearance

−

Diameter of piston = $3\frac{5}{8}$ − _______

= _______ − 0.007

= _______ in *Ans.*

0.007
3.625
3.618

Problems

1. Write the following fractions as decimal fractions:
   a. $\frac{17}{1,000}$
   b. $\frac{245}{1,000}$
   c. $\frac{75}{100}$
   d. $\frac{9}{10}$
   e. $\frac{3}{1,000}$
2. Change the following mixed numbers to decimals:
   a. $3\frac{9}{100}$
   b. $5\frac{7}{1,000}$
   c. $15\frac{56}{100}$
   d. $1\frac{34}{1,000}$
3. Change each of the following fractions to a decimal, rounded off as directed:
   a. $\frac{5}{8}$ to the nearest hundredth
   b. $\frac{15}{32}$ to the nearest hundredth
   c. $\frac{3}{16}$ to the nearest thousandth
   d. $\frac{1}{3}$ to the nearest thousandth
   e. $\frac{5}{64}$ to the nearest thousandth
   f. $\frac{9}{32}$ to the nearest thousandth
   g. $\frac{5}{9}$ to the nearest thousandth
   h. $\frac{13}{24}$ to the nearest hundredth
4. (*E*) An electric generator delivers 223.8 V. If 4.95 V is lost in the line wires, find the voltage delivered at the end of the line.
5. (*M*) Find the depth *D* of the flat that has been ground on the shaft shown in Fig. 8-3.

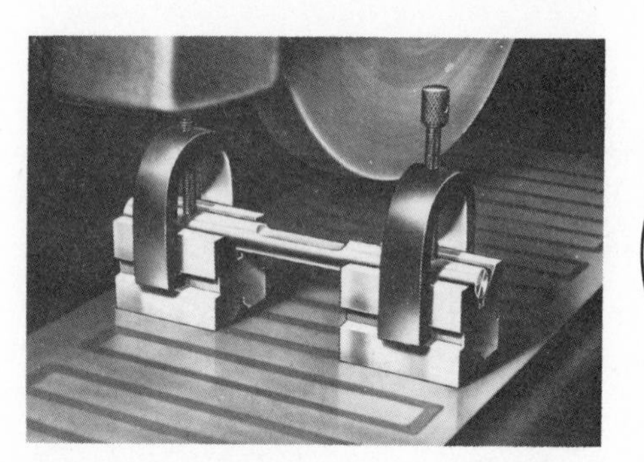

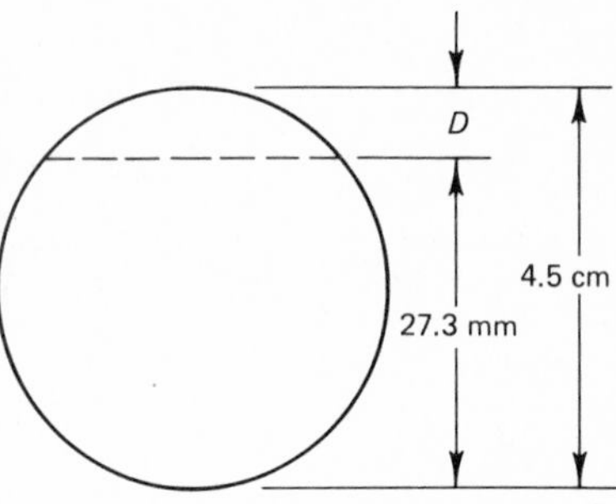

Fig. 8-3 Grinding a flat on a shaft. (Courtesy The L. S. Starrett Company)

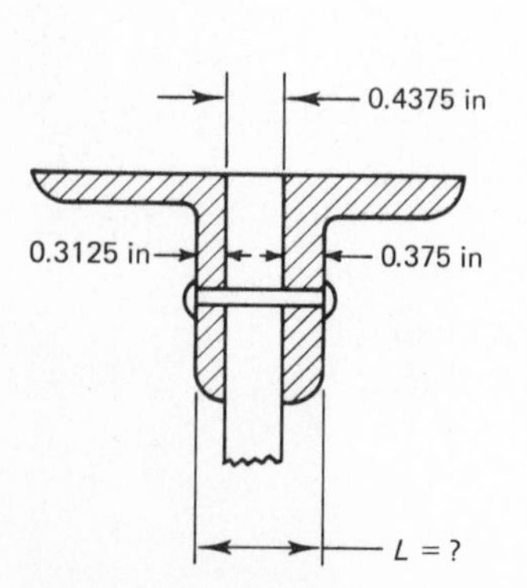

Fig. 8-4 Find the length under the heads of the rivet.

6. (*C*) In Fig. 8-4, find the length *L* of the rivet holding the flanges to the plate.
7. (*A*) The 1968 Cadillac engine 8-472 has a recommended top compression ring side clearance between 0.0022 in and 0.0035 in. Find the total permissible tolerance.
8. (*F*) If a bushel of wheat weighs 60.5 lb, find the weight of 1,000,000 bushels.
9. (*F*) In a recent year in the United States, 700,000,000 bushels of wheat were harvested by 100,000 combines (Fig. 8-5). How many bushels were harvested per combine?
10. (*G*) Change:
    a. 185 mm to centimeters
    b. 0.2 m to millimeters
    c. 50 dam to kilometers
    d. 0.8 cm to millimeters

Fig. 8-5 International 915 combine. (Courtesy International Harvester Company)

11. (*G*) Multiply:
    a. 0.865 × 1,000
    b. 92.6 × 1,000,000
    c. 0.0035 × 1,000
    d. 0.0005 × 100
12. (*G*) Divide:
    a. 0.09 ÷ 100
    b. 8.2 ÷ 1,000
    c. 4,683 ÷ 100
    d. 386 ÷ 1,000,000
13. (*M*) Set up gage blocks for the following dimensions:
    a. 0.5711 in
    b. 2.7962 in
14. (*A, M*) What will be the new bore diameter after a $3\frac{5}{8}$-in-diameter cylinder is rebored 0.038 in oversize?

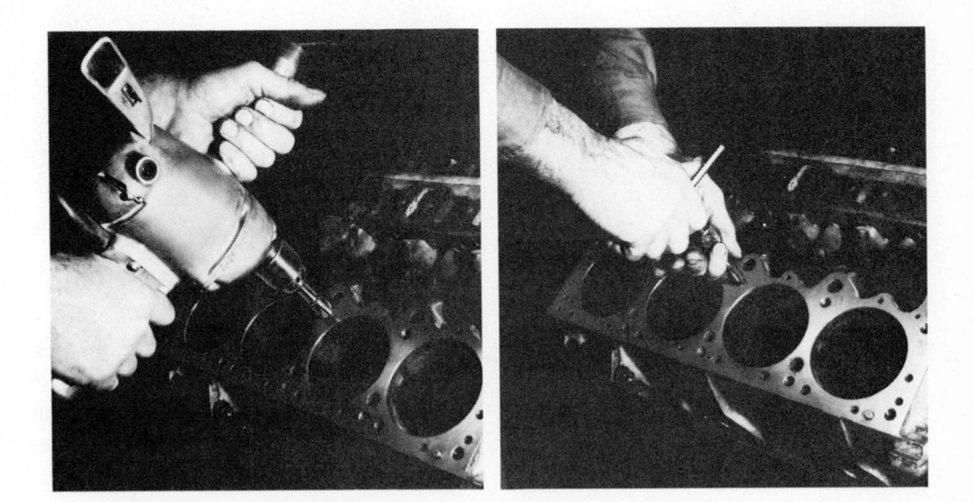

Fig. 8-6 After reboring the cylinders, the bolt holes are chamfered and retapped before a new gasket is installed. (Courtesy Fel-Pro Inc.)

15. (*B, E*) An electrician figured the cost to rewind a motor as follows: #17 wire at $1.75, #28 wire at $0.95, #00 top sticks at $0.20, armalac at $0.83, and 2 hr of labor at $7.75/hour. Find the total cost.

16. (*G*) In Fig. 8-7, read the distances indicated in centimeters and millimeters.

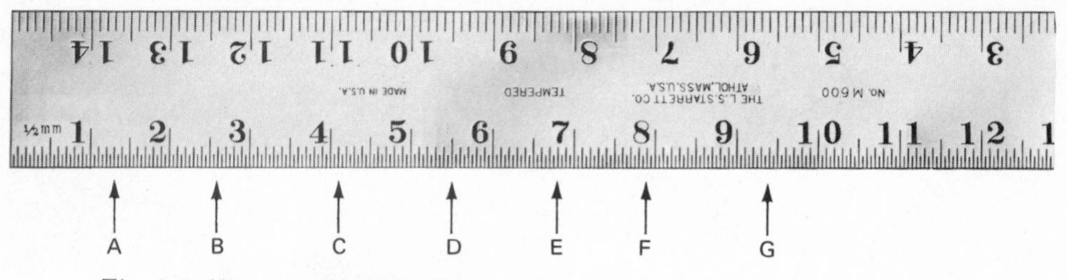

Fig. 8-7 (Courtesy The L.S. Starrett Company)

17. (*C*) If the bucket capacity of the Melroe Bobcat loader shown in Fig. 8-8 is 1.5 tons, how many pounds of rock can be stockpiled in 100 trips? (1 ton = 2,000 lb)

Fig. 8-8 Stockpiling landscaping rock. (Courtesy Clark Equipment Company)

18. (*M*) A piece of work 1.6875 in in diameter is being turned in a lathe. What will be the diameter after a cut 0.03125 in deep is taken?
19. (*M*) A milling machine arbor is 3.25 in long. What length of arbor remains uncovered after the following collars are placed on the arbor: 0.75 in, 0.625 in, 0.3125 in, and 0.4375 in?
20. (*E*) A voltage divider in a small radio is essentially as shown in Fig. 8-9. Find the total resistance by adding the resistances. Remember that the prefix "k" means "kilo," or 1,000.
21. (*F, C*) If 1 hp = 33,000 ft-lb of work per minute, how many foot-pounds of work will be delivered by a 100-hp tractor in 100 min?
22. (*A*) The cylinder lining in an engine has an inside diameter of 5.125 in. The lining should be rebored when it is worn to an ID larger than 5.130 in. Would you rebore the cylinder if the diameter were worn to 5.132 in? Why?
23. (*F*) A sample of milk was found to contain 3.78 percent protein, 3.58 percent fat, 4.83 percent lactose, 9.79 percent ash, and water. What percent was water?
24. (*M*) A 10.5-cm-diameter piece of work is turned down in a lathe in three cuts. The first cut takes off 40 mm, the second takes off 2.4 mm, and the third takes off 1.7 mm. What is the finished diameter of the work?

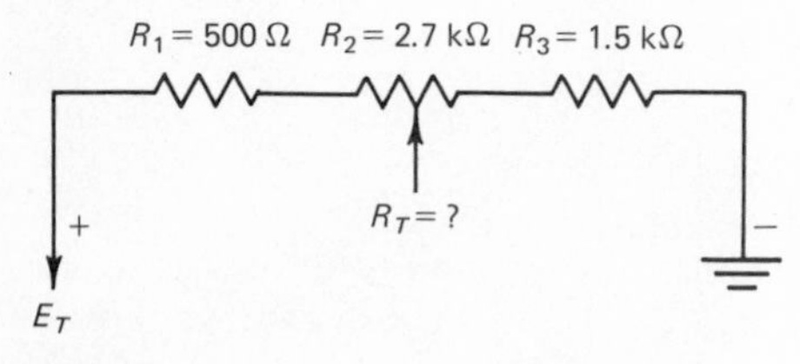

Fig. 8-9

Fig. 8-10 Clark planetary axles give this tractor added drawbar pull. (Courtesy Clark Equipment Company)

25. (*G, M*) Using Fig. 8-11, find the missing dimensions *A* and *B*.

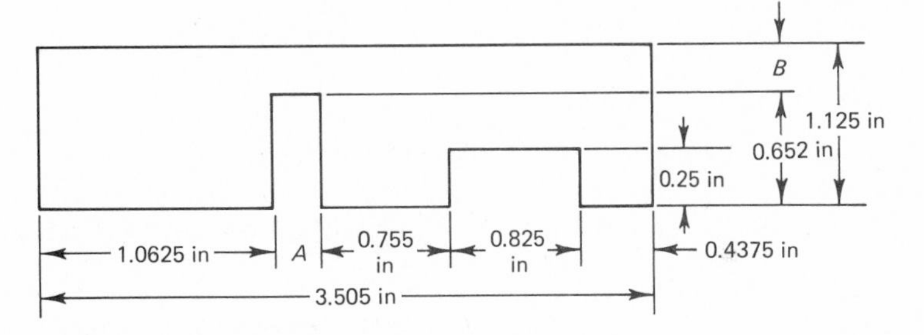

Fig. 8-11

26. (*M*) A machinist used a set of Jo blocks to check a measurement of 0.4955 in. She has already wrung together the blocks measuring 0.250 in and 0.145 in. What additional block is needed to total 0.4955 in?
27. (*A*) An engine cylinder has a diameter of 4.125 in. Find the clearance between the cylinder and a 4.120-in-diameter piston.
28. (*F, B*) Salt costs \$0.006/lb. Find the cost per year for salt for 2,000 head of range cattle if each animal needs 20 lb/yr.
29. (*M*) Find the depth of cut required to mill a square end on the round shaft shown in Fig. 8-12.
30. (*E*) The heaters of the tubes shown in Fig. 8-13 are connected in series. Find the total voltage used by all 5 tubes by adding the individual voltages. If they are all supplied from a 117-V outlet (E_T), how many volts are used by the dropping resistor (E_R)?

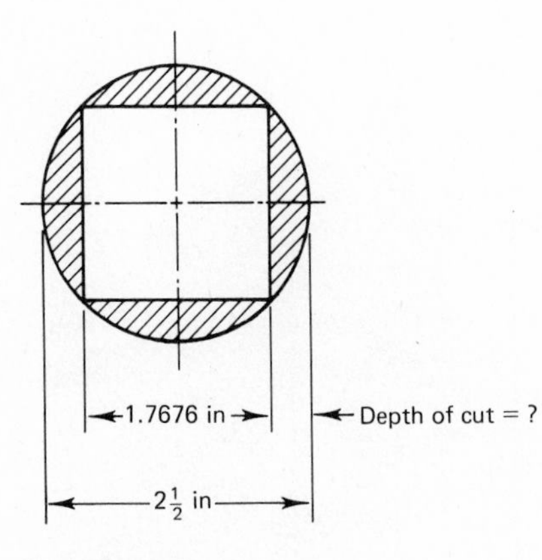

Fig. 8-12

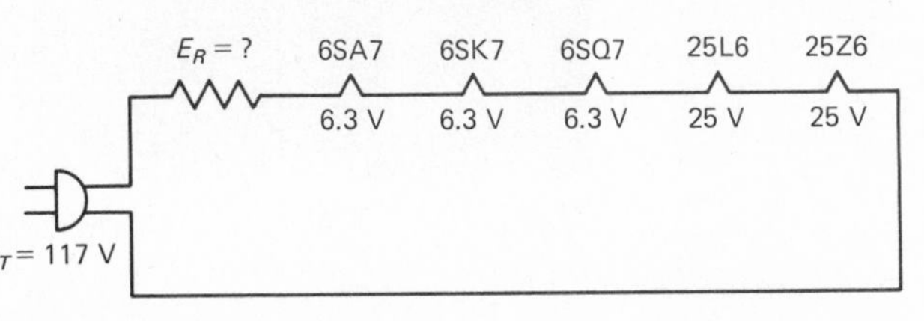

Fig. 8-13

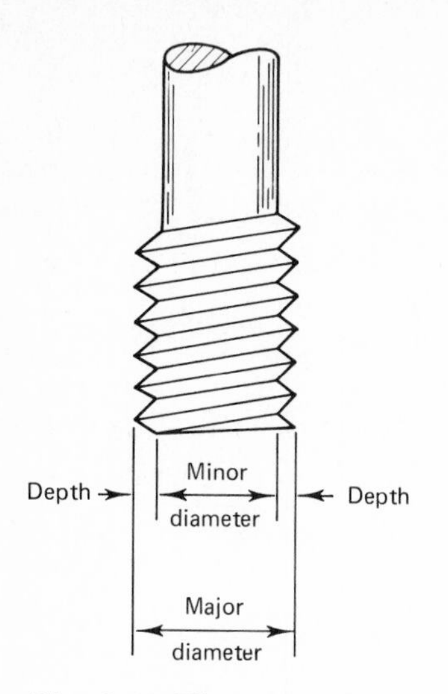

Fig. 8-14 The minor diameter is equal to the major diameter minus twice the depth of the thread.

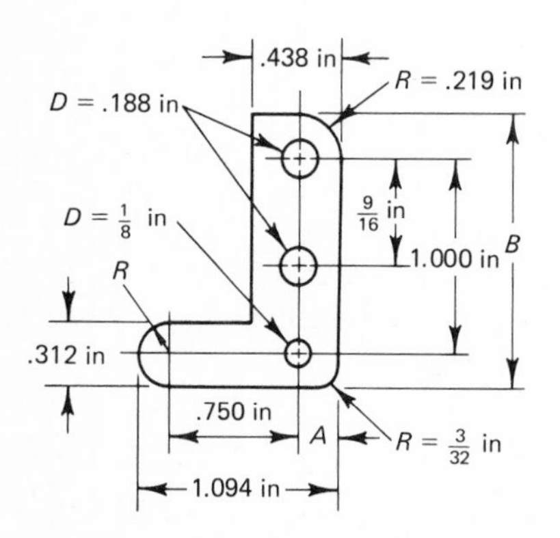

Fig. 8-15

31. (*E, A, M*) The diameter of the commutator of a motor armature was 3.125 in. It was turned down in a lathe to remove an imperfection. If a cut 0.017 in deep was taken, find the finished diameter.
32. (*M*) In Fig. 8-14, find the minor diameter of a $\frac{3}{8}''$-16 sharp V thread if the major diameter is 0.375 in and the depth is 0.0541 in.
33. (*G*) Find the missing dimensions *A* and *B* in the part shown in Fig. 8-15.
34. (*A, G*) A car was customized as follows: the upper and lower steel steering control arms weighing a total of 7.58 lb were replaced with aluminum arms weighing a total of 4.69 lb; a steel manifold weighing 32.84 lb was replaced with an aluminum alloy manifold weighing 16.29 lb; and decorative accessories weighing 12.57 lb were added to the car. Does the car weigh more or less than originally, and by how much?
35. (*M, G*) Find the missing dimensions *X* and *Y* in the lathe slide shown in Fig. 8-16.

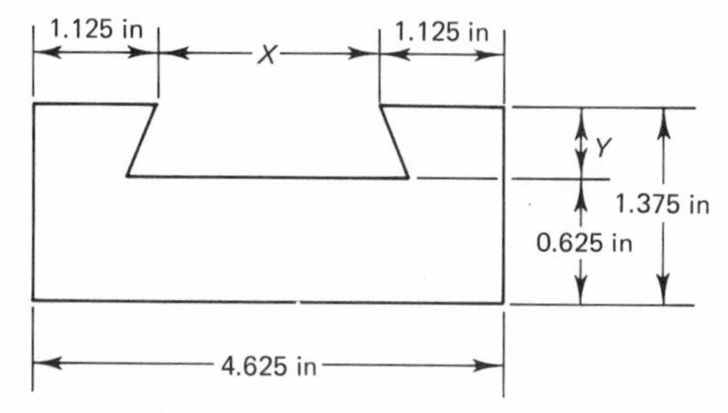

Fig. 8-16 A lathe slide.

36. (*M, A, G*) A bushing has an outside diameter of 2.813 in and an inside diameter of 2.375 in. Find the thickness of the wall.
37. (*M*) The inside diameter of the piece shown in Fig. 8-17 is 12.5 cm. What will be the new ID after a cut 2.5 mm deep has been taken?

Fig. 8-17 A boring operation on a lathe. (Courtesy LeBlond Inc.)

(See Answer Key for Test 2—Review of Jobs 1 to 7)

JOB 9 Multiplying Decimals

Decimals are multiplied in exactly the same way that ordinary numbers are multiplied. However, in addition to the normal multiplication, the decimal point must be correctly placed in the answer.

RULE **The number of decimal places in a product is equal to the sum of the number of decimal places in the numbers being multiplied.**

Example 9-1 Multiply 0.62 by 0.3.

Solution

0.62	(multiplicand—2 places)	
× 0.3	(multiplier—1 place)	
0.186	(product = 2 + 1 = 3 places)	*Ans.*

Example 9-2 Multiply 0.35 by 0.004.

Solution

0.35	(multiplicand—2 places)	
× 0.004	(multiplier—3 places)	
0.00140	(product = 2 + 3 = 5 places)	*Ans.*

In this problem, extra zeros must be inserted between the decimal point and the digits of the answer to make up the required number of decimal places.

Self-Test 9-3 A carpenter describes the area of 100 sq ft as 1 *square.* If it takes 3.3 hr to finish flooring 1 square, (a) how long will it take to install a floor with an area of 870 sq ft? (b) At $7.50 per hr, what will be the total cost?

Solution

a. **1.** The number of squares (of 100 sq ft each) in 870 sq ft is obtained by (multiplying)(dividing) 870 by ______.
Thus, 870 ÷ 100 = ______ squares.

dividing
100
8.7

2. At 3.3 hr per square, 8.7 squares require 8.7 × ______ hr.
When 8.7 is multiplied by 3.3, the number of digits after the decimal point is ______.
If 8.7 × 3.3 = 2871, then the decimal point should be placed after the digit ______.
The answer is ______ hr.

3.3
2
8
28.71

b. At $7.50 per hr, the total cost = 7.50 × ______
The total cost = $______ *Ans.*

28.71
215.33
(Money values are always rounded up.)

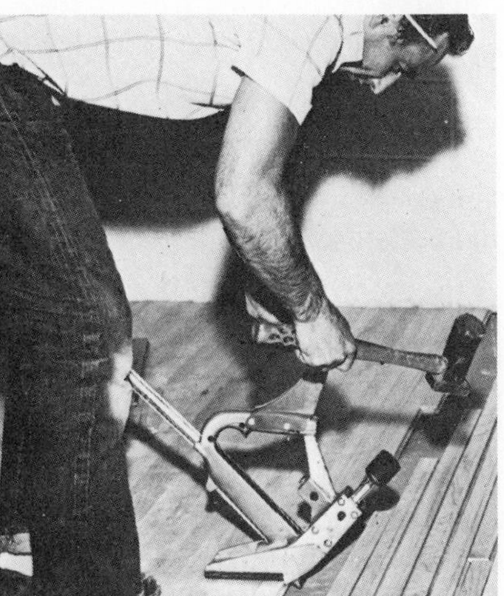

Fig. 9-1 Installing a floor with a Rockwell Portanailer. (Courtesy Rockwell International)

Self-Test 9-4 What is the total material cost for a job that uses 1,475 ft of two-conductor BX cable at $0.035/ft and 32 boxes at $0.082 each?

Solution

The price of $0.035/ft means that the cable costs $0.035 for (how many) ft? For 2 ft of this cable, the cost is ______ × $0.035. For 5 ft of this cable, the cost is 5 × ______. For 1,475 ft of this cable, the cost is 1,475 × ______.	1 2 $0.035 $0.035
When 1,475 is multiplied by 0.035, the number of digits after the decimal point is ______. If 1,475 × 0.035 = 51,625, then the decimal point should be placed after the digit ______. The answer $51.625 should be rounded off to ______. The cost for 32 boxes at $0.082 each means that the total cost for the boxes is 32 × ______. 32 × 0.082 = ______. Since this represents dollars, $2.624 should be rounded off to $______.	3 1 $51.63 $0.082, 2.624 2.62
The total material cost is obtained by adding $51.63 and ______.	$2.62
When adding decimals, we must remember to keep the decimal points in a(n) ______ line.	vertical
$51.63 + $ 2.62 $______ *Ans.* ← Lined up decimal points	$54.25

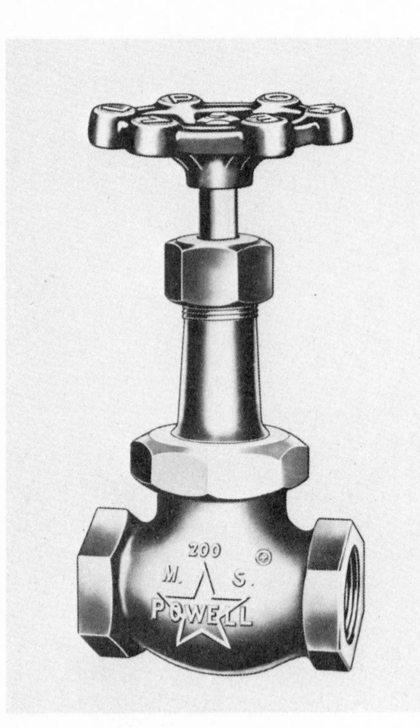

Fig. 9-2 A globe valve. (Courtesy The William Powell Company)

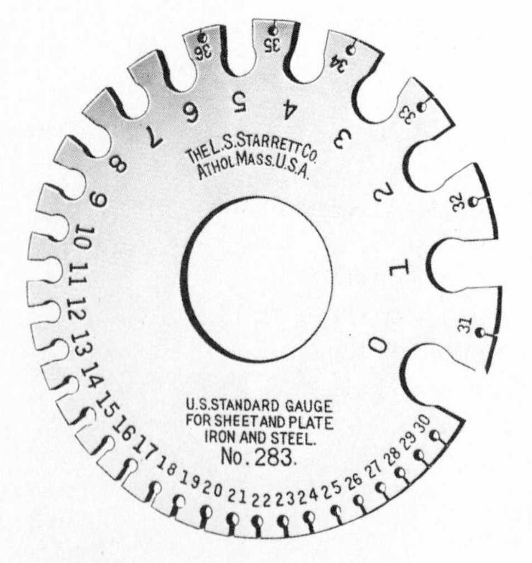

Fig. 9-3 U.S. Standard Gauge for sheet and plate iron and steel. (Courtesy The L.S. Starrett Company)

Problems

1. 0.005 × 82
2. 1.732 × 40
3. 1.13 × 0.41
4. 0.9 × 0.09
5. 0.866 × 35
6. 44.6 × 805
7. 7.63 × 0.029
8. 0.354 × 0.008
9. 6.2 × 0.003
10. 0.033 × 0.0025
11. 106 × 0.045
12. 73.8 × 1.09
13. (*E*) If a 100-W lamp uses 0.91 A, how much current would be used by 5 such lamps connected in parallel?
14. (*B, C*) If seed costs $0.35/lb and a lawn requires 1 lb/1,000 sq ft, what is the cost of seeding a lawn with an area of 7,500 sq ft?
15. (*B, C*) What is the total cost for 2,000 bricks at $0.095 per brick if the delivery charge is an additional $12.50?
16. (*C*) The bronze globe valve shown in Fig. 9-2 weighs 4.85 lb. Find the weight of 12 of these valves.
17. (*B*) A piece worker averages 192 items per day. If she is paid at the rate of $0.13 per item, how much does she earn in 5 days?
18. (*F, B*) At $0.036/lb, what are the freight charges to ship 500 lb of fertilizer?
19. (*M*) The screw thread in a micrometer permits the spindle to advance 0.025 in in one complete turn of the thimble. How far will it advance in 11 complete turns?
20. (*C, A*) The gage shown in Fig. 9-3 measures No. 8 gage sheet steel as 0.1285 in thick. Find the thickness of a pile of 32 of these sheets.
21. (*C, B*) A contractor bought 230 cu yd of ready-mixed concrete at $15.00/cu yd, 50 barrels of lime at $9.50/barrel, and 45 cu yd of sand at $3.00/cu yd. Find the total cost of the materials.
22. (*C*) A pile of lumber contains 190 boards 12 ft long, 75 boards 10.5 ft long, 112 boards 8.75 ft long, and 30 boards 15.5 ft long. How many linear feet of lumber are in the pile?

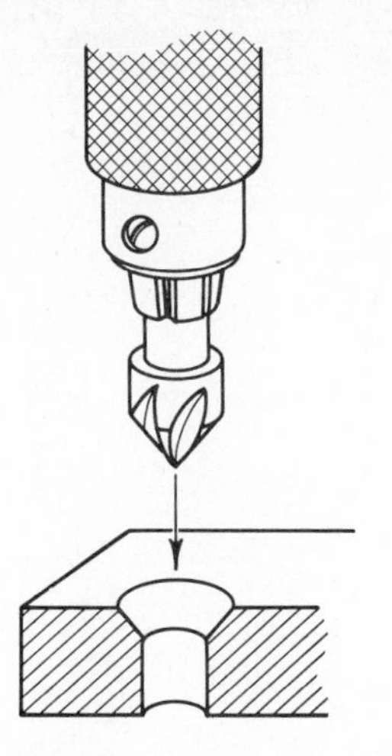

Fig. 9-4 *Using a countersink drill.*

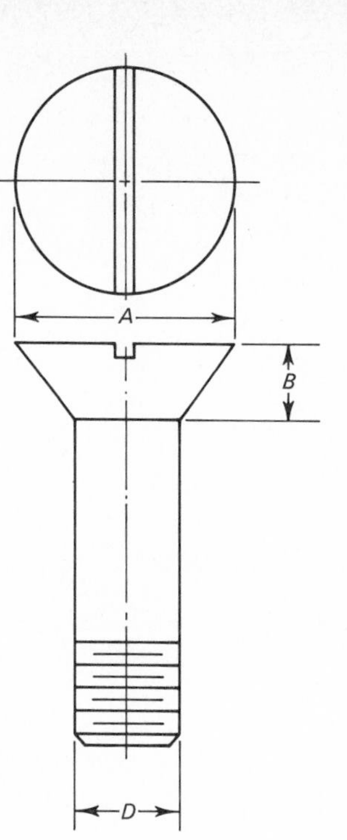

Fig. 9-5 *A bolt with a countersunk head.*

Fig. 9-6 *Model 7778 Rockwell Portable Magnetic Drill Press.*

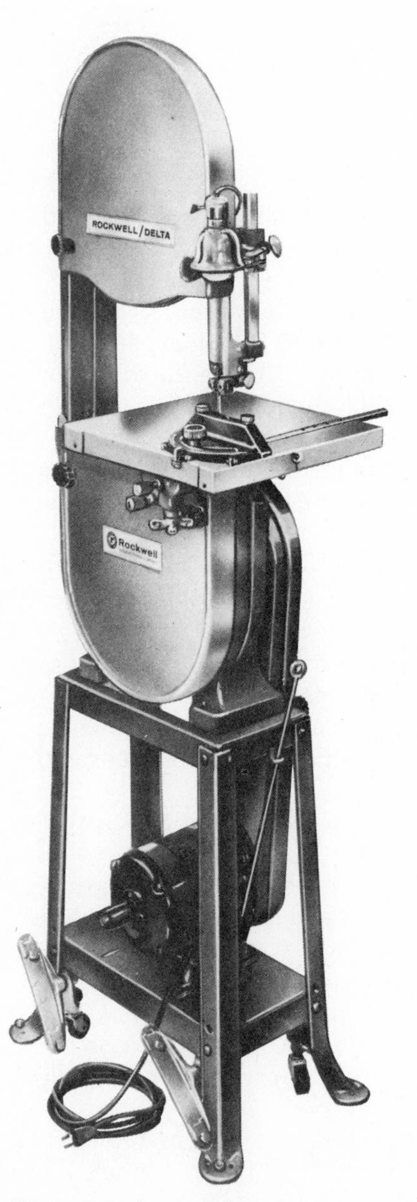

Fig. 9-7 *A wood-cutting bandsaw. (Courtesy Rockwell International)*

23. (*E*) The number of milliamperes of current is found by multiplying the number of amperes by 1,000. Find the number of milliamperes equal to (a) 0.25 A, (b) 0.025 A, and (c) 2.5 A.
24. (*B*, *M*) A countersink (Fig. 9-4) is used to drill a conelike depression at the outer end of a hole to receive the head of a screw. Find the cost of 6 countersink drills at $2.35 each.
25. (*F*, *B*) A feed lot operator figured that it required 6.7 lb of feed to produce 1 lb of gain. How many tons of feed were needed to put 500 lb of gain on each of 200 head of cattle? (1 ton = 2,000 lb.)
26. (*E*) The laminated core of a transformer is made of 32 thicknesses of metal. If each lamination is 0.03 in thick, find the total thickness of the core.
27. (*F*, *B*) At 60 lb/bushel, find the weight of 134.5 bushels of wheat.
28. (*F*, *B*) Find the cost of seeding 85 acres of logged timberland with crested wheatgrass at 6 lb/acre and $0.52/lb.
29. (*C*, *B*) If one person can lay 120 sq ft of flooring per hour, how many square feet can a 2-person crew lay in a 7-hr day?
30. (*C*, *B*) If nails cost $22.35/100-lb keg, how much would 350 lb cost?
31. (*M*) Figure 9-5 shows a bolt with a countersunk head. The distance *A* is 1.85 times the diameter. If the diameter is 0.91 cm, find the distance *A*.
32. (*M*) In Fig. 9-5, the distance *B* is 0.424 times the diameter. If the diameter is 12.5 mm, find the distance *B*.
33. (*E*) The power required by the motor of the drill shown in Fig. 9-6 is equal to the voltage supplied multiplied by the current. Find the power used (in watts) if the voltage is 115 V and the current drawn is 3.28 A.
34. (*E*) Find the power used by the bandsaw motor shown in Fig. 9-7 if the voltage is 120 V and the current is 4.15 A.
35. (*B*, *G*) Newspapers define a column inch as an area 1 column wide and 1 in deep. Find the cost for an advertisement 2 columns wide and $3\frac{1}{2}$ in deep at $4.25/column in.

Fig. 9-8 An Austin-Western grader pioneers a new forest road. (Courtesy Clark Equipment Company)

Fig. 9-9 The operator of this Trailing Plow may swing the hitch over so that the plow centers behind the tractor for transport. (Courtesy International Harvester Company)

36. (*C, B*) A contractor estimated that his cost to build a road would be \$23,500/mi (Fig. 9-8). What would be the estimated cost for 3.65 mi of road?
37. (*G, C*) If a steel tape expands 0.00012 in for each inch of its length when heat-treated, how much will a tape 20 ft long expand?
38. (*C, M, G*) The diameter of a rivet head is 1.7 times the diameter of the rivet. If a rivet has a diameter of 9.5 cm, find the diameter of the rivet head.
39. (*E, B*) An electrician worked 1.5 hr to install a junction box, make connections, and test the wiring. If her rate of pay is \$7.50 per hour, how much did she earn?
40. (*A*) An automobile engine uses 15 times as much air as gasoline. How much air is needed for each 3.5 cubic centimeters (cc) of gasoline?
41. (*A*) A certain car uses 0.06 gal of gas per mile at 50 mph. If it uses 1.25 times as much at 70 mph, find the gas consumption at 70 mph.
42. (*E*) If the allowable current-carrying capacity of a copper bus bar is 1,200 A/sq in, find the current permitted in a copper bus bar with a cross-sectional area of 0.52 sq in.
43. (*F*) A plow similar to that shown in Fig. 9-9 plowed 8.35 acres/hr. How many acres can it plow in 7.5 hr?
44. (*F, B*) An egg farmer bought 330 lb of mash and 500 lb of grain. If mash costs \$12.00/hundredweight (cwt) and grain costs \$10.25/cwt, find the total cost.
45. (*C, B*) Plastic pipe costs \$0.36/ft and copper tubing costs \$0.47/ft. Find the difference in cost for runs totaling 150 ft.
46. (*M*) The taper (decrease in diameter) per inch of length of a Brown & Sharpe taper is 0.04167 in. Find the total taper in a length of (a) 4 in, (b) 6 in, and (c) 12 in.
47. (*C, B*) A contractor estimates that it will take 0.47 hr/100 sq ft to place deadening felt over rough flooring. Find the cost to felt 1,260 sq ft of flooring at \$4.75 per hour.
48. (*M, G*) The forging press shown in Fig. 9-10 exerts a force of 150,000 lb/sq in. Find the total force exerted on an area of 20.5 sq in.

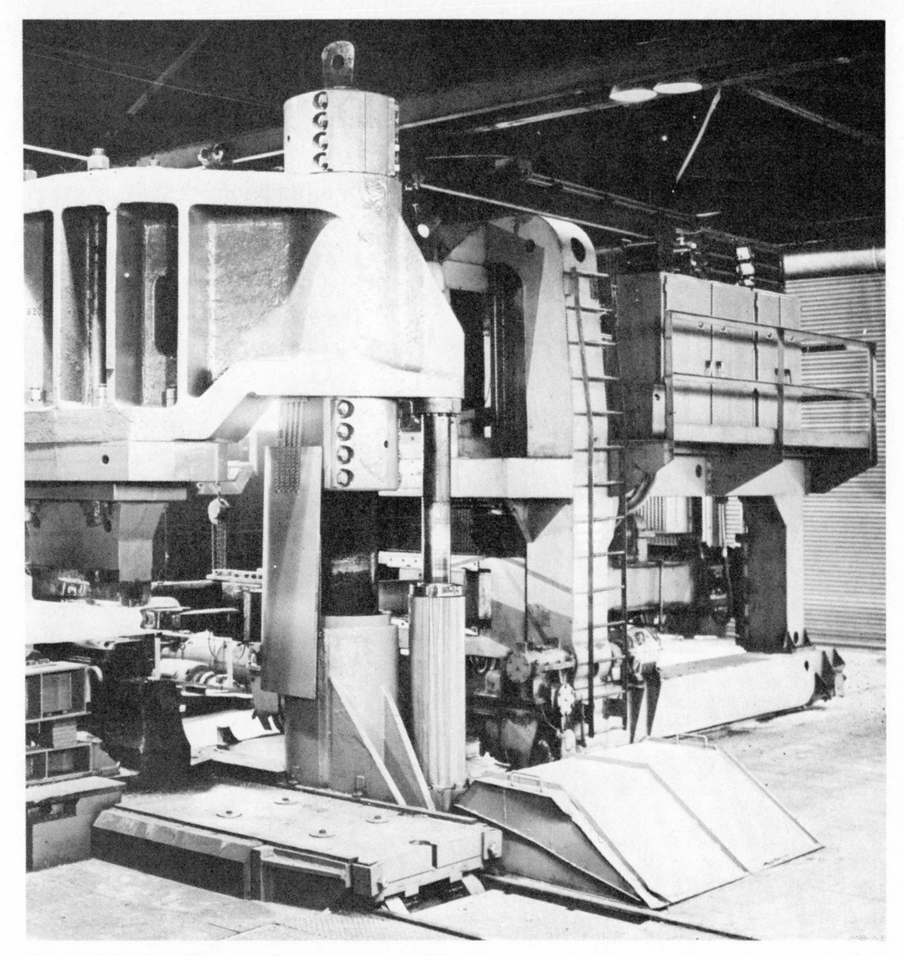

Fig. 9-10 A 2,000-ton forging press. (Courtesy Allegheny Ludlum Industries, Inc.)

49. (*C, B*) A contractor estimated the labor cost for lathing and plastering as follows:

OPERATION	HR/100 SQ FT
Applying metal lath	1.6
Scratch-coat plastering	0.7
Finish-coat plastering	1.2

At an average rate of \$7.25/hr, find the cost to lath and plaster an area of 475 sq ft.

50. (*M*) How far will a drill penetrate a piece of work in $\frac{1}{2}$ min if the drill rotates at 100 revolutions per minute (rpm) and penetrates at the rate of 1.5 mm/rev?
51. (*E*) The standard unit of electrical resistance is equal to the resistance of a column of mercury 106.3 cm high. If 1 cm equals 0.3937 in, what is the height of the mercury column correct to the nearest thousandth of an inch?
52. (*C*) In plumbing work, a "head" of 1 ft of water is equal to a pressure of 0.434 lb/sq in. What will be the pressure exerted by a head of 34 ft of water?
53. (*B, A*) The following rates were advertised by a Rent-A-Car company:
    Mustang: \$7.50/day + \$0.11/mi
    Fairlane: \$11.50/day + \$0.07/mi
    a. Find the cost of renting a Mustang for a 4-day, 586-mi trip.
    b. Find the cost of renting a Fairlane for the same trip.
    c. Find the difference in the costs.

54. (*A, G*) The distance around a circle (its circumference) is found by multiplying its diameter by 3.14. Find the circumference to the nearest thousandth of an inch of a compressed piston ring whose diameter is 3.125 in.
55. (*B*) The currency conversion rates on a certain day were as follows:
    Japan: 1 yen = 0.36 cents
    Spain: 1 peseta = 1.8 cents
    Portugal: 1 escudo = 4 cents
    France: 1 franc = 22.1 cents
    Brazil: 1 cruziero = 16.5 cents
    Italy: 1 lira = 0.166 cents
    Find the value of the following items in U.S. currency:
    a. A camera costing 120,000 yen
    b. A book costing 357 pesetas
    c. A vase costing 775 escudos
    d. A TV set costing 1,525 francs
    e. A watch costing 56 cruzieros
    f. A dress costing 40,500 lira
56. (*B*) Which hat is more expensive, one costing 12,650 lira in Italy or one costing 100 francs in France?
57. (*F, B*) A logged area is replanted with oatgrass at the rate of 5 lb/acre. Find the cost of reseeding 155 acres if oatgrass seed costs $0.95/lb.
58. (*C, B*) In the construction trades, the letter C means hundred, and the letter M means thousand. Find the cost of the following:
    a. 7,500 ft of siding @ $13.75/C.
    b. 4,190 ft of flooring @ $95.50/M.
    c. 17,500 bricks @ $106.00/M.
    d. 14,750 sq ft of shingles @ $9.45/C.

Example 10-1 Divide 4.788 by 14.

Solution The division is accomplished in the same manner as in changing fractions to decimals. See Examples 2-1 to 2-5.

```
     0.342     Ans.
14)4.788
    4 2↓
      58
      56↓
       28
       28
        0
```

Example 10-2 Divide 1.38 by 0.06.

Solution When dividing by a decimal, it is best to move the decimal point all the way over to the right so as to bring it to the end of the divisor.

```
.06,)1.38
```

If this is done, the decimal point in the dividend must also be moved to the right *for the same number of places*. Then we can divide as before.

```
          23.     Ans.
 .06.)1.38.
      1 2
        18
        18
         0
```

Example 10-3 Divide 3.6 by 0.08.

Solution A zero must be added after the 6 to provide the two places that the decimal point must be moved to the right.

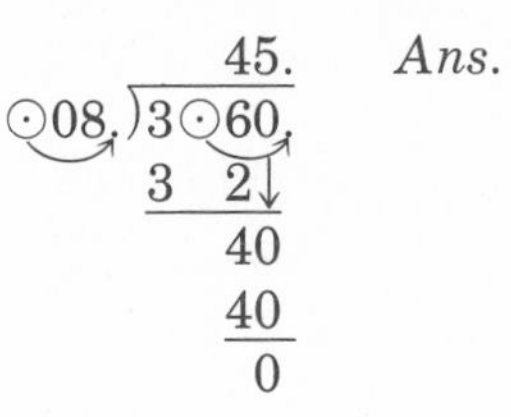

Example 10-4 Divide 0.0007 by 0.125.

Solution

```
              0.0056     Ans.
 .125.).000.7000
            625
            750
            750
              0
```

Example 10-5 The formula for the depth of an American Standard thread is

$$D = \frac{0.6495}{\text{no. of threads per inch}}$$

Find the depth of cut needed to cut the 12 threads per inch as shown in Fig. 10-1.

Fig. 10-1 Cutting a screw thread on a lathe. (Courtesy LeBlond Inc.)

Solution $D = \frac{0.6495}{N} = \frac{0.6495}{12}$

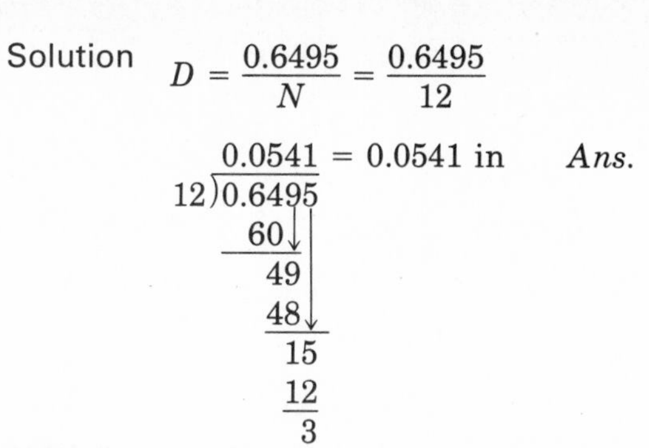

$0.0541 = 0.0541$ in *Ans.*

Self-Test 10-6 In qualifying for the "Indy 500," a racing car covered 200 mi in 1.47 hr. Find its average speed correct to the nearest hundredth of a mile.

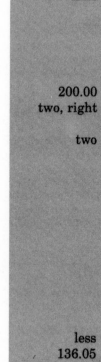

Solution The average speed is obtained by dividing the ______ by the ______. In our problem, the distance (200) divided by the time (1.47) may be written as

$200 \div 1.47$ or $\frac{200}{1.47}$ or $1.47\overline{)200}$

The decimal point in the number 200 is placed so that it appears as ______. Before we divide, the decimal point in 1.47 must be moved ______ places to the ______. Then the decimal point in 200.00 must also be moved ______ places to the right. The problem will now appear as

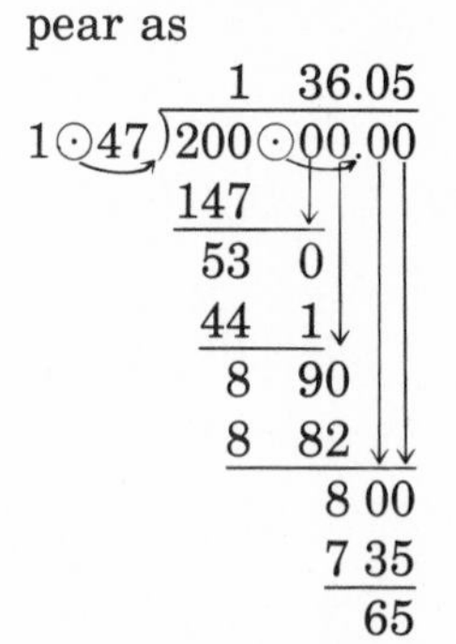

The remainder (65) is (more)(less) than half of 147. The final answer is ______ mph.

Problems

1. $3.9 \div 0.3$
2. $12.56 \div 0.4$
3. $80.5 \div 0.5$
4. $51 \div 0.06$
5. $38.54 \div 8.2$
6. $1{,}591 \div 0.43$
7. $2.8296 \div 0.0036$
8. $140.7 \div 0.021$
9. $9.1408 \div 3.94$
10. (*E*) A 50-ft-long wire has a resistance of 10.35 Ω. What is the resistance of 1 ft of this wire?
11. (*E*) If #30 annealed copper wire has an electrical resistance of 103.2 Ω/1,000 ft, find the resistance per foot.
12. (*M*) A stack of 35 sheets of spring steel is 2.17 in high. Find the thickness of each sheet.
13. (*C*) A gallon of a certain paint will cover 325 sq ft of surface. How many gallons are needed to cover 2,745 sq ft?

14. (*E*) Shielded rubber-jacketed microphone cable weighs 0.075 lb/ft. How many feet of cable are there in a coil weighing 15 lb?
15. (*A*) The compression ratio of an engine cylinder is 9.5:1. Find the volume at full compression if the initial volume is 315 cu in.
16. (*F*) If 1.25 acres were sprayed with a total of 3 gal, 1 qt of insecticide, how many gallons were used per acre?
17. (*E*) If a certain diameter of nichrome wire has a resistance of 1.5 Ω/ft, how many feet of this wire are needed to make a resistance of 12 Ω?
18. (*B*) After a man made a down payment on a car, he still owed $747.00. If he wants to pay off the balance in 12 equal monthly payments, find the amount of each payment.
19. (*M*) In grinding the flutes on a tap, the tap feeds under the stone at the rate of 0.008 in/rev of the wheel (Fig. 10-2). How many revolutions of the wheel are required to grind a flute 2 in long?

Fig. 10-2 Grinding the flutes on a tap.
(Courtesy The K.O. Lee Company)

20. (*C, E*) A 400-ft-long coil of wire weighs 10.5 lb. How many feet of wire are there in 1 lb?
21. (*C*) If each strip of tongue-and-groove flooring covers a width of 3.5 in, how many strips are needed to cover a floor that is 10 ft wide?
22. (*C, B*) A contractor paid $1,037.50 for 250 cu yd of topsoil. Find the cost per cubic yard.
23. (*M*) The Morse No. 5 tapered reamer shown in Fig. 10-3 has a taper of 0.631 in/ft of length. Find the taper in 1 in of its length.
24. (*B, F*) A merchant buys potatoes at $9.00/cwt and sells them in 5-lb bags at $0.89/bag. Find her profit (a) per lb, and (b) per cwt.

Fig. 10-3 A Morse Standard tapered reamer.

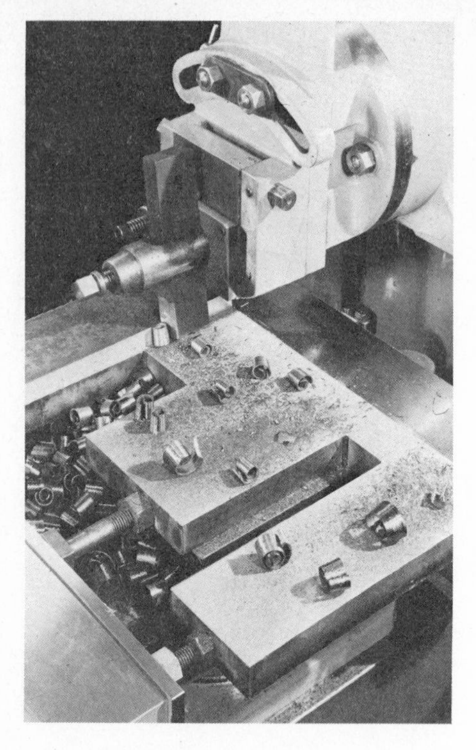

Fig. 10-4 The reciprocating action of the ram on this shaper cuts off a chip on each stroke. (Courtesy Rockford Machine Tool Company)

Fig. 10-5 Slotting a screwdriver handle on a horizontal milling machine. (Courtesy Hamilton Associates)

25. (*M*) A shaper removes a chip 0.012 in thick on each stroke of the ram, as shown in Fig. 10-4. How many strokes are needed to shave off 1.766 in of metal?
26. (*F*) If 15,800 lb of cotton was produced on 35.8 acres, find the yield per acre to the nearest hundredth of a pound.
27. (*M*) How long did it take to mill the 4.5-cm-deep slot in the screwdriver handle shown in Fig. 10-5 if the handle advanced into the cutter at the rate of 20 mm/min?
28. (*A*) Find the pressure in pounds per square inch acting on a piston of 153.9 sq in if the total force on the piston is 72,000 lb.
29. (*C*) A contractor must locate 9 equally spaced parking meters on a street. The total distance from the first to the last meter is 117.5 ft. Find the distance between two adjacent meters.
30. (*C*) A bundle of wood laths spaced $\frac{1}{4}$ in apart will cover 6.48 sq yd of wall surface. How many bundles are needed to cover 150 sq yd?
31. (*C*) After allowing for waste, a room requires 517 sq ft of wallpaper. How many single rolls of paper, each covering 29.5 sq ft of surface, will be needed for the job?
32. (*A*) If the mechanical advantage of a clutch link mechanism is 42.5:1, find the force needed to disengage a pressure plate being held under a force of 952 lb.
33. (*A*) The outside diameter of a piston ring is 4.275 in. If the inside diameter is 4.125 in, how thick is the ring?
34. (*M*) The centers of two holes 3.719 in apart are to be located equally distant from both ends, as shown in Fig. 10-6. Find the distance *X*.
35. (*M*) The table of a planer is 24 in long. Eight equally spaced holes are to be drilled in the table, the center line of the end holes being located $2\frac{1}{4}$ in from the ends. Find the distance between the centers of two adjacent holes.
36. (*F, B*) On a certain day, corn was selling at $3.60/bushel and grain at $7.20/cwt. Find the number of bushels of corn equal in value to 250 lb of grain.
37. (*C*) If 1 ft of water exerts a pressure of 0.434 lb/sq in, how many feet of water are needed to exert a pressure of 5 lb/sq in?
38. (*A, B*) Find the cost per mile for the following tires:
    a. $26.95 for a guaranteed 20,000 mi
    b. $45.95 for a guaranteed 35,000 mi

    Which tire is the better buy, and what is the saving per mile?
39. (*M*) The 12-in-diameter saw blade shown in Fig. 10-7 has a circumference of 37.68 in. Find the distance between the teeth if there are 40 teeth.
40. (*A*) The compression ratio of the cylinder shown in Fig. 10-8 is obtained by dividing the volume at bottom dead center (350.2 cu in) by the volume at top dead center (38.4 cu in). Find the compression ratio.
41. (*B*) The IBM Model 10 copier-duplicator can produce 1 copy every 0.8 sec. How many copies can be made per minute?

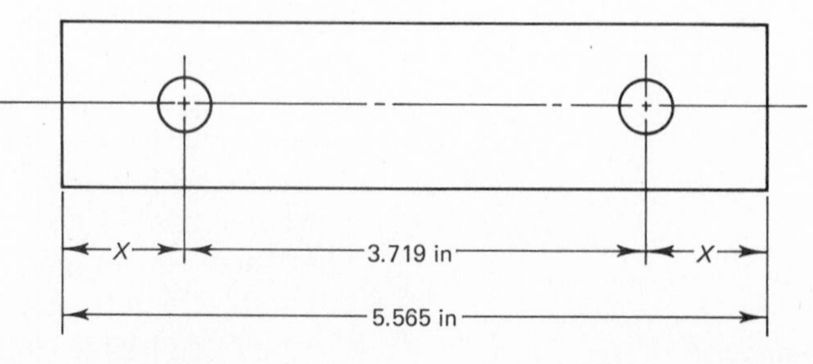

Fig. 10-6

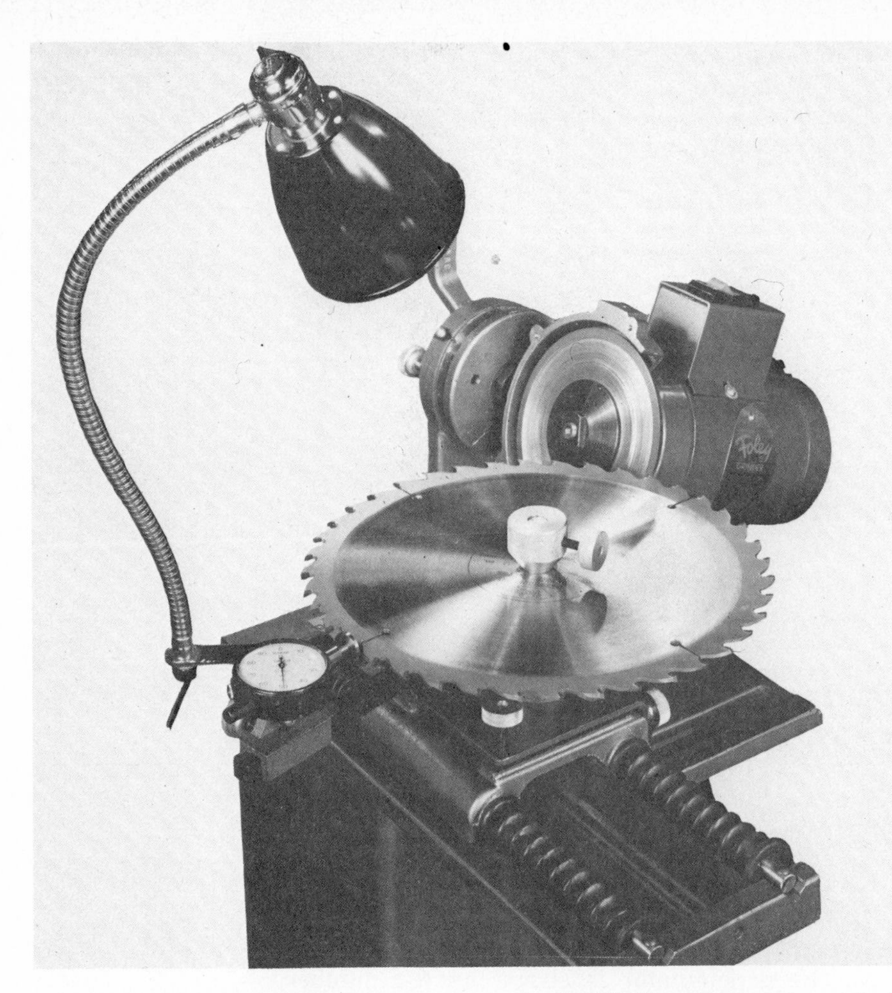

Fig. 10-7 *Sharpening a carbide-tipped circular saw with a Spartan grinder. (Courtesy Foley Manufacturing Company)*

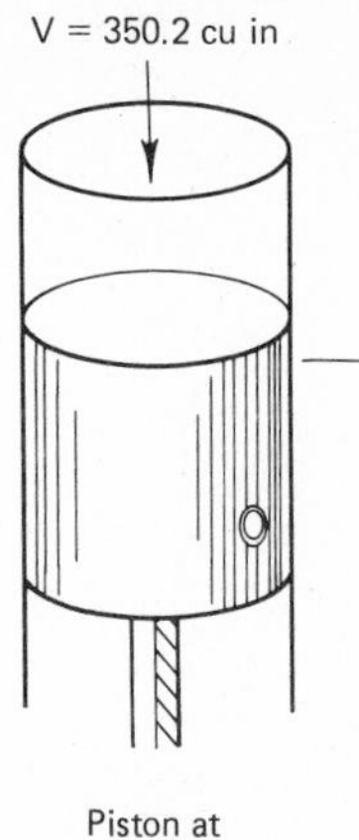

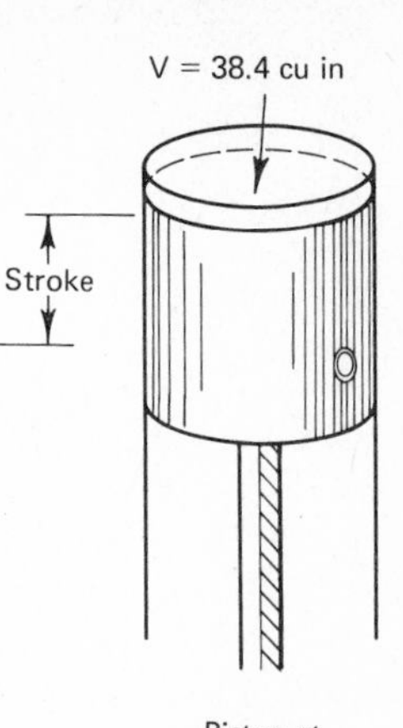

Fig. 10-8

42. (*G, F*) In a barometer, a height of 1 mm of mercury represents an atmospheric pressure of 0.0193 lb/sq in. How many millimeters of mercury is equivalent to a normal atmospheric pressure of 14.67 lb/sq in?
43. (*M*) In Fig. 10-9, if the drill feeds into the work at the rate of 0.6 mm/rev, how many revolutions are needed to drill a hole 6.24 cm deep? How long will it take if the shaft turns at a speed of 80 rpm?

Fig. 10-9 Drilling operation using a steady rest. (Courtesy LeBlond Inc.)

44. (*M*) A piece of work in a lathe makes 20 rpm, as shown in Fig. 10-10. How far will the tool advance along the work in 1 min if the feed is 0.0625 in/rev? How long will it take to turn a piece 18 in long under these conditions?

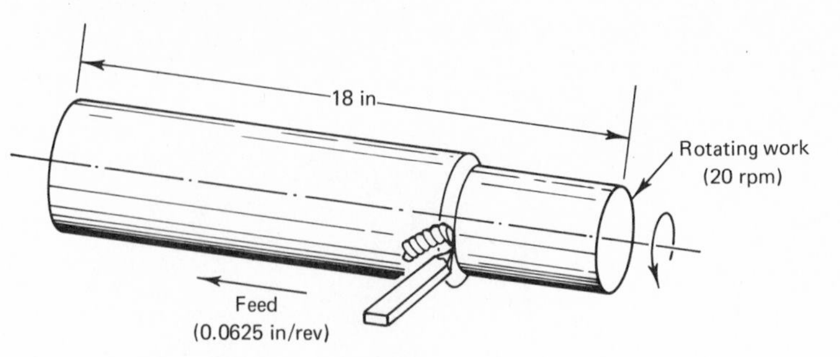

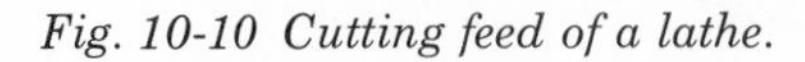

Fig. 10-10 Cutting feed of a lathe.

JOB 11 Cumulative Review of Decimals

SUMMARY—WORKING WITH DECIMALS

1. Decimal fractions are fractions whose _______ are numbers like 10, _______, 1,000, etc.	denominators 100
2. The denominator is shown by the number of digits to the _______ of the decimal point. Thus, 0.6 represents six _______. 0.57 represents fifty-seven _______. 0.123 represents one hundred-twenty-three _______. 3.09 represents three and nine-_______.	right tenths hundredths thousandths hundredths
3. Decimals can be compared only when they have the _______ number of decimal places.	same
4. The word "and" in a mixed number such as five and three-hundredths is written as a(n) _______ point. This number would be written as _______.	decimal 5.03
5. Fractions are changed to decimals by _______ the numerator by the _______.	dividing denominator
6. A whole number always has a decimal point understood to be at the (beginning)(end) of the number.	end
7. When dividing decimals, if a remainder is more than _______ of the divisor, drop it and add a full unit to the last _______ of the answer. If the remainder is _______ than half, drop it completely.	half digit less
8. When adding or subtracting decimals, line up the decimal points in a(n) _______ column.	vertical

9. The product of two decimals has as many decimal places as the ______ of the places in the numbers being multiplied.

10. When you divide decimals, move the decimal point in the divisor to the ______ as many places as is necessary to bring the point behind the last digit. Then move the point in the dividend to the right for the ______ number of places.

Self-Test 11-1 The tightening specification for the cylinder head bolts on a 1974 Cadillac Eldorado 500 is 115 lb-ft. If the wrench available has an effective length of 15 in, find the force needed to satisfactorily tighten the bolts.

Fig. 11-1 The twisting effort needed to loosen a cap from a jar is called the torque.

Solution *Torque* is a twisting or turning effort of the type used in twisting open the cap on a jar (Fig. 11-1). This turning effort (torque) is measured in pound-feet, and is equal to the force applied (measured in pounds) multiplied by the distance (in feet) from the force to the center of the turning item, as shown in Fig. 11-2.

In formula form, Torque = force F × distance L. This formula may be rearranged by algebraic methods to find the force F.

$$\text{Force} = \frac{\text{torque (lb-ft)}}{\text{distance } L \text{ (ft)}}$$

Now to solve our problem. Since the distance L must be measured in feet, we must change 15 in to ft by (multiplying)(dividing) the 15 in by the ______ in in a foot.

Thus, 15 ÷ 12 = ______ ft.

dividing, 12

1.25

1. Write the formula.

$$\text{Force} = \frac{\text{torque}}{\text{distance}}$$

2. Substitute numbers.

$$\text{Force} = \frac{?}{1.25}$$

115

3. Divide the numbers.

Force = ______ lb *Ans.*

92

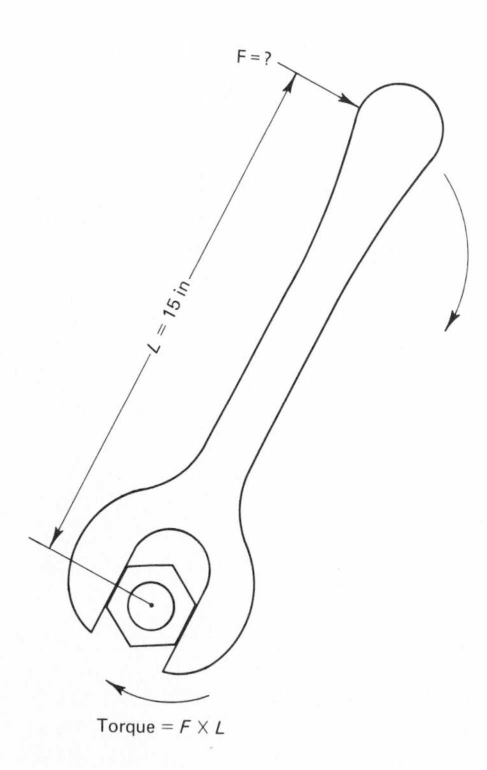

Fig. 11-2 Applying a torque to turn a nut.

Note: the torque in an automobile means the effort available to turn the wheels of the car. The torque increases as the rpm of the engine increases up to a certain point. Beyond this critical rpm, even though the horsepower increases, the torque will *decrease*. Therefore, on ice, or in a rut, do not race the engine, as this will *decrease* the available torque.

Problems

1. (*M*) Find the dimensions *A* and *B* in the spindle shaft shown in Fig. 11-3.

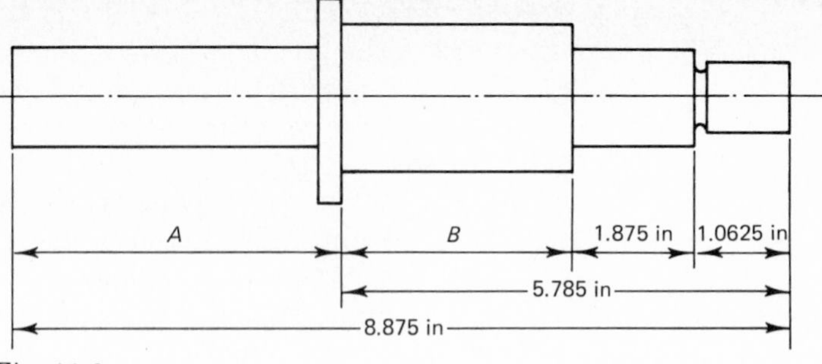

Fig. 11-3

2. (*F*) If there are 43,560 sq ft to an acre of ground, what decimal part of an acre is covered by 2,178 sq ft?
3. (*M*) The single depth of an American Standard thread is found by dividing 0.6495 by the number of threads per inch. Find the single depth of screws with (a) 24, (b) 16, (c) 10, (d) 8, and (e) 6 threads per inch.
4. (*C*) As shown in Fig. 11-4, a plumbing plan calls for the condensate pipe to have a drop of 2.25 in along a length of 108 in. Find the drop per inch of length.

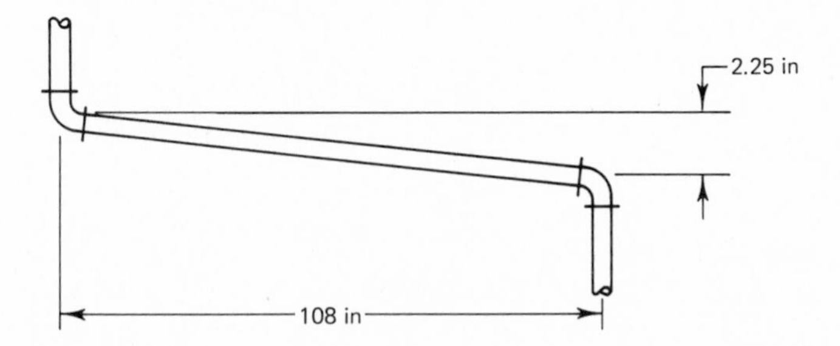

Fig. 11-4 A condensate pipe is pitched downward.

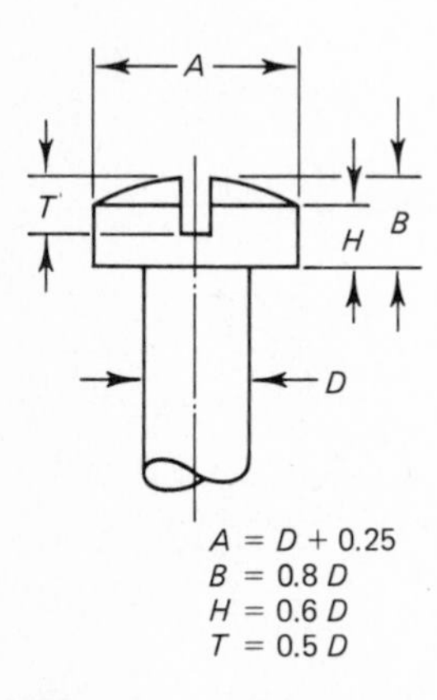

Fig. 11-5 Dimensions for a fillister-head machine screw.

5. (*M*) The formulas for the dimensions of a fillister-head machine screw are given in Fig. 11-5. Find the dimensions A, B, H, and T for a screw whose diameter $D = 0.375$ in.
6. (*E*) In an ac voltage wave, the maximum voltage is equal to 1.414 times the effective voltage as read on a meter. Find the maximum voltage of a wave whose meter reading is 117 V.
7. (*M*) On a lathe making 160 rpm, the cutting tool advances 0.031 in/rev of the work. How far does it travel in (a) 2 min, and (b) 3.5 min?
8. (*M*) A shaft is to be turned to a diameter of 0.240 in. A machinist "mikes" it, as shown in Fig. 11-6, as 0.258 in. How deep a cut should be taken to turn the shaft to the correct diameter?
9. (*E*) The power factor of an ac circuit is found by dividing its resistance (R) by its impedance (Z). Find the power factor (pf) of a circuit if $R = 2.46\ \Omega$ and $Z = 30\ \Omega$.
10. (*A, B*) If 1 cu ft = 7.48 gal, find the number of gallons of gas in a tank with a volume of 2.67 cu ft.
11. (*G, C*) How many $\frac{3}{4}$-in-diameter bolts are in a 75-lb keg if each bolt weighs 0.52 lb?
12. (*F, B*) The cost of seeding a tract of timberland by airplane is \$7.65 per acre. Find the cost of seeding 145 acres.
13. (*M*) A drill operates at a feed of 2 mm/rev. How many revolutions of the drill are needed to drill a hole 11.25 cm deep?

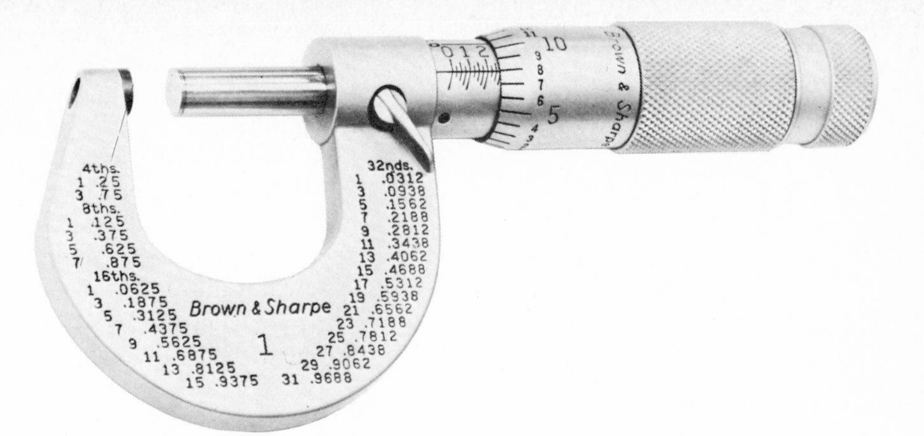

Fig. 11-6 The micrometer registers 0.258 in. (Courtesy Brown & Sharpe Manufacturing Company)

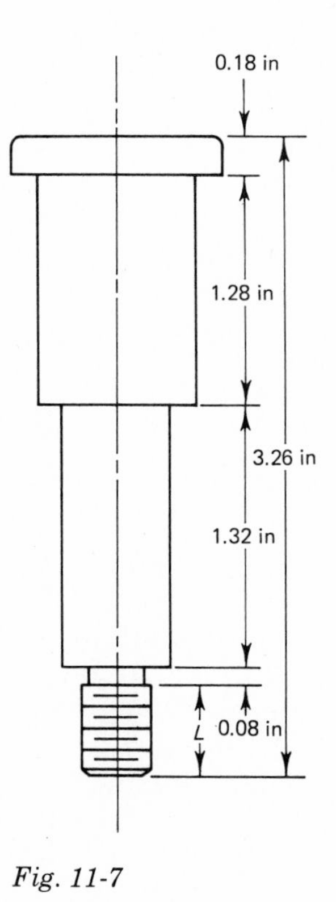

Fig. 11-7

14. (*C*) A strip of hardwood flooring is 9.5 cm wide. How many strips are needed to cover a floor 3.5 m wide?
15. (*M*) Two hardened pins 0.562 in and 0.406 in in diameter are spaced 2.875 in center-to-center. Find the distance from the outside of one pin to the outside of the other.
16. (*A, B*) A rebuilt alternator that cost \$47.50 was sold for \$63.65. (a) Find the profit. (b) What decimal part of the cost was the profit?
17. (*F*) If there are 1.2445 cu ft/bushel, change 150 bushels to cubic feet.
18. (*M*) Find dimension *L* in the stud shown in Fig. 11-7.
19. (*F, B*) If 120 boxes of fancy apples (average weight 14.6 lb per box) were shipped at a cost of \$0.355/100 lb, find the total shipping cost.
20. (*M*) A keyway must be milled to a depth of 1.0625 in. If the cutter has already cut a depth of 0.719 in, find the depth that remains to be cut.
21. (*F, B*) Fuel and maintenance costs for the combine shown in Fig. 11-8 are estimated to be \$3.50/hr. Find the cost for a working life of 14,500 hr.

Fig. 11-8 Quick-attach feeder on this combine permits fast changeover of grain header to corn head, or from cutterbar to windrow pickup. (Courtesy International Harvester Company)

22. (*B, A*) Mr. Rhodes drove his car 13,500 mi last year. He averaged 15 mi/gal of gas and 1 qt of oil for each 500 mi. (a) At 53.9 cents/gal, find the total cost for gas. (b) At $1.15/qt, find the total cost for oil. (c) Find the total cost for gas and oil.
23. (*G*) The weight of 1 cu in of pure water at 62 degrees Fahrenheit (°F) is 0.0361 lb. If gold weighs 19.32 times as much, find the weight of an equal quantity of gold.
24. (*C*) A plate is to be riveted to an angle bracket, as shown in Fig. 11-9. If 6 equally spaced rivet holes are to be drilled, find the center-to-center distance between the holes.

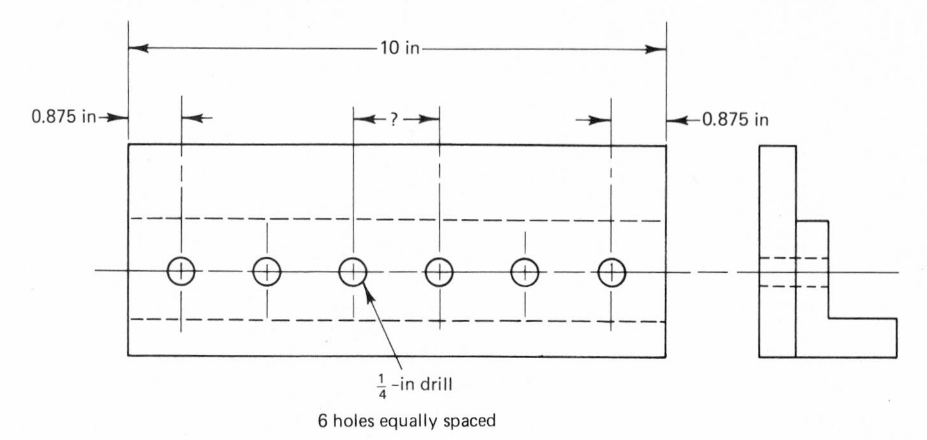

Fig. 11-9

25. (*M*) A $1\frac{5}{8}$-in-diameter rod is to be turned down in a lathe. If one cut $\frac{1}{64}$ in deep is taken and then another cut 0.003 in deep is taken, what will be the final diameter?
26. (*E, B*) If BX cable costs $0.235/ft, what would be the cost of 52.5 ft?
27. (*F*) A steer dressed out at 735 lb from a live weight of 1,290 lb. Expressed to the nearest hundredth, what part of the live weight was the dressed weight?
28. (*M*) A 6-in-long reamer is 1.2575 in in diameter at the small end and 1.5625 in at the large end. (a) Find the total increase in diameter. (b) Find the increase in diameter for 1 in of the reamer's length.
29. (*E*) What is the total current supplied to a parallel circuit consisting of a broiler (9.25 A), a radio (0.93 A), and 2 lamps, each drawing 0.91 A?
30. (*M*) A shaft is 1.0625 in in diameter. What should be the inside diameter of the thrust bearing in the pillow block shown in Fig. 11-10 if 0.003 in is allowed for clearance on all sides?
31. (*A, B*) Mr. Graham received an increase from $5.65/hr to $5.90/hr for his work as an auto mechanic. Find the amount of his increase for a 40-hr week.
32. (*M*) A pattern for an aluminum casting must allow an additional 0.0078 in/in of length for shrinkage. How long should a pattern be to provide for shrinkage in a 9-in-long aluminum casting?
33. (*C*) An I-beam weighs 15.25 lb/ft. Find the weight of 4.5 ft of this beam.
34. (*E*) A 60-W lamp uses about 0.625 A when connected in an ordinary house line. How many amperes would 3 such lamps use when connected in parallel?
35. (*M*) In Fig. 11-11, the work advances into the milling cutter at the rate of 2 mm/rev of the cutter. How many revolutions of the cutter are needed to cut a tang 1.8 cm long?

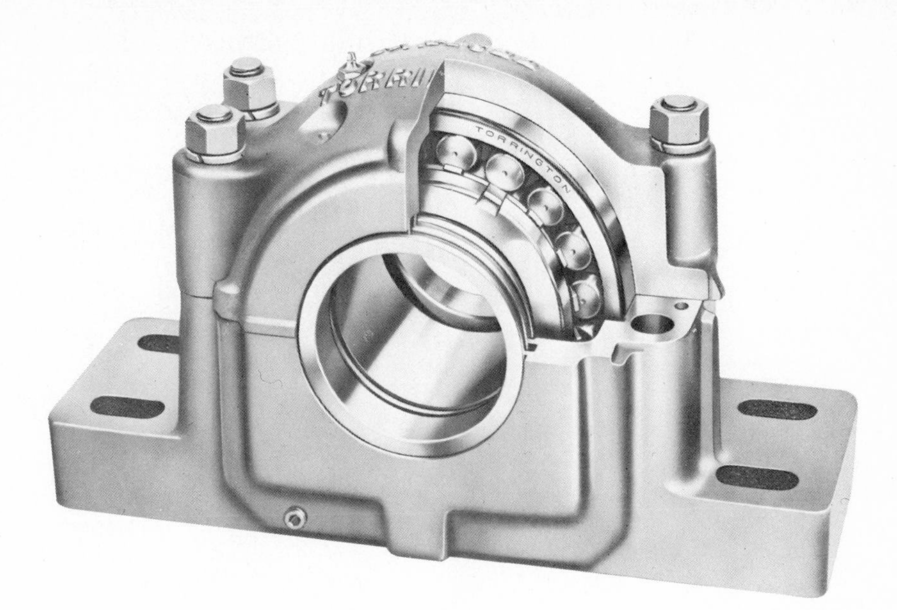

Fig. 11-10 A pillow block. (Courtesy The Torrington Company)

Fig. 11-11 Milling a screwdriver tang. (Courtesy Hamilton Associates)

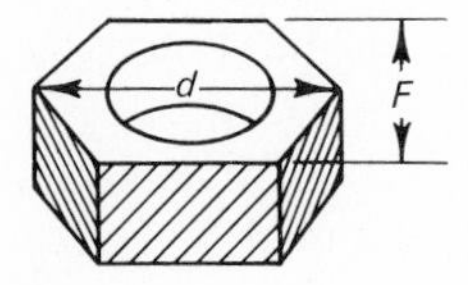

Fig. 11-12 A hexagon nut.

36. (*E*) A neon electric sign uses 4.75 W of power for each foot of tubing. Find the power consumption for a sign which uses 34.5 ft of tubing.
37. (*C*) Cast iron can safely sustain a compressive force of 15,000 lb/sq in. What is the maximum weight that a column of 1.35 sq in cross-sectional area can safely sustain?
38. (*A, M, B*) If $\frac{1}{8}$-in welding electrodes cost $0.08 each, and each electrode can weld a length of 10 in, find the cost of welding a length of 15 ft.
39. (*M*) In Fig. 11-12, the distance F is given by the formula $F = 0.866 \times d$. Find F if d = (a) 0.75 in, (b) 1.5 cm, (c) 0.625 in, and (d) 17.3 mm.
40. (*A, M*) If No. 8 (B & S) gage sheet metal is 0.1285 in thick, find the thickness of a pile of 32 of these sheets.
41. (*F, B*) A chicken farmer made a baby chick mixture of the following nutrients:

300 lb ground yellow corn @	$0.07/lb
90 lb ground oats @	$0.094/lb
50 lb cottonseed meal @	$0.081/lb
37 lb bran @	$0.058/lb
15 lb ground shell @	$0.02/lb
8 lb salt @	$0.015/lb

Find the cost per 100 lb.

42. (*M*) The automatic turret lathe shown in Fig. 11-13 is programmed to cut pins of cold-rolled steel 9.1 cm long. If the width of the cutting tool is 2.5 mm, how many pins can be cut from a rod 4 m long?
43. (*M*) Four locating pins are required for a drill jig. The pins are 2.625 in long. Allowing 0.125 in for cutting off each piece and 0.200 in for facing and finishing each end, find the total length of stock required to make the pins. (See Fig. 11-14.)
44. (*M*) The feed per revolution of a drill is 0.008 in. How many revolutions of the drill are needed to drill a hole $4\frac{1}{2}$ in deep?
45. (*M*) If the drill in Prob. 44 makes 100 rpm, how long will it take to drill the hole?

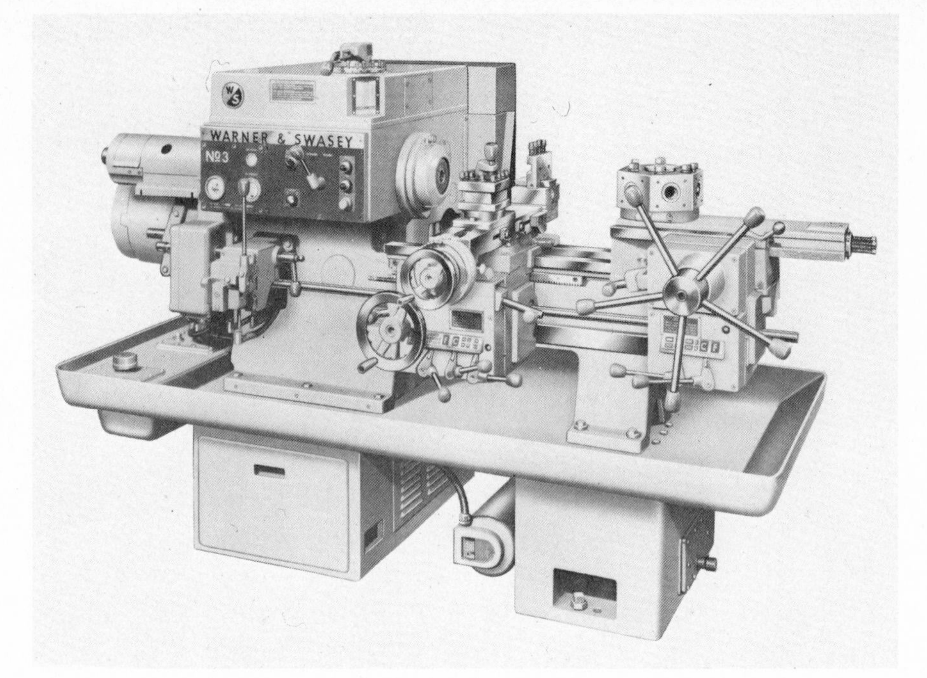

Fig. 11-13 An automatic turret lathe. (Courtesy Warner & Swasey Company)

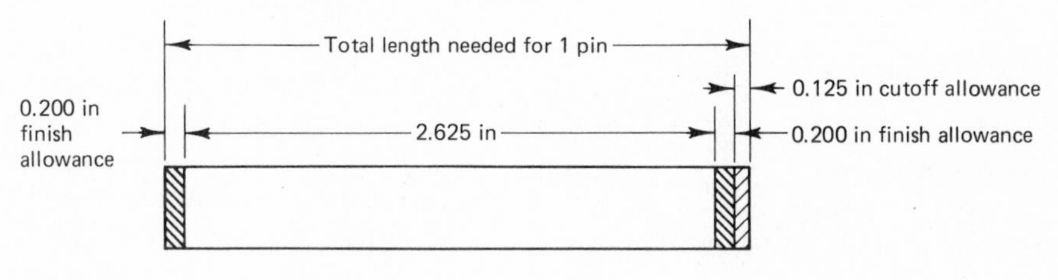

Fig. 11-14

46. (*M*) How many cuts are needed to turn down 1.150-in-diameter stock to 1 in on a lathe if each cut is 0.025 in deep?
47. (*B*) An outboard motor (Fig. 11-15) can be purchased for $485.95 or $45.75 a month for 12 months. Find the finance charge.
48. (*A*) Using a 15-in wrench, find the force needed to tighten the cylinder head bolts on the following Buick Special automobiles.

YEAR	ENGINE	TORQUE (lb-ft)
a. 1967	V6	70
b. 1968	8-350	65
c. 1969	8-400	100
d. 1970	L6	95
e. 1971	8-350	75

49. (*E, A*) An electrician is to wire a light switch to operate the following light bulbs on a truck.

NUMBER	TYPE	CURRENT (EACH)
2	Marker	0.3 A
3	Clearance	0.4 A
2	Stop/tail	2.4 A
2	Front park	1.5 A
2	Headlamp	7.6 A

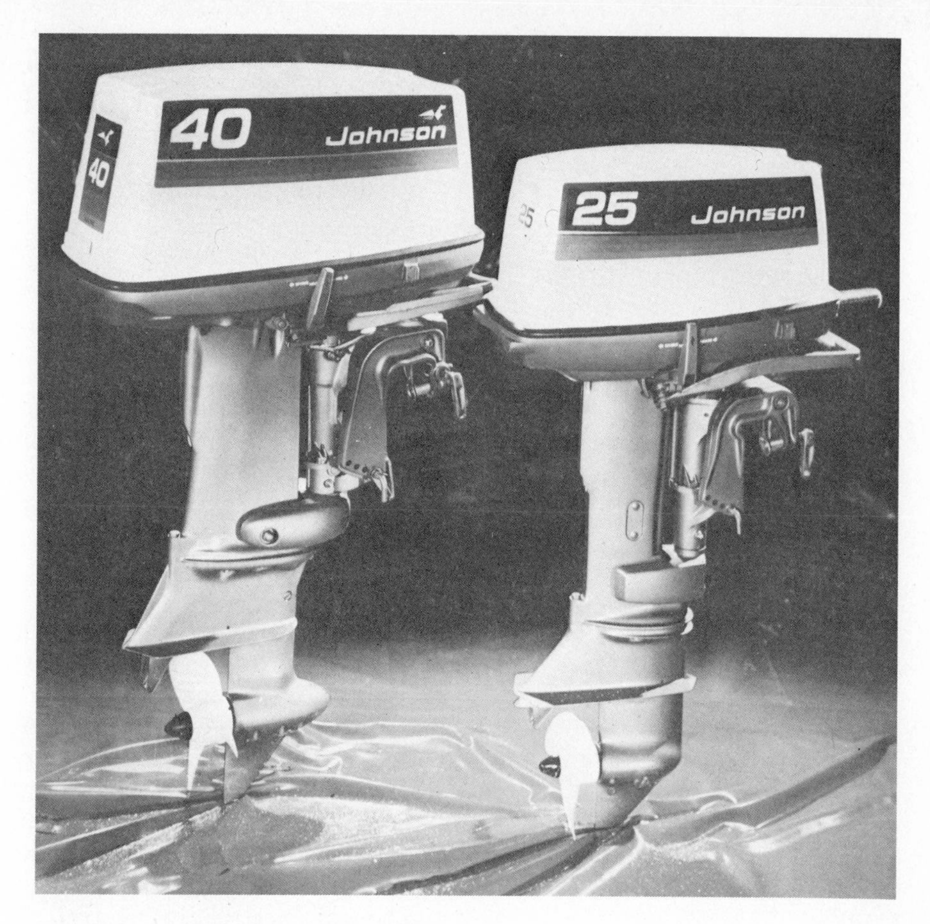

Fig. 11-15 Outboard motors. (Courtesy Johnson Outboards)

The maximum allowable current-carrying capacities of wire are:

16 gage	6 A
14 gage	15 A
12 gage	20 A
10 gage	30 A

Find the smallest gage wire that may be safely used if all lights are operating.

50. (*A, B*) Find the total cost of the following items used on an engine repair job.

8 spark plugs	\$ 1.75 each
1 set of rings	\$35.50
1 set of gaskets	\$23.50
8 valves	\$ 4.80 each
16 valve springs	\$ 1.40 each
8 rod bearings	\$ 2.90 each

(See Answer Key for Test 3—Cumulative Review)

JOB 12 Applying Decimals—Electricity and Ohm's Law

BASIC ELECTRICAL CONCEPTS

We shall start our study of electricity with an examination of the materials from which electric energy is produced.

ELEMENTS

Science has discovered more than 100 different kinds of material called *elements*. These cannot be made from other materials and cannot be broken up to form other materials by ordinary methods. Gold, copper, iron, oxygen, and carbon are some examples of elements.

ATOMS

An *atom* is the smallest particle of an element which has all the properties of the element. An atom may be broken down into smaller pieces, but these pieces have none of the properties of the element. It is now believed that all material is electrical in nature. All matter is made of combinations of elements. All elements are made of atoms. The atoms themselves are merely combinations of different kinds of electric energy. The presently accepted theory is that an atom, because it is electrically neutral, is made of a number of positive charges of electricity called *protons* and an equal number of negative charges called *electrons*. There are also a number of electrically neutral particles called *neutrons*. Similarly charged particles will repel each other. Particles of opposite charge will attract each other. For example, two electrons will repel each other, but an electron will be attracted to a proton (Fig. 12-1).

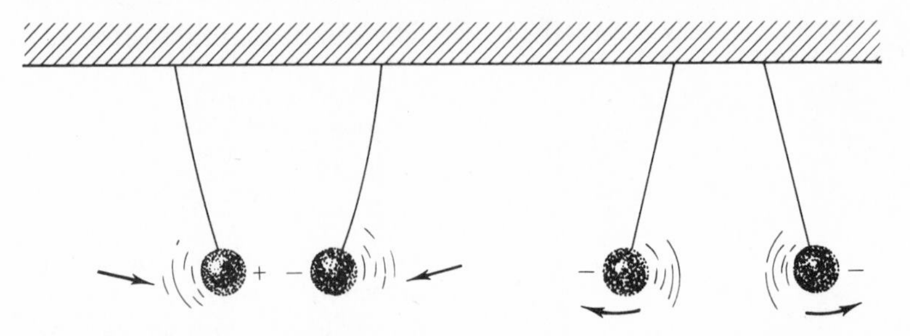

Fig. 12-1 Opposite charges attract; like charges repel each other.

STRUCTURE OF THE ATOM

An atom is believed to resemble our solar system, with the sun at the center and the planets revolving around it (Fig. 12-2*a*). An atom consists of electrons revolving around a nucleus which contains the protons and neutrons, as shown in Fig. 12-2*b*. The number of revolving, or *planetary*, electrons is equal to the number of positively charged protons in the nucleus.

FREE ELECTRONS

The electrons farthest from the nucleus are called *free electrons* because they are bound very loosely to the nucleus. The word *electricity* comes from *elektron,* the Greek word for amber. Thales, a Greek philosopher, observed that if amber were rubbed, it would attract small objects. You

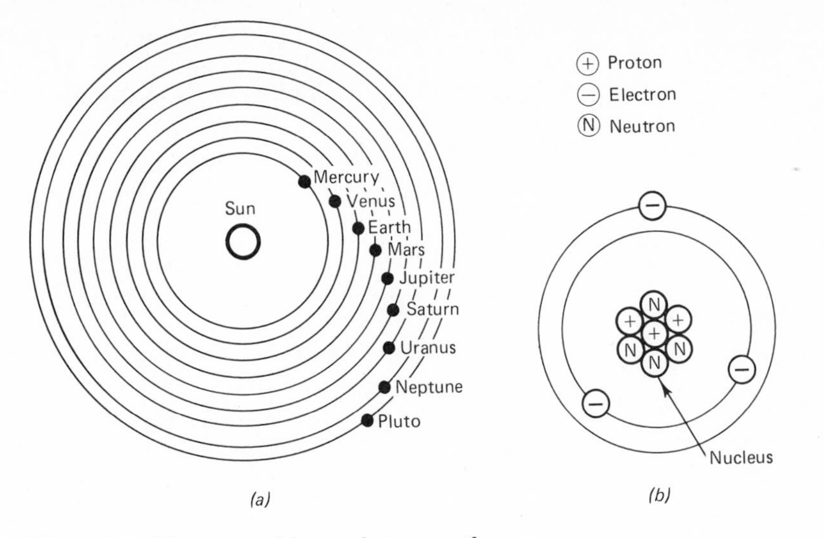

Fig. 12-2 The resemblance between the solar system (a) *and the theoretical diagram of an atom of lithium* (b).

can observe the same effect by running a hard rubber comb through your hair or by rubbing the comb on your coat sleeve. The comb will then attract small pieces of paper. As the comb is rubbed on the sleeve, some of the free electrons are rubbed off the coat and deposited on the comb. The comb is now considered to be *negatively charged* because it has accumulated an excess of electrons. On the other hand, the coat sleeve is *positively charged* because of the *shortage* of electrons. A small bit of paper will now be attracted and held by the charged comb. A similar transfer of electrons occurs if, on a dry day, you shuffle your feet over a rug. Electrons are rubbed off the rug and are accumulated on your body. Touching a metal doorknob or some other conductor will produce a slight shock as the electrons *flow* through your body to the ground.

ELECTRON FLOW

Notice that no shock will be experienced while the body is accumulating the electrons. The shock will occur only when the electrons *flow* through your body in one concerted surge. This directed flow, or movement, of electrons is one of the most important ideas in electricity. In order to understand this better, let us compare the particles of matter with a brick wall, as in Fig. 12-3. A brick in a wall is like an atom, since it is the smallest part of the wall that has the characteristics of the wall. If an individual brick were to be powdered into dust, we could compare a single grain of dust with an electron. The grains of brick cannot do any useful work by themselves, but if a powerful pressure like a blast of compressed air were allowed to hit the particles (as would occur in a sandblasting machine), enormous energies would be available.

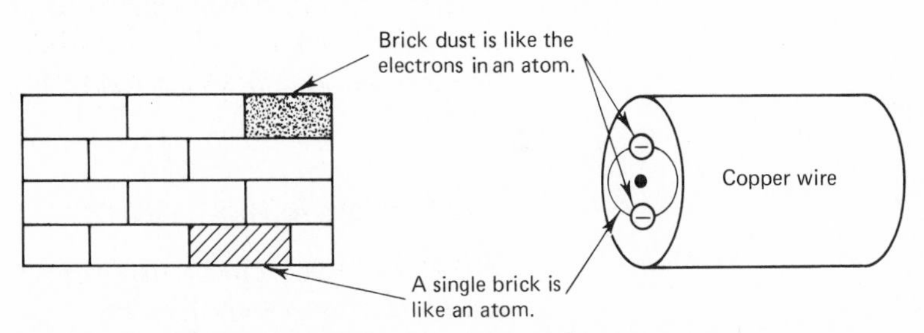

Fig. 12-3 Comparison of brick dust with electrons.

We can also consider the drops of water in a stream to be like the electrons in a piece of matter. If the drops of water move aimlessly, as in a water sprinkler, their energy is small. But if all the drops are forced to move *in the same direction,* as in a high-pressure fire hose, then their energy is large. We can see, then, that if electrons are to do useful work, they must be moving *under pressure* like the grains of sand or drops of water.

ELECTRICAL PRESSURE: VOLTAGE

In order to move the drops of water or the grains of sand, a mechanical pressure supplied by a water pump or an air compressor is required. Similarly, an electron pressure is required to move the electrons along a wire. This electrical pressure is called the *voltage* and is measured in units of *volts* (V). This unit of measurement was named after Count Allesandro Volta, an Italian physicist (1745–1827).

PRODUCING ELECTRICAL PRESSURES

A working pressure is considered to exist when there is a *difference* in energies. Water in a high tank will exert a pressure because of the *difference* in the height of the water levels, as shown in Fig. 12-4. Elements differ not only in the number of electrons which make up their individual "solar systems" but also in the energies of their electrons. Therefore, if two different materials are brought together under suitable conditions, there will be a *difference* of electron energies. This difference produces the electrical pressure that we call *voltage.*

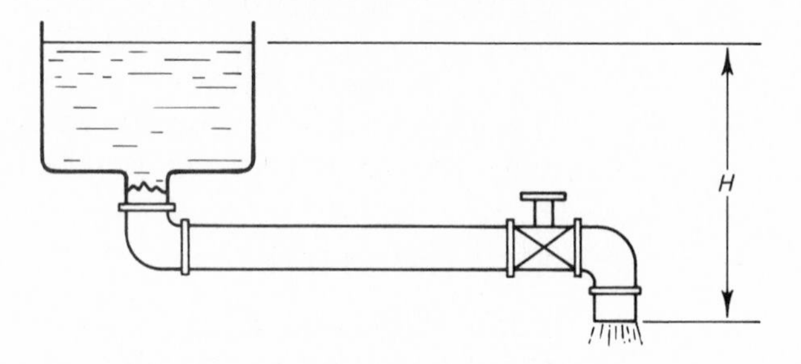

Fig. 12-4 The difference between the heights of the water levels causes pressure.

Different combinations of materials will produce different voltages. A dry cell of carbon and zinc produces 1.5 V, a cell of lead and sulfuric acid produces 2.1 V, and an Edison storage cell of nickel and iron produces 1.2 V. These combinations are called *batteries.* A voltage pressure may be produced in several other ways. For example, the 110–120 V supplied by an ordinary house outlet is produced by a machine called a *generator.* An automobile spark coil delivers about 1,500 V. The purpose of all voltages, however they are produced, is to provide a force to *move* the electrons. It is appropriately called an *electron-moving* or *electromotive force,* which is abbreviated as emf. Very often the simple symbol E is used to indicate voltage.

QUANTITY OF ELECTRICITY

The amount of electricity represented by a single electron is very small—much too small to be used as a measure of quantity in practical electrical work. The electron has already been compared with a drop of water. It is obviously ridiculous to measure quantities of water by the number of drops. Instead, we use quantities like a gallon or a quart, each of which represents a certain number of drops. Similarly, the practical unit of electrical quantity represents a certain number of electrons. This unit is called a *coulomb* (C). It was named after Charles A. de Coulomb, a French physicist (1736–1806). The coulomb is equal to about 6 billion billion electrons (6,000,000,000,000,000,000). The symbol for the amount of electricity measured in coulombs is Q, since this represents a definite quantity of electrons.

CURRENT

When a voltage is applied across the ends of a conductor, the electrons, which up to now have been moving in many different directions (Fig. 12-5*a*) are forced to move in the *same direction* along the wire (Fig. 12-5*b*). Individual electrons all along the path are forced to leave the

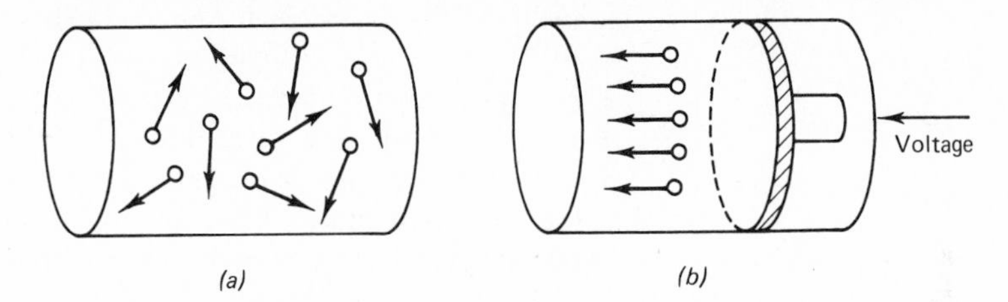

Fig. 12-5 (a) *The electrons in the wire move in many different directions.* (b) *When a voltage pushes the electrons, they all move in the same direction and make an electric current.*

atoms to which they are attached. They travel only a short distance until they find an atom that needs an electron. This motion is transmitted along the path from atom to atom, as motion in a whip is transmitted from one end to the other. This *flow* of electrons is called an *electric current* (Fig. 12-6). The speed of this flow is very nearly equal to the speed of light, which is about 186,000 miles per second. The actual speed of the individual electrons is much slower, but the effect of a pressure at one end of a wire is felt almost instantaneously at the other end.

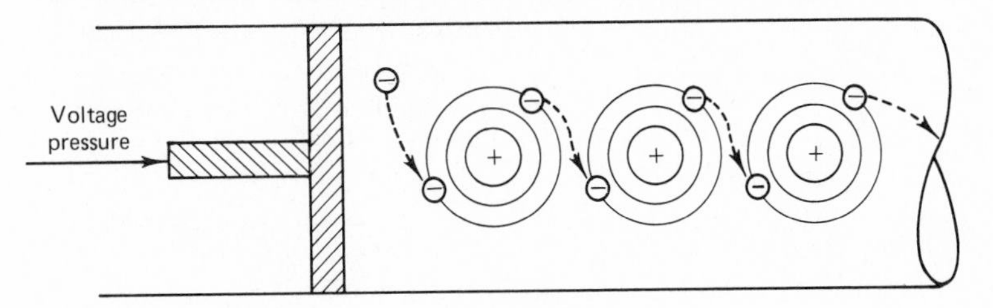

Fig. 12-6 An electric current is produced by billions of electrons moving through a wire.

The flow of water is measured as the number of gallons per minute, barrels per hour, etc. Similarly, the flow of electricity is measured by the number of electrons that pass a point in a wire in 1 sec. We do not have special names for the flow of water, but we do have a special name for the flow of electrons. This name is *ampere* (A) of current. The ampere was named after André Marie Ampère, a French physicist and chemist (1775–1836). A current of 1 ampere represents a flow of 1 coulomb of electricity (6 billion billion electrons) past a point in a wire in 1 sec. The symbol for current is I.

Since an electron is so small (about 25 trillion to an inch) and since there are so many of them, it is impossible to count them as they go by. However, when electrons are moving as an electric current, they can do useful work such as lighting lamps, running motors, producing heat, and plating metals. We can make use of this last ability of an electric current to measure and define it accurately. An international commission has defined an ampere as that number of electrons which can deposit a definite amount of silver (0.001118 gram) from a silver solution in 1 second. A 100-W 110-V lamp uses about 1 A. A 600-W 110-V electric iron uses about 5.5 A. The current required by a radio or television tube may be as low as 0.001 A.

RESISTANCE

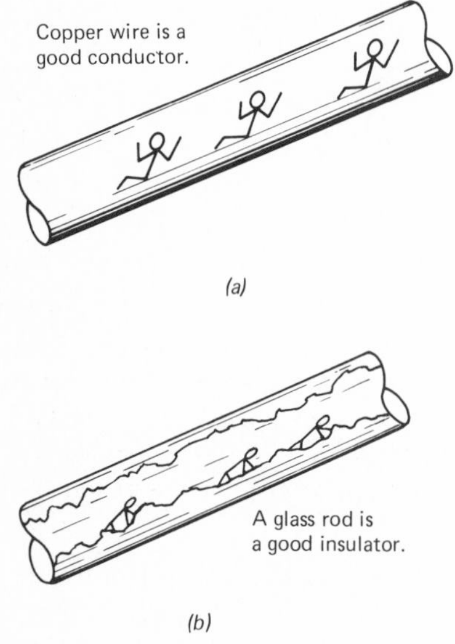

Fig. 12-7 (a) *Electrons travel freely through conductors.* (b) *Insulators prevent electrons from flowing freely.*

We have learned that "free" electrons may be forced to move from atom to atom when a voltage pressure is applied. Different materials vary in their number of "free" electrons and in the ease with which electrons may be transferred between atoms. A *conductor* (Fig. 12-7*a*) is a material through which electrons may travel freely. Most metals are good conductors. An *insulator* (Fig. 12-7*b*) is a material which prevents electrons from traveling through it easily. Nonmetallic materials like glass, mica, porcelain, rubber, and textiles are good insulators. No material is a perfect insulator or a perfect conductor.

The ability of a material to resist the flow of electrons is called its *resistance* and is measured in units called *ohms* (Ω). This unit of measurement was named after Georg Simon Ohm, a German physicist (1787–1854). The symbol for resistance is R. An international agreement defines the ohm as the resistance offered by a column of mercury of uniform cross section, 106.3 centimeters long (about 41.8 in), and weighing 14.45 grams (about $\frac{1}{2}$ oz). For example, 1,000 ft of #10 copper wire has a resistance of almost exactly 1 Ω. The resistance of a 40-W electric light is 300 Ω when hot. A 150-V voltmeter has a resistance of 15,000 Ω.

ELECTRICAL MEASUREMENTS AND CIRCUITS

A *voltmeter* measures the voltage E in units of volts and is always connected across the ends of the part in order to measure the difference in pressure. See Fig. 12-8. An *ammeter* measures the current I in units of amperes and is always connected into the circuit as shown in Fig. 12-8 so as to measure the flow of electrons through it.

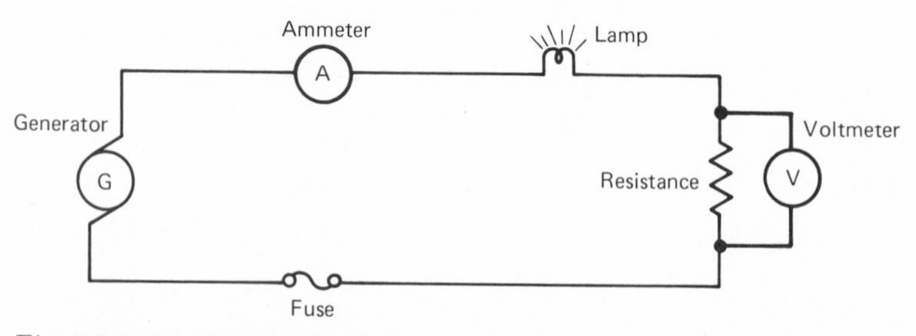

Fig. 12-8 A voltmeter is always connected across the ends of the part. An ammeter is always inserted into the line of the circuit.

ELECTRIC CIRCUITS

A *circuit* is simply a *complete* path along which the electrons can flow. A complete circuit must have an *unbroken* path, as shown in Fig. 12-9*a*. The source of energy acts as an electron pump to force the electrons through the conductor (usually a copper wire) against the resistance of the device to be operated. When the switch is opened, as in Fig. 12-9*b*, the electrons cannot leave one side of the switch to enter the other side. They cannot return to their source because of the very high resistance of the air gap. This is an *open circuit,* and no current will flow.

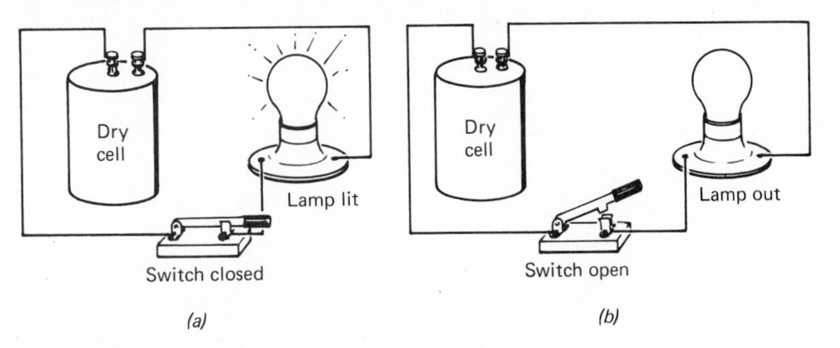

Fig. 12-9 (a) *With the switch closed, current flows through this complete circuit.* (b) *With the switch open, no current flows through this broken or open circuit.*

SYMBOLS

Up to this point, most of our diagrams have used pictures of the various electrical devices. However, not everybody can draw pictures quickly and accurately, and so we shall substitute special simple diagrams to illustrate the various parts of any circuit. Each circuit element is represented by a simple diagram called the *symbol* for the part. The standard symbols for the commonly used electrical and electronic components are given in Fig. 12-10.

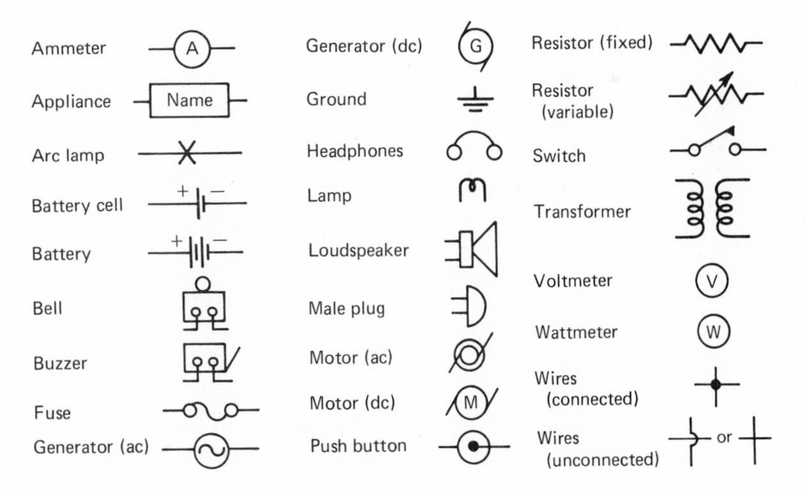

Fig. 12-10 Standard circuit symbols.

CIRCUIT DIAGRAMS

Every electric circuit must contain the following:

1. A source of electrical pressure, or voltage E, measured in volts.

2. An unbroken conductor through which the electrons may flow easily. The amount of electron flow per unit of time is the current I, measured in amperes.

3. A load, or resistance R, measured in ohms.

Example 12-1 Draw a circuit containing 2 battery cells, an ammeter, a fuse, a switch, a resistor, and a lamp. Label each part, using E for voltage, R for resistance, and I for current.

Solution See Fig. 12-11. The numbers 1 and 2 under the letter R are a convenient way to indicate the first resistance R_1 and the second resistance R_2.

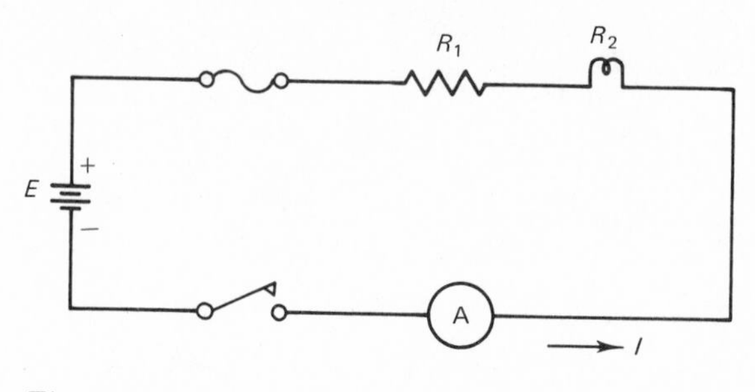

Fig. 12-11

Problems

Using the circuit symbols shown in Fig. 12-10, draw a circuit for each problem, including the elements indicated. Label the diagram completely, using the subscripts 1, 2, 3, etc., to indicate the parts of the circuit.

1. A battery of 2 cells, a fuse, and 2 resistors
2. An ac generator, a switch, an ammeter, a bell, and a buzzer
3. A battery, a switch, an ammeter, 2 resistors, and a lamp with a voltmeter across it
4. A battery, a switch, 2 bells, and an ammeter
5. A dc generator, a switch, a variable resistor, and a dc motor
6. A dc generator, a switch, a fuse, an arc lamp, and a resistor
7. A male plug, an electric iron, a variable resistor, and a switch
8. A dc generator, a switch, an arc lamp, and 2 resistors

Summary

1. All material is made of electric charges.
2. Negative charges are called *electrons*.
3. Positive charges are called *protons*.
4. Similar charges repel each other, and opposite charges attract each other.
5. A *coulomb* (C) represents a quantity of 6 billion billion electrons.
6. An *ampere* (A) is a current of 1 coulomb per second. The symbol for current is I.
7. *Current* is measured by an ammeter, which is always inserted directly into the circuit and reads amperes.
8. Electrons move because of the pressure of an electromotive force, or *voltage*. The symbol for voltage is E.
9. *Voltage* is measured by a voltmeter, which is always connected across the ends of the circuit element and reads volts (V).

10. The *resistance* of a circuit is the resistance offered by the circuit to the flow of electrons. The symbol for resistance is R.

11. *Resistance* is measured by an ohmmeter, which is always connected across the ends of the element and reads ohms (Ω).

12. A *conductor* is a material which allows electrons to flow easily.

13. An *insulator* is a material which prevents current from flowing easily.

14. A *fuse* is a thin strip of easily melted material. It protects a circuit from large currents by melting quickly and thereby breaking the circuit.

OHM'S LAW

There are three quantities which are present in every complete operating electric circuit.

1. The electromotive force E, expressed in volts (V), which causes the flow of current.

2. The resistance R of the circuit, expressed in ohms (Ω), which attempts to stop the flow of current.

3. The current I, expressed in amperes (A), which flows as a result of the voltage pressure exceeding the resistance.

A definite relationship exists among these three quantities. It is known as *Ohm's law*. This relationship is very important, since it is the basis for most of the calculations in electrical and electronics work. In the three circuits shown in Fig. 12-12, the voltage E of the battery is measured by a voltmeter. The current I that flows is measured by an ammeter. The resistance R of the resistor is indicated by the manufacturer by distinctive markings on the resistor.

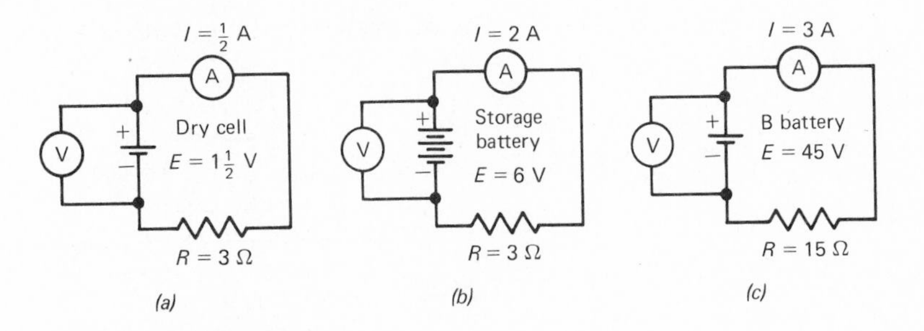

Fig. 12-12 The voltage equals the current multiplied by the resistance in a simple circuit.

Let us put the information from Fig. 12-12 into a table.

FIGURE	E	I	R	$I \times R$
a	$1\frac{1}{2}$	$\frac{1}{2}$	3	$\frac{1}{2} \times 3 = 1\frac{1}{2}$
b	6	2	3	$2 \times 3 = 6$
c	45	3	15	$3 \times 15 = 45$

In the last column of the table, we have multiplied the current I by the resistance R for each circuit. Evidently, the product of current I and resistance R is always equal to the voltage E of the circuit. This is true for all circuits; it was first discovered by Georg S. Ohm. This simple relationship is called *Ohm's law.*

RULE **Voltage equals current multiplied by resistance.**

FORMULA $E = I \times R$

In this formula, E must always be expressed in volts, I must always be expressed in amperes, and R must always be expressed in ohms.

SOLVING PROBLEMS

1. Read the problem carefully.
2. Draw a simple diagram of the circuit.
3. Record the given information directly on the diagram. Indicate the values to be found by question marks.
4. Write the formula.
5. Substitute the given numbers for the letters in the formula. If the number to be substituted for the letter is unknown, merely write the letter again. Be sure to include all mathematical signs like ×, =, etc.
6. Do the indicated arithmetic at the side so as not to interrupt the continued progress of the solution.
7. In the answer, indicate the letter, its numerical value, and the units of measurement.

Example 12-2 A doorbell requires $\frac{1}{4}$ A in order to ring. The resistance of the coils in the bell is 24 Ω. What voltage must be supplied in order to ring the bell?

Solution The diagram for the circuit is shown in Fig. 12-13.

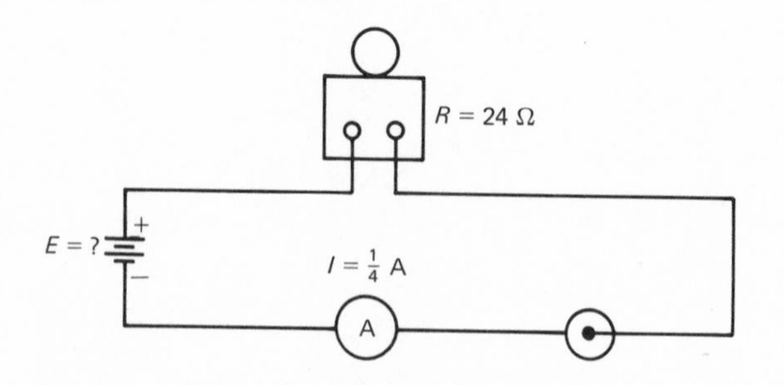

Fig. 12-13 The bell is operated by the push button.

1. Write the formula.

$$E = I \times R$$

2. Substitute numbers.

$$E = \frac{1}{4} \times 24$$

3. Multiply the numbers.

$E = 6$ V *Ans.*

Example 12-3 A relay coil is used to control the current from a power source to a motor. The resistance of the relay is 28 Ω. What voltage is required to operate the relay if it requires a current of 0.05 A?

Solution The diagram for the circuit is shown in Fig. 12-14.

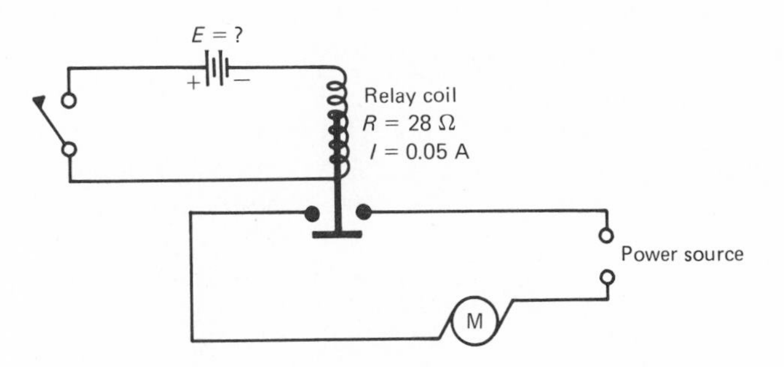

Fig. 12-14 The relay controls the large current from the power source.

1. Write the formula.

$E = I \times R$

2. Substitute numbers.

$E = 0.05 \times 28$

3. Multiply the numbers.

$E = 1.4$ V *Ans.*

Self-Test 12-4 An automobile battery supplies a current of 7.5 A to a headlamp whose resistance is 0.84 Ω. Find the voltage delivered by the battery.

Solution The diagram for the circuit is shown in Fig. 12-15.

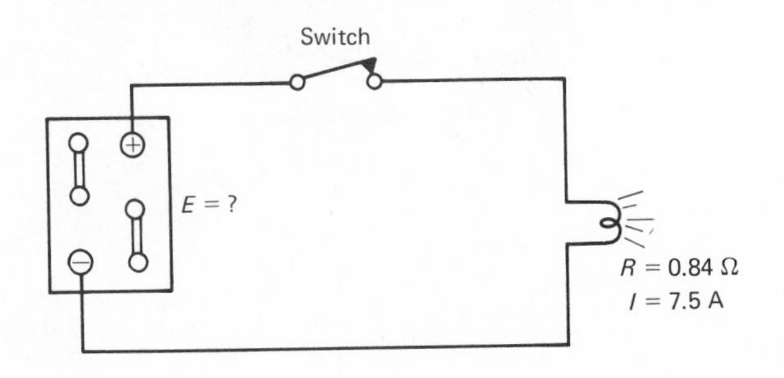

Fig. 12-15

1. Write the formula.

$E =$ ______ × ______

2. Substitute numbers.

$E =$ ______ × 0.84

3. Multiply the numbers.

$E =$ ______ V *Ans.*

I, R

7.5

6.3

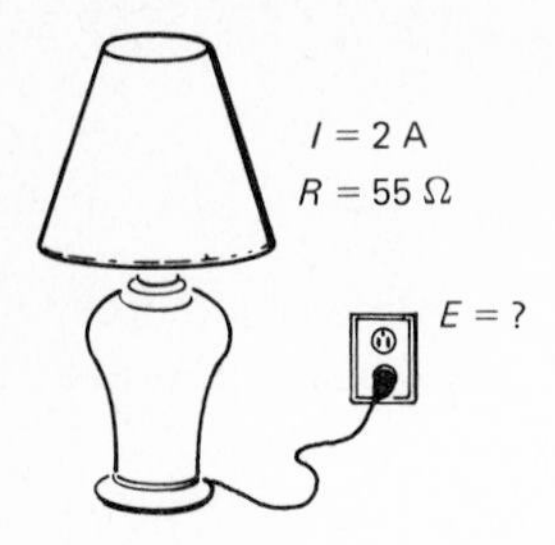

Fig. 12-16

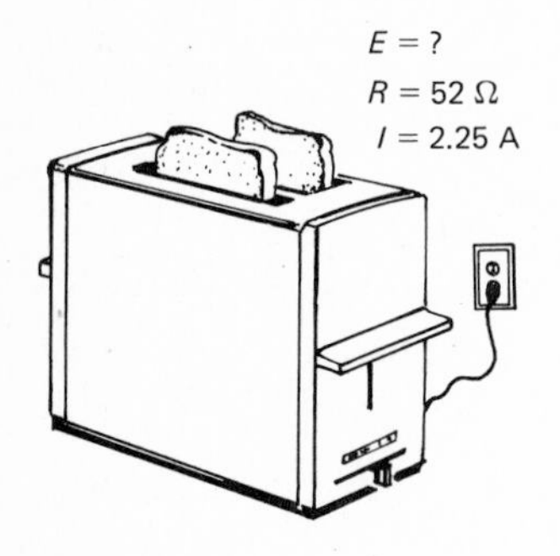

Fig. 12-17

Problems

1. What voltage is required to light the lamp shown in Fig. 12-16 if the current required is 2 A and the resistance of the lamp is 55 Ω?
2. A 20-Ω heating element resistor draws 3 A from a line. Find the voltage across the resistor.
3. If the total resistance (impedance) of a radio receiver is 240 Ω and it draws 0.5 A, what voltage is needed?
4. A certain television tube takes 0.15 A. Its resistance is 80 Ω. What voltage does it need?
5. An arc lamp whose hot resistance is 9 Ω draws 6.2 A. What voltage is required?
6. What voltage is required to operate a 5,500-Ω electric clock which draws 0.02 A?
7. The 52-Ω electric toaster shown in Fig. 12-17 uses 2.25 A. Find the required voltage.
8. What voltage is needed to energize the field coil of a loudspeaker if its resistance is 1,100 Ω and it draws 0.04 A?
9. The coils of a washing machine motor have a resistance of 21 Ω. What voltage is required if the motor draws 5.3 A?
10. What is the voltage required to operate the electric impact wrench shown in Fig. 12-18 if its resistance is 24.5 Ω and it draws 4.8 A?

Fig. 12-18 An electric impact wrench. (Courtesy Rockwell International)

11. What is the voltage required for an electroplating tank whose resistance is 0.35 Ω and which requires a current of 80 A?
12. A neon electric sign draws 1.07 A. If its resistance is 98 Ω, find the voltage needed.
13. An electric bell has a resistance of 25 Ω and will not operate on a current of less than 0.25 A. What is the smallest voltage that will ring the bell?
14. The resistance of a telephone receiver is 1,000 Ω. If the current is 0.032 A, what is the voltage across the receiver?

15. What voltage is supplied to a 0.015-Ω dc arc welder drawing 650 A?
16. A 125-Ω relay coil needs 0.15 A to operate. What is the lowest voltage needed to operate the relay?
17. The resistance of a spark plug air gap is 2,200 Ω. What voltage is needed to force 0.16 A across the gap?
18. A 250,000-Ω resistor in the plate circuit of a 6CL6 video amplifier tube in a TV set draws 0.0003 A. Find the voltage across the resistor.
19. The line from an automobile battery to a distant transmitter must not use more than 0.25 V when the transmitter is operating. If the current from the battery is 18 A, will a 0.015-Ω line be satisfactory?
20. In the voltage divider circuit shown in Fig. 12-19, find the voltage across the resistors R_1, R_2, and R_3:

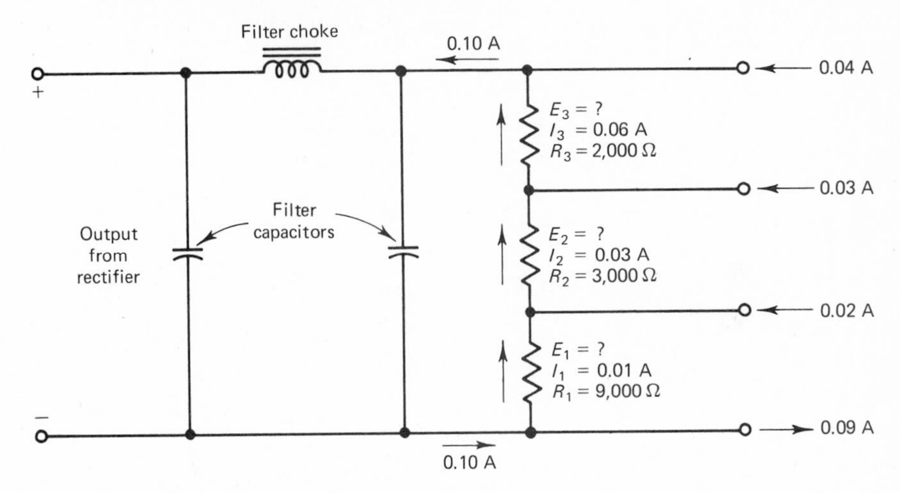

Fig. 12-19 A voltage divider provides different voltages from a single source.

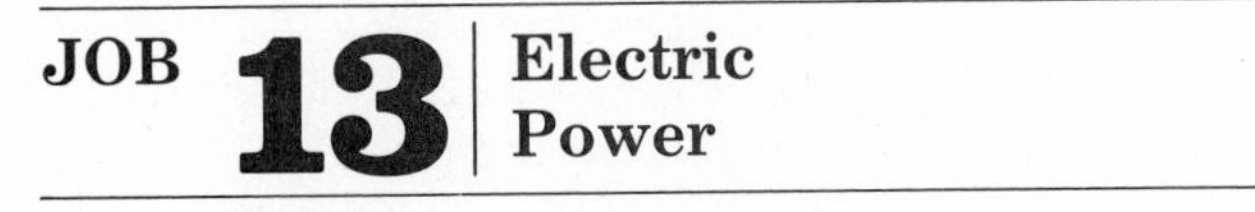

JOB 13 Electric Power

RULE **The electric power in any part of a circuit is equal to the current in that part multiplied by the voltage across that part of the circuit.**

FORMULA $P = I \times E$

where: P = power in watts
I = current in amperes
E = voltage in volts

Note: a watt (W) of electric power is the power used when 1 volt causes 1 ampere of current to flow in a circuit. A *kilowatt* (kW) of power is equal to 1,000 W. See Table 13-1.

Table 13-1. Power Used by Electrical Devices (in Watts)

Device	Watts	Device	Watts
Electric clocks	1–3	Vacuum cleaners	300–700
Door chimes	15	Washers	350–450
Electric fans	50–300	Toasters	600–1,100
Lamp bulbs	15–200	Air conditioners	800–1,500
Sewing machines	40–80	Broilers	800–1,500
Radios	50–100	Electric ironers	1,000–1,500
Televisions	150–250	Clothes driers	4,000–4,700
Refrigerators	200–300	Electric ranges	Up to 23,000

Example 13-1 What is the power consumed by an automobile headlight if it takes 2.8 A at 6 V?

Solution Given: $E = 6$ V Find: $P = ?$
$I = 2.8$ A

1. Write the formula.

$P = I \times E$

2. Substitute numbers.

$P = 2.8 \times 6$

3. Multiply the numbers.

$P = 16.8$ W *Ans.*

Example 13-2 A 1,000-Ω resistor is used in the power-supply filter circuit of many small radio receivers. If the resistor carries 0.06 A at 60 V, how many watts of power are developed in the resistor? What must be the wattage rating of the resistor in order to dissipate this power safely as heat?

Solution Given: $I = 0.06$ A Find: $P = ?$
$E = 60$ V Wattage rating = ?

$P = I \times E$

$= 0.06 \times 60 = 3.6$ W *Ans.*

The wattage rating of a resistor describes its ability to dissipate the heat produced in it by the passage of an electric current without overheating. For example, a 2-W resistor could dissipate 2 W of heat energy without overheating. However, if 3 W of power were to be developed in it, it would overheat because it could not dissipate the extra 1 W of power. Owing to the lack of ventilation in the close quarters of most television receivers, the wattage rating of these resistors is usually at least twice the wattage developed in them.

Wattage rating $= 2 \times P$

$= 2 \times 3.6 = 7.2$ W *Ans.*

Self-Test 13-3 A room air conditioner draws 7.5 A from the 120-V house line. (a) Find the power consumed. (b) At $0.06 for each kilowatt used per hour, what is the cost of operating the air conditioner for 8 hr?

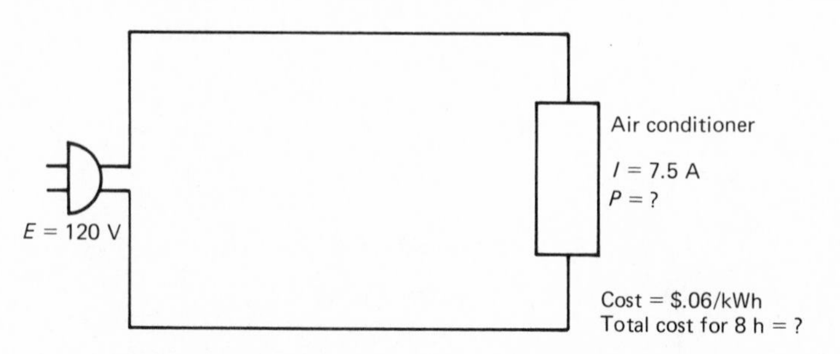

Fig. 13-1

Solution **a.** Find the power consumed.

1. Write the formula.

$P =$ ______ × ______

2. Substitute numbers.

   $P = 7.5 \times$ ______ — **120**

3. Multiply the numbers.

   $P =$ ______ W *Ans.* — **900**

b. Find the cost.

1. Since 1 kW = ______ W, we can change 900 W into kilowatts by (multiplying)(dividing) 900 by 1,000. — **1,000**, **dividing**
   Therefore, 900 W = 900 ÷ 1,000 = ______ kW. — **0.9**

2. The number of kilowatthours (kWhr) equals

   kilowatts × ______ — **hours**

   or kWhr = 0.9 × 8 = ______ kWhr — **7.2**

3. The total cost = kilowatthours × cost per ______. — **kilowatthour**

   = 7.2 × ______ — **$0.06**
   = ______ *Ans.* — **$0.43**

Problems

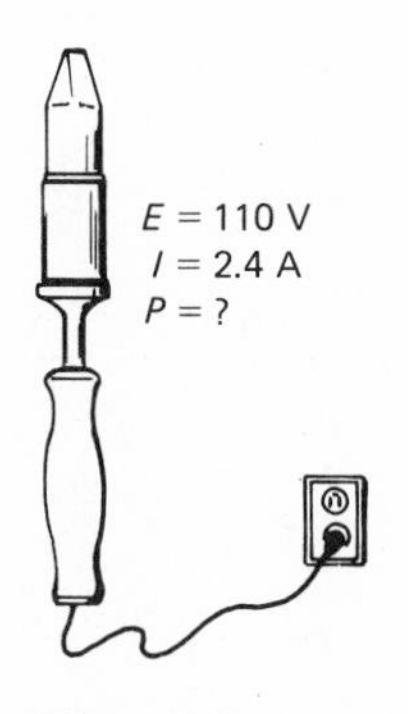

Fig. 13-2

Fig. 13-3 A portable saw. (Courtesy Rockwell International)

1. An automobile starting motor draws 80 A at 6 V. How much power is drawn from the battery?
2. A 20-hp motor takes 74 A at 230 V when operating at full load. Find the power used.
3. What is the wattage dissipated as heat by an electric heater operating at 110 V and 0.2 A?
4. What is the power consumed by a truck headlight if it takes 5.8 A at 12 V?
5. Find the power used by the 2.4-A 110-V soldering iron shown in Fig. 13-2.
6. The heater of a 12B4A vertical-deflection-amplifier tube used 12.6 V and 0.3 A. Find the power consumed.
7. The high contact resistance of a poorly wired electric plug reduced the current by 1 A on a 110-V line. Find the power wasted in the plug.
8. How much power is consumed by an electric clock using 0.02 A at 110 V?
9. If the voltage drop across a spark plug air gap is 30 V, find the power consumed in sending 0.002 A across the gap.
10. Find the power consumed by the portable saw shown in Fig. 13-3 if it draws 7.8 A at 115 V.
11. A power supply delivers 0.16 A at 250 V to a public address amplifier. Find the watts of power delivered.
12. A 180-Ω line cord resistor carrying 0.15 A causes a voltage drop of 27 V. How much power must be dissipated as heat? What must be the wattage rating of the line cord?
13. A window air conditioner is rated at 6.9 A and operated on a 117-V line. Find the wattage used by the conditioner.
14. An electric oven uses 36.3 A at 117 V. Find the wattage used by the oven.

Fig. 13-4 Tapping an electric arc furnace. (Courtesy Allegheny Ludlum Industries, Inc.)

15. The emitter resistor in a transistor circuit carries 0.045 A at 10 V. What must be its wattage rating?
16. An electric enameling kiln takes 9.2 A from a 117-V line. Find the power used.
17. How much power is used by a $\frac{3}{4}$-ton air conditioner drawing 11.4 A from a 220-V line?
18. The electric arc furnace shown in Fig. 13-4 uses 94 A at 440 V. Find the power used.
19. The cathode bias resistor for a television tube causes a drop of 16.4 V when 0.45 A flows through it. Find the wattage rating of the resistor.
20. How many watts are dissipated as heat by a line cord resistor if it carries 0.22 A at 30 V?
21. A Tungar battery charger uses 15 A at 110 V. What is the cost of charging a bank of storage batteries at $0.05/kWhr if it requires 1 hr to charge the batteries?
22. If 15 lamps in parallel each take 1.67 A when connected across a 120-V line, find the wattage of each lamp and the total wattage used.
23. A certain transistor can pass a maximum of 0.3 A. If it is rated at 0.15 W maximum, will it be able to withstand a 0.45-V collector-emitter voltage?
24. What is the total power used by a 4.5-A electric iron, a 0.85-A fan, and a 2.2-A refrigerator motor if they are all connected in parallel across a 115-V line?
    Hint: add the currents to get the total current used at 115 V.
25. An electric percolator drawing 9 A and an electric toaster drawing 10.2 A are connected in parallel to the 117-V house line. Find the total power consumed.
26. Find the power drawn from a 6-V battery by a parallel circuit of 2 headlights (4 A each) and 2 taillights (0.9 A each).
27. An automobile headlight lamp draws 3 A at 6 V. If it is used for 3 hr/day for 300 days, find the cost at $0.05/kWhr.

JOB 14 Solving Equations and Electrical Formulas

As you may have noticed, every formula contains the sign of equality. The statement that a combination of quantities is *equal* to another combination of quantities is called an *equation*. In this sense, every formula is an equation. Examples of some equations are

$$3 \times 4 = 12$$

$$2 \times R = 10$$

$$E = I \times R$$

$$3y = 12$$

Note: no mathematical sign between quantities means that they are to be multiplied. Thus $3y$ means $3 \times y$. For this reason, Ohm's law is often written as $E = IR$.

For a statement to be termed an equation, it is necessary only that the value on the left side of the equality sign be *truly equal* to the value on the right side.

WORKING WITH EQUATIONS

Many mathematical operations may be performed on an equation. Whatever is done, however, the basic equality of the statements on each side of the equality sign must not be destroyed. This equality must be maintained if the equation is to remain an equation. This is accomplished by applying the following basic principle.

PRINCIPLE **Any mathematical operation performed on one side of an equality sign must also be performed on the other side.**

For example, let us subject the equation $3 \times 4 = 12$ to different mathematical operations.

RULE 1 **The same number may be added to both sides of an equality sign without destroying the equality.**

1. Write the equation.

$3 \times 4 = 12$

2. Add **3** to both sides.

$(3 \times 4) + \mathbf{3} = 12 + \mathbf{3}$

3. Do the arithmetic.

$12 + \mathbf{3} = 12 + \mathbf{3}$

$15 = 15$

which is a true equation, since the left side is still equal to the right side.

RULE 2 **The same number may be subtracted from both sides of an equality sign without destroying the equality.**

1. Write the equation.

$3 \times 4 = 12$

2. Subtract **3** from both sides.

$(3 \times 4) - \mathbf{3} = 12 - \mathbf{3}$

3. Do the arithmetic.

$12 - \mathbf{3} = 12 - \mathbf{3}$

$9 = 9$

which is a true equation, since the left side is still equal to the right side.

RULE 3 **Both sides of an equality sign may be multiplied by the same number without destroying the equality.**

1. Write the equation.

$3 \times 4 = 12$

2. Multiply both sides by **3**.

$(3 \times 4) \times \mathbf{3} = 12 \times \mathbf{3}$

3. Do the arithmetic.

$$12 \times 3 = 12 \times 3$$

$$36 = 36$$

which is a true equation, since the left side is still equal to the right side.

RULE 4 **Both sides of an equality sign may be divided by the same number without destroying the equality.**

1. Write the equation.

$$3 \times 4 = 12$$

2. Divide both sides by **3**.

$$\frac{3 \times 4}{\mathbf{3}} = \frac{12}{\mathbf{3}}$$

3. Do the arithmetic.

$$\frac{12}{\mathbf{3}} = \frac{12}{\mathbf{3}}$$

$$4 = 4$$

which is a true equation, since the left side is still equal to the right side.

These examples show that if the same mathematical operation is performed on *both* sides of an equation, it does not destroy the basic equality of the equation.

SOLVING EQUATIONS

Consider the equation $2R = 10$. To solve this equation means to find the value of the unknown letter R that will make the equation a true statement. This value is found *when the letter stands all alone on one side of the equality sign*. When this occurs, the equation has the form

R = some number

This number will obviously be the value of the letter R, and the equation will be solved.

How do we get the letter all alone? In the equation $2R = 10$, the letter will be alone on the left side of the equality sign if we can somehow eliminate the number 2 on that side. The number 2 will actually be eliminated if we can change it to the number 1, since $1R$ means the same as R. This can be done by applying the basic principle that permits us to perform the same operation on both sides of the equality sign. But which mathematical operation will eliminate the number 2? In general, the operation to be performed on both sides of the equality sign will be the *opposite* of that used in the equation. Since $2R$ means 2 *multiplied* by R, we shall use rule 4 above, and *divide both sides* of the equation by that same number 2.

1. Write the equation.

$$2R = 10$$

2. Divide both sides by 2.

$$\frac{2R}{2} = \frac{10}{2}$$

3. Simplify each side separately.

$$\frac{\overset{1}{\cancel{2}}R}{\underset{1}{\cancel{2}}} = \frac{\overset{5}{\cancel{10}}}{\underset{1}{\cancel{2}}} \quad \text{or} \quad 1R = 5 \quad \text{or} \quad R = 5 \quad \textit{Ans.}$$

Only those numbers that appear on the *same* side of the equality sign may be canceled. Do *not* cancel across the equality sign.

Suppose the unknown letter appears on the right side of the equality sign, as in the equation $12 = 3R$. In this situation, we proceed exactly as before. To solve the equation means to get the letter *all alone* on one side of the equality sign. It is not important which side is chosen, as long as the letter is *all alone* on that side. We can get the letter R all alone on the right side by changing the $3R$ to $1R$ by applying the *opposite* operation of *dividing both sides* of the equation by the number 3.

1. Write the equation.

$$12 = 3R$$

2. Divide both sides by 3.

$$\frac{12}{3} = \frac{3R}{3}$$

3. Simplify each side separately.

$$\frac{\overset{4}{\cancel{12}}}{\underset{1}{\cancel{3}}} = \frac{\overset{1}{\cancel{3}}R}{\underset{1}{\cancel{3}}}$$

$$4 = 1R \quad \text{or} \quad 4 = R \quad \text{or} \quad R = 4 \quad \textit{Ans.}$$

Notice that the number that is multiplied by the unknown letter will be canceled out *only* if we divide both sides of the equality sign *by that same number.* Dividing both sides by any other number will *not* eliminate this number.

RULE 5 **To eliminate the number which is multiplied by the unknown letter, divide both sides of the equality sign by the multiplier of the letter.**

Example 14-1 Solve the following equations for the values of the unknown letters.

Solution 1. Write the equations.

$$2R = 10 \qquad 14 = 7E$$

2. Divide both sides of each equation by the multiplier of the letter.

$$\frac{2R}{2} = \frac{10}{2} \qquad \frac{14}{7} = \frac{7E}{7}$$

3. Cancel out the multiplier of the letters.

$$R = \frac{10}{2} \qquad \frac{14}{7} = E$$

4. Divide,

$R = 5$ *Ans.* $2 = E$ *Ans.*

We are now in a position to shorten our work. Notice that in each example, the effect of dividing both sides of the equality sign by the multiplier of the letter has been to *move* the multiplier *across the equality sign* into the position shown in step 3. Since this will always occur, we can eliminate step 2 and proceed as shown in Example 14-2.

Example 14-2 Solve the following equations for the values of the unknown letters: $3R = 12$ and $15 = 5E$.

Solution

1. Write the equations.

$3R = 12$ $15 = 5E$

2. Divide the quantity all alone on one side of the equality sign by the multiplier of the letter.

$$R = \frac{12}{3} \qquad \frac{15}{5} = E$$

3. Divide.

$R = 4$ *Ans.* $3 = E$ *Ans.*

RULE 6 **To solve a simple equation of the form "a number multiplied by a letter equals a number," divide the number all alone on one side of the equality sign by the multiplier of the letter.**

Example 14-3 Solve the equation $4R = 23$ for the value of R.

Solution $4R = 23$

$$R = \frac{23}{4}$$

$R = 5.75$ *Ans.*

Example 14-4 Solve the equation $18 = 0.3Z$ for the value of Z.

Solution $18 = 0.3Z$

$$\frac{18}{0.3} = Z$$

$60 = Z$ or $Z = 60$ *Ans.*

Self-Test 14-5 Solve the equation $0.04E = 4.68$ for the value of E.

Solution

1. Write the equation.

$0.04E = 4.68$

2. Solve for E.

$$E = \frac{?}{?}$$

$$\frac{4.68}{0.04}$$

3. Divide the numbers.

$E =$ ______ *Ans.* 117

Problems

Solve the following equations for the value of the unknown letter:

1. $3I = 15$	8. $16R = 4$	15. $40 = 0.2Z$
2. $5R = 20$	9. $20I = 117$	16. $8 = 0.4Z$
3. $2E = 12$	10. $19 = 2E$	17. $0.15R = 120$
4. $48 = 8R$	11. $\frac{1}{2}W = 20$	18. $0.003R = 78$
5. $7I = 63$	12. $20 = 100R$	19. $117 = 0.3Z$
6. $4L = 21$	13. $0.3R = 120$	20. $\frac{3}{5}T = 12$
7. $3R = 41$	14. $0.04Z = 60$	21. $8E = \frac{1}{2}$

SOLVING THE FORMULA FOR OHM'S LAW

The formula for Ohm's law is actually an equation. By applying Rule 6, we can solve Ohm's law for any unknown value of current or resistance.

Example 14-6 The total resistance of a relay coil is 50 Ω. What current will it draw from a 20-V source?

Solution The diagram for the circuit is shown in Fig. 14-1.

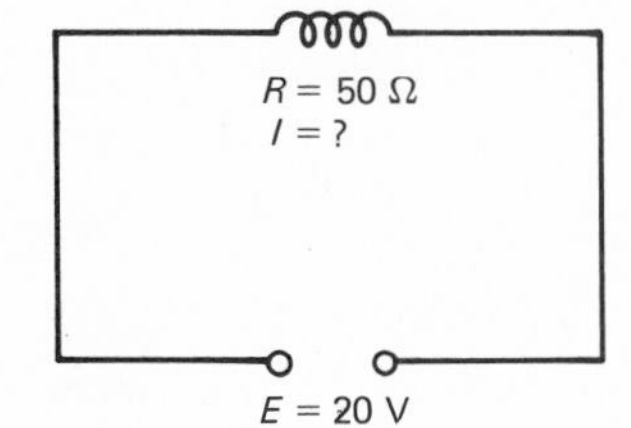

Fig. 14-1

1. Write the formula.

$E = IR$

2. Substitute numbers.

$20 = I \times 50$

3. Solve for I.

$$\frac{20}{50} = I$$

4. Divide the numbers.

$0.4 = I$ or $I = 0.4$ A *Ans.*

Example 14-7 Find the total resistance of a telegraph coil if it draws 0.015 A from a 6.6-V source.

Solution The diagram for the circuit is shown in Fig. 14-2.

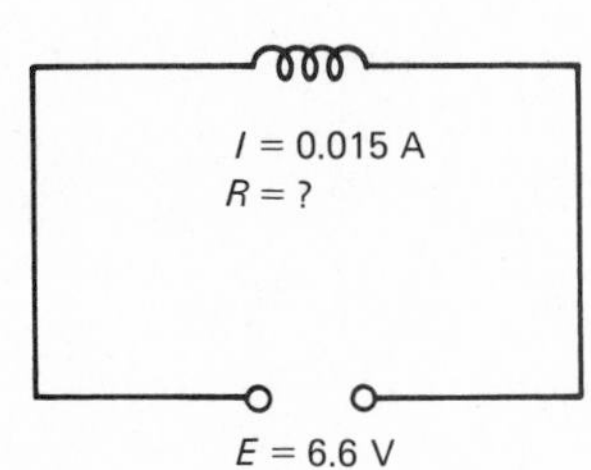

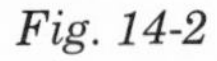
Fig. 14-2

1. Write the formula.

$E = IR$

2. Substitute numbers.

$6.6 = 0.015 \times R$

3. Solve for R.

$$\frac{6.6}{0.015} = R$$

4. Divide the numbers.

$440 = R$ or $R = 440$ Ω *Ans.*

Self-Test 14-8 Find the current drawn by a 90-Ω subway-car heater from the 550-V electrical system.

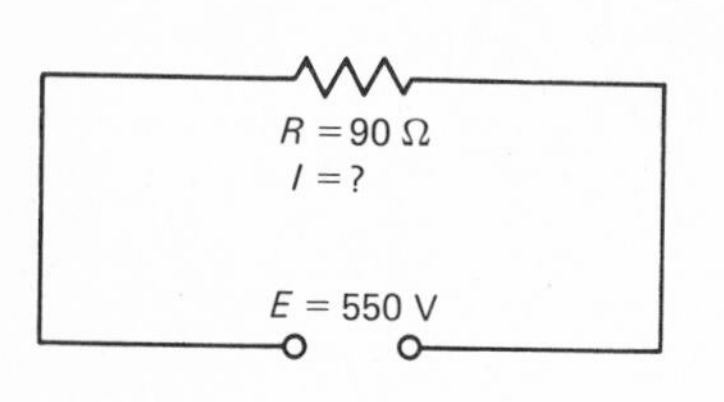

Fig. 14-3

Solution The diagram for the circuit is shown in Fig. 14-3.

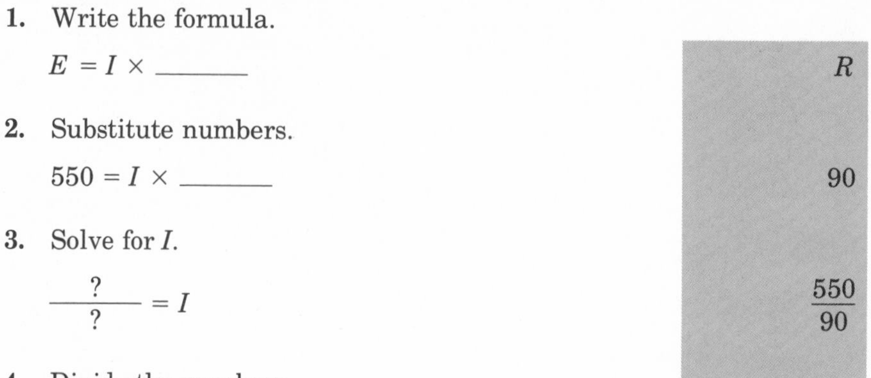

1. Write the formula.

   $E = I \times$ ______ | R

2. Substitute numbers.

   $550 = I \times$ ______ | 90

3. Solve for I.

   $\frac{?}{?} = I$ | $\frac{550}{90}$

4. Divide the numbers.

   ______ $= I$ or $I =$ ______ A *Ans.* | 6.11, 6.11

Problems

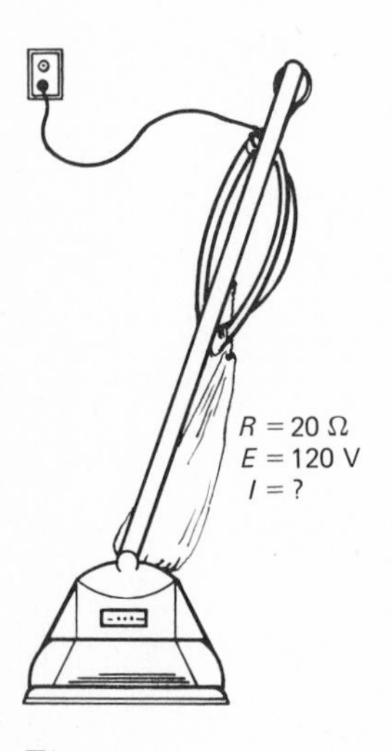

Fig. 14-4

1. What is the hot resistance of an arc lamp if it draws 15 A from a 30-V line?
2. The resistance of the motor windings of the electric vacuum cleaner shown in Fig. 14-4 is 20 Ω. If the voltage is 120 V, find the current drawn.
3. An electric enameling kiln draws 9 A from a 117-V line. Find the resistance of the coils.
4. The field magnet of a loudspeaker carries 0.4 A when connected to a 40-V supply. Find its resistance.
5. How much current is drawn from a 12-V battery when it is operating an automobile horn of 8 Ω resistance?
6. What is the hot resistance of a tungsten lamp if it draws 0.25 A from a 110-V line?
7. What current would flow in a 0.3-Ω short circuit of a 6-V automobile ignition system?
8. A dry cell indicates a terminal voltage of 1.2 V when a wire of 0.2 Ω resistance is connected across it. What current flows in the wire?
9. Find the resistance of an automobile starting motor if it draws 90 A from the 12-V battery.
10. A meter registers 0.0002 A when the voltage across it is 3 V. Find the total resistance of the meter circuit.
11. What is the resistance of a telephone receiver if there is a voltage drop of 24 V across it when the current is 0.02 A?
12. What current is drawn by the 75-Ω electric drill shown in Fig. 14-5 when it is operated from a 120-V line?
13. Find the current drawn by a 52-Ω toaster from a 117-V line.
14. Find the resistance of an electric furnace drawing 41 A from a 230-V line.
15. The resistance of the field coils of a shunt motor is 60 Ω. What is the field current when the voltage across the coils is 220 V?
16. A 12-V air-conditioning unit in an automobile has a resistance of 0.82 Ω. What is the size of fuse needed to protect the circuit?
17. The large copper leads on switchboards are called *bus bars*. What is the resistance of a bus bar carrying 400 A if the voltage across its ends is 0.6 V?

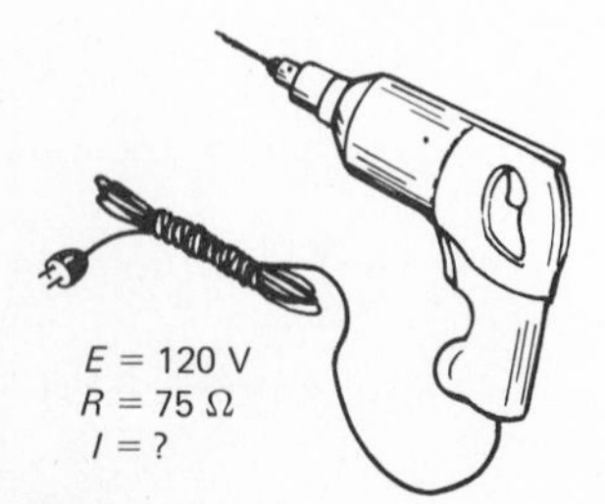

Fig. 14-5

18. If a radio receiver draws 0.85 A from a 110-V line, what is the total resistance (impedance) of the receiver?
19. A 32-candlepower lamp in a truck headlight draws 3.4 A from the 6-V battery. What is the resistance of the lamp?
20. If the resistance of the air gap in an automobile spark plug is 2,500 Ω, what voltage is needed to force 0.16 A through it?
21. A 2N525 transistor is used as a transistor switch to control a 20-V source across a 100-Ω load. Find the current that flows when the switch is conducting.
22. A voltage of 5 V appears across the 2,500-Ω load resistor in a self-biased transistor circuit. Find the current in the resistor.
23. What current flows through an automobile headlight lamp of 1.2 Ω resistance if it is operated from the 6-V battery?
24. What is the resistance of a buzzer if it draws 0.14 A from a 3-V source?
25. Find the resistance of an iron if it draws 3.8 A from a 110-V line.
26. A 160-Ω telegraph relay coil operates on a voltage of 9.6 V. What is the current drawn by the relay?

SOLVING THE POWER FORMULA FOR CURRENT OR VOLTAGE

We've just learned how to solve Ohm's law for the current or the resistance. The power formula is the same general type of equation, and we can use the same methods to solve it for the values of current or voltage.

Example 14-9 What is the operating voltage of an electric toaster rated at 600 W if it draws 5 A?

Solution Given: $P = 600$ W Find: $E = ?$
$I = 5$ A

$$P = I \times E$$

$$600 = 5 \times E$$

$$E = \frac{600}{5} = 120 \text{ V} \quad \textit{Ans.}$$

Self-Test 14-10 A stabilizing resistor in the emitter circuit of a 2N109 output transistor develops a voltage drop of 1.2 V while consuming 0.06 W of power. Find the current through the resistor.

Solution Given: $P = 0.06$ W Find: _______ = ?
$E =$ _______ V

$$P = I \times E$$

_______ $= I \times 1.2$

$$I = \frac{0.06}{?}$$

$I =$ _______ A *Ans.*

I
1.2
0.06
1.2
0.05

Problems

1. A 550-W neon sign operates on a 110-V line. Find the current drawn.
2. What current is drawn by a 480-W steam iron from a 120-V line? (Fig. 14-6)

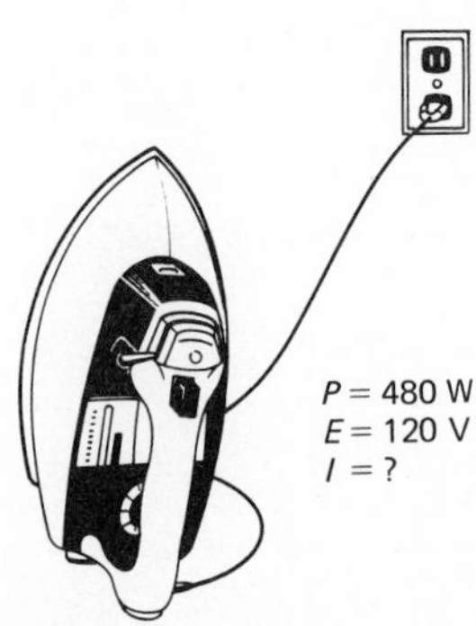

Fig. 14-6

Fig. 14-7 Model 311 sander/grinder. (Courtesy Foley Manufacturing Company)

3. Find the current drawn by a 1,200-W aircraft system from a 24-V source.
4. What current is drawn by a 1,500-W electric ironing machine from a 120-V line?
5. What current is drawn by a 55-candlepower 110-V lamp if it uses 1 W/candlepower?
6. The motor of the sander/grinder shown in Fig. 14-7 consumes 550 W at 115 V. Find the current drawn.
7. Twenty 60-W lamps are connected in parallel to light a stage. Find the current drawn from a 220-V source.
8. A washing machine motor requires 350 W of power. If it draws $3\frac{1}{8}$ A, find the operating voltage.
9. An electric broiler rated at 1,550 W operates from a 117-V line. The available fuses are rated at 20 A, 30 A, 50 A, and 60 A. Which fuse should be used in the circuit to protect the broiler?
10. A 6EM5 tube is used in the vertical output stage of the Emerson No. 120969 television receiver and delivers 3.4 W. If the current is 0.032 A, what is the voltage across the circuit?

USING OHM'S LAW TO SOLVE POWER PROBLEMS

In some problems, the power cannot be found because either the voltage or the current is unknown. In these instances, the unknown value is found by Ohm's law.

Example 14-11 Find the power taken by a soldering iron with 60 Ω resistance if it draws a current of 2 A.

Solution Given: $R = 60\ \Omega$ Find: $P = ?$
$I = 2$ A

1. Find the voltage.

$E = I \times R$

$E = 2 \times 60 = 120$ V

2. Find the power.

$P = I \times E$

$P = 2 \times 120 = 240$ W *Ans.*

Self-Test 14-12 Find the power used by the 11-Ω resistance element of an electric furnace if the voltage is 110 V.

Solution Given:

$R = 11\ \Omega$ Find: $P = ?$
$E =$ ______ V

1. Find the current.

$E = I \times R$

$110 = I \times$ ______

$I = \frac{110}{11} =$ ______ A

110

11

10

2. Find the power.

$P = \_\_\_\_\_\_ \times E$

$P = 10 \times 110 = \_\_\_\_\_\_$ W *Ans.*

Problems

1. A 20-Ω neon sign operates on a 120-V line. Find the power used by the sign.
2. Find the power used by a 55-Ω electric light which draws 2 A.
3. What is the wattage dissipated by a 10,000-Ω voltage divider if the voltage across it is 250 V? What is its wattage rating?
4. What is the maximum power obtainable from a Grenet cell of 2 V which has an internal resistance of 0.02 Ω?
5. A 240-Ω resistor in the emitter circuit of the output stage of a transistor radio carries 0.005 A. Find the wattage developed in the resistor.
6. The resistance of an ammeter is 0.025 Ω. Find the power used by the meter when it reads 4 A.
7. What is the power consumed by a 90-Ω subway-car heater if the operating voltage is 550 V?
8. A 40-Ω pilot light is to be operated in a 0.15-A circuit. How many watts are developed in the lamp?
9. A 20-Ω toaster operates on a 115-V line. Find the power used.
10. A 12FX5 tube is used as an audio output tube in the RCA model KCS 176 television receiver. If it has a cathode resistor of 180 Ω which develops a bias of 5 V, find the power used by the bias resistor.
11. A voltmeter has an internal resistance of 220,000 Ω. How much power does it use when the meter reads 110 V?
12. In the Magnavox TV chassis T928 series, a 4,000-Ω dropping resistor is used to feed voltage to the screen of the 6BQ6 horizontal output tube. If the screen current is 0.03 A, how many watts are developed in the resistor?

JOB 15 Review of Ohm's Law and Power

In any electric circuit,

1. The voltage forces the ______ through a conductor against its resistance. — current

2. The ______ tries to stop the current from flowing. — resistance

3. The current that flows in a circuit depends on the ______ and the resistance. The relationship among these three quantities is described by Ohm's law. Ohm's law applies to an entire circuit or to any component part of a circuit. — voltage

The formula for Ohm's law is

$E = I \times \_\_\_\_\_\_$ — R

where E = ______, measured in ______ — voltage, volts

I = ______, measured in ______ — current, amperes

R = ______, measured in ______ — resistance, ohms

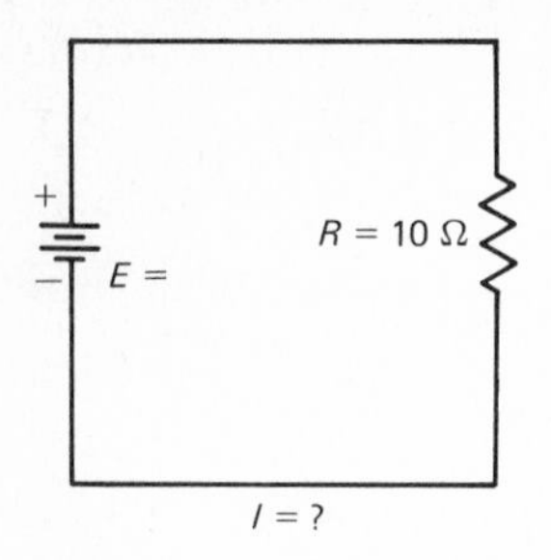

Fig. 15-1 When the resistance remains constant, the larger the voltage, the larger the current.

The formula for Ohm's law may be used to find the value of any one of the quantities in the formula. The student should also be able to determine the relative values of each quantity as one of the other quantities is changed in amount.

In Fig. 15-1, if the resistance $R = 10\ \Omega$ remains unchanged:

When the voltage $E = 10$ V, the current $I =$ ______.	1 A
When the voltage $E = 20$ V, the current $I =$ ______.	2 A
When the voltage $E = 50$ V, the current $I =$ ______.	5 A
When the voltage $E = 100$ V, the current $I =$ ______.	10 A
As you can see, when the resistance remains constant:	
The larger the voltage, the larger the current.	
The smaller the voltage, the ______ the current.	smaller
In Fig. 15-2, if the voltage $E = 100$ V remains unchanged:	
When the resistance $R = 1\ \Omega$, the current $I =$ ______.	100 A
When the resistance $R = 10\ \Omega$, the current $I =$ ______.	10 A
When the resistance $R = 50\ \Omega$, the current $I =$ ______.	2 A
When the resistance $R = 100\ \Omega$, the current $I =$ ______.	1 A
As you can see, when the voltage remains constant:	
The larger the resistance, the smaller the current.	
The smaller the resistance, the ______ the current.	larger
The power used by any part of a circuit is equal to the ______ in that part multiplied by the ______ across that part.	current, voltage
The formula for power is	
$P = I \times$ ______	E
where $P =$ ______, measured in ______	power, watts
$I =$ ______, measured in ______	current, amperes
$E =$ ______, measured in ______	voltage, volts
The formula may be used to find	
P if ______ and E are known	I
I if ______ and E are known	P
E if P and ______ are known	I
The wattage rating of a resistor is equal to ______ times the wattage developed in the resistor.	2

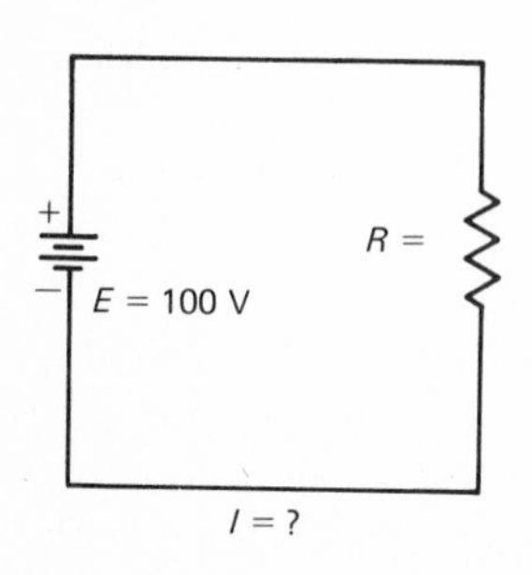

Fig. 15-2 When the voltage remains constant, the larger the resistance, the smaller the current.

FORMULAS IN ELECTRICAL WORK

A formula is a shorthand method for writing a rule. A number may be substituted for each letter in a formula. The signs of operation tell us what to do with these numbers.

STEPS IN SOLVING PROBLEMS

1. Read the problem carefully.
2. Draw a simple diagram of the circuit.
3. Record the given information directly on the diagram. Indicate the values to be found by question marks.
4. Write the formula.
5. Substitute the given numbers for the letters in the formula. If the number for the letter is unknown, merely write the letter again. Include all mathematical signs.
6. Do the indicated arithmetic. If after substitution the unknown letter is multiplied by some number, divide the number all alone on one side of the equality sign by the multiplier of the unknown letter.
7. In the answer, indicate the letter, its numerical value, and the unit of measurement.

Example 15-1 Solve the equation $6.3 = 0.3R$ for the letter R.

Solution 1. Write the equation.

$6.3 = 0.3R$

2. Solve for R.

$$\frac{6.3}{0.3} = R$$

3. Divide the numbers.

$21 = R$ or $R = 21$ *Ans.*

Self-Test 15-2 Find the current drawn by an 18-W automobile headlight from a 12-V battery.

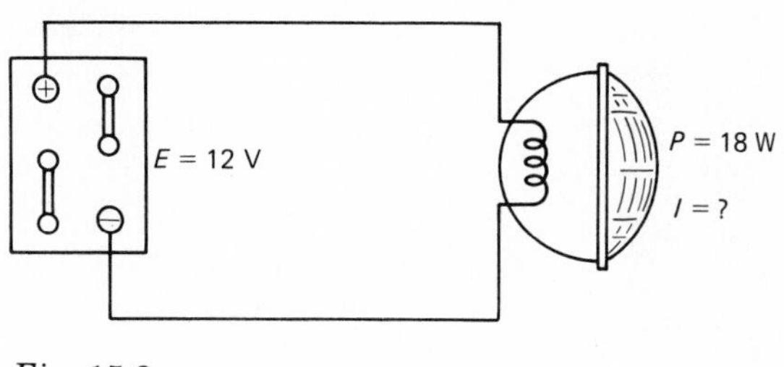

Fig. 15-3

Solution 1. The diagram for the circuit is shown in Fig. 15-3.

2. Write the formula.

$P = I \times$ _______

3. Substitute numbers.

_______ $= I \times 12$

E
18

4. Solve the equation.

$$\frac{18}{?} = I$$

12

5. Divide the numbers.

_______ $= I$ or $I = 1.5$ _______ *Ans.*

1.5, A

Self-Test 15-3 A poorly soldered joint has a contact resistance of 100 Ω. What is the power lost in the joint if the current is 0.5 A?

Solution Given: $R = 100\ \Omega$ Find: $P = ?$
$I =$ _______ A

0.5

1. Find the voltage.

$E =$ _______ $\times R$

$= 0.5 \times$ _______

$=$ _______ V

I

100

50

2. Find the power.

$P = I \times$ _______

$=$ _______ $\times 50$

$=$ _______ W *Ans.*

E

0.5

25

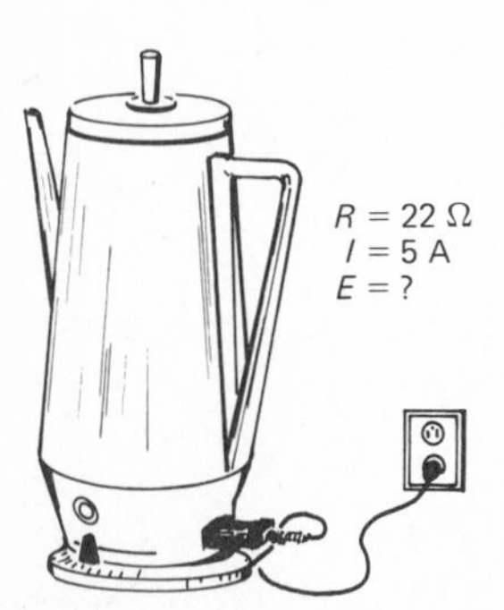

Fig. 15-4

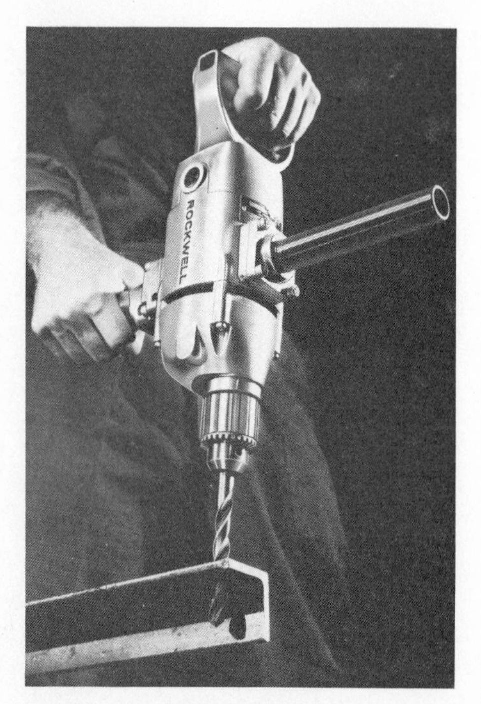

Fig. 15-5 A heavy duty electric drill. (Courtesy Rockwell International)

Problems

1. The resistance of the electric percolator shown in Fig. 15-4 is 22 Ω. If it draws 5 A, what is the operating voltage?
2. An electric heater whose coil is wound with #18 iron wire is connected across 110 V. If it draws a current of 10 A, what is the value of its resistance?
3. According to the National Electrical Code, #14 asbestos-covered type A wire should never carry more than 32 A. Is it safe to use this wire to carry power to a 10-Ω 230-V motor?
4. A washing-machine motor has a total resistance of 39 Ω and operates on 117 V. Find the current taken by the motor.
5. What is the voltage drop across an Allied model BK relay of 12,000 Ω resistance if it carries 0.0015 A?
6. A 30-Ω electric toaster draws 4 A. Find the power used.
7. Find the power used by the electric drill shown in Fig. 15-5 if it draws 6 A from a 117-V line.
8. A 6-V pilot light is to be operated in a 0.15-A circuit. How many watts does the lamp use?
9. A 20-Ω neon sign drawing 6 A operates on a 120-V line. Find the power consumed.
10. What is the resistance of a cathode bias resistor which causes a drop of 20 V when 0.05 A flows through it?
11. An electric cream separator requires 6.25 A at 120 V to process 100 gal of milk per hour. (a) How many watts of power are used per hour? (b) How many kilowatts are used per hour? (c) If electric energy costs $0.065/kWhr, find the cost to process 800 gal of milk.
12. What current is drawn by the 250-W motor of the scroll saw shown in Fig. 15-6 when it is operated from a 117-V line?
13. What is the voltage across a telephone receiver of 800 Ω resistance if the current flowing is 0.03 A?

Fig. 15-6 A scroll saw. (Courtesy Rockwell International)

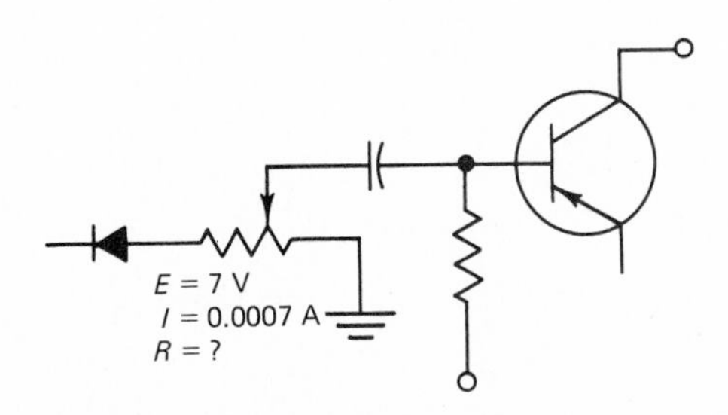

Fig. 15-7 A volume control in a simple transistor circuit.

14. Find the power consumed by a 100-Ω electric iron operating on a 115-V line.
15. A spot welder delivers 70 A when the voltage is 4.9 V. What is the resistance of the piece being welded?
16. What is the resistance of a bus bar carrying 300 A if the voltage drop across it is 1.2 V?
17. A volume control similar to that used in the Panasonic RF 738 transistor radio is shown in Fig. 15-7. How many ohms of resistance are engaged in the potentiometer if the voltage drop is 7 V and it passes a current of 0.0007 A?
18. A sensitive dc meter takes 0.009 A from a line when the voltage is 108 V. What is the resistance of the meter?
19. What is the power consumed by a voltage divider carrying 0.0025 A with a voltage drop of 250 V?
20. An electromagnet draws 5 A from a 110-V line. What current will it draw from a 220-V line?
21. If the voltage drop across a 10,000-Ω voltage divider is 90 V, find the power used.
22. A power pack delivers 4.2 A to a flash lamp through a cord with a resistance of 0.65 Ω. Find the voltage drop along the cord.
23. A 5,000-Ω resistor in a voltage divider reduces the voltage across it by 150 V. What current flows through the resistor?
24. The combined resistance of a coffee percolator and toaster in parallel is 22 Ω. Find the total power used if the line voltage is 110 V.
25. A series resistor is used to reduce the voltage to a motor by 45 V. What must be the resistance of the resistor if the motor draws 0.52 A?
26. What voltage is required to operate a 25-W automobile headlight bulb if the current drawn is 4 A?
27. A 250,000-Ω resistor in the plate circuit of a 6CL6 video amplifier tube draws 0.0003 A. Find the voltage across the resistor.
28. What should be the wattage rating of a 125-Ω resistor if it must carry 0.2 A?
29. The winding of a transformer has a resistance of 63 Ω. What is the voltage drop in the winding when it carries a current of 0.069 A?

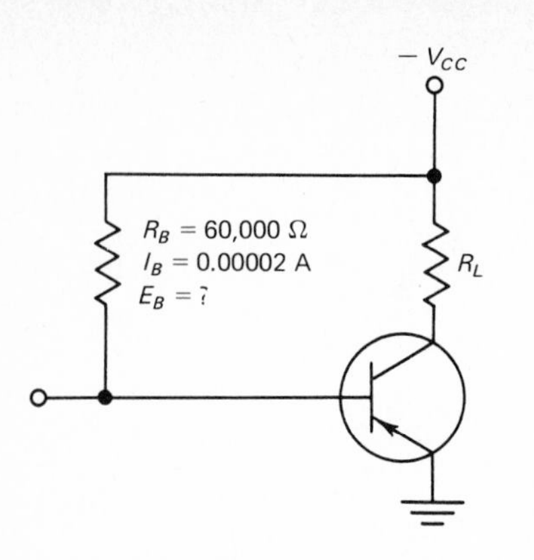

Fig. 15-8 A fixed-bias transistor circuit.

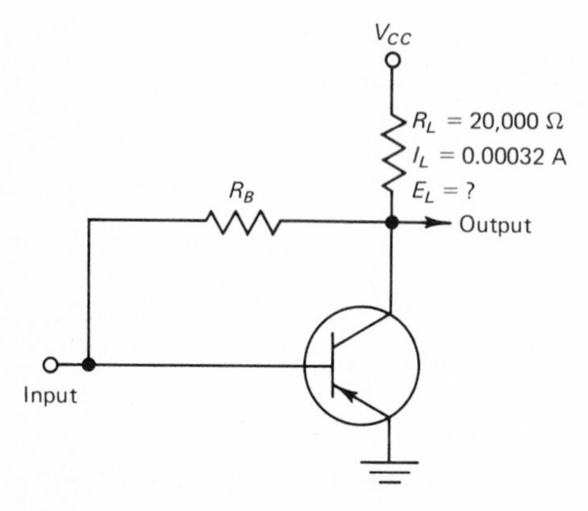

Fig. 15-9 A self-bias transistor circuit.

30. A 110-V line is protected with a 15-A fuse. Will the fuse "carry" a 5.5-Ω load?
31. In the fixed-bias transistor circuit shown in Fig. 15-8, the base resistor $R_B = 60{,}000\ \Omega$, and the base current $I_B = 0.00002$ A. Find the voltage drop across the base resistor.
32. A voltage of 28.8 V is required to send 7.2 A of current through a wire 5 mi long. What is the resistance of the wire? What is the resistance per mile of the wire?
33. A series of insulators leak 0.00003 A at 9,000 V. Find the resistance of the insulator string.
34. What bias voltage is developed across a grid leak resistor of 2,000,000 Ω resistance if the current through it is 0.0000002 A?
35. In the self-bias transistor circuit shown in Fig. 15-9, the load resistor $R_L = 20{,}000\ \Omega$ and carries a current $I_L = 0.00032$ A. Find the voltage drop across the load.

(See Answer Key for Test 4—Ohm's Law and Power)

JOB 16 Applying Decimals—The Micrometer

Each inch of the top scale shown in Fig. 16-1 is divided into 10 parts. Each division represents $\frac{1}{10}$ in or 0.1 in. Each inch on the bottom scale is divided into 100 parts. Each division represents $\frac{1}{100}$ in, or 0.01 in. These machinists' scales are used to provide an accuracy which cannot be obtained with the ordinary ruler. However, the machinist and auto mechanic require devices which can measure even more accurately.

Fig. 16-1 A decimal rule. (Courtesy The L.S. Starrett Company)

THE MICROMETER

The micrometer, or "mike," is a very sturdy tool that can measure distances to the nearest thousandth of an inch. In Fig. 16-2, the spindle C is attached to the thimble E so that they may turn as a single unit. The part

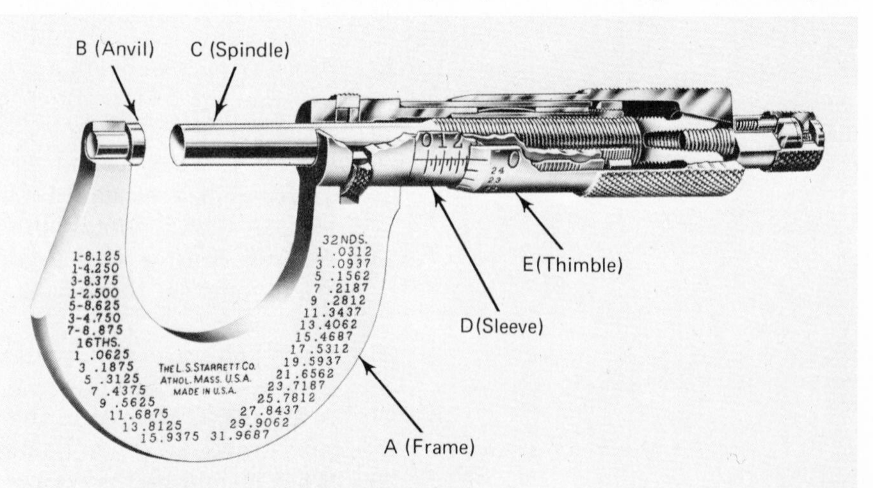

Fig. 16-2 A micrometer can measure to the nearest thousandth of an inch. (Courtesy The L.S. Starrett Company)

of the spindle concealed inside the sleeve D and thimble is a precisely machined thread of 40 threads to the inch. This thread is free to turn inside a nut which is fixed in the sleeve. When the thimble is revolved, the thread on the spindle moves through the nut and changes the distance between the anvil (B) and the spindle as shown in Fig. 16-3. A piece to be measured is enclosed between the anvil and the spindle. The measurement of this space is obtained by reading the lines and figures that are exposed as the thimble turns.

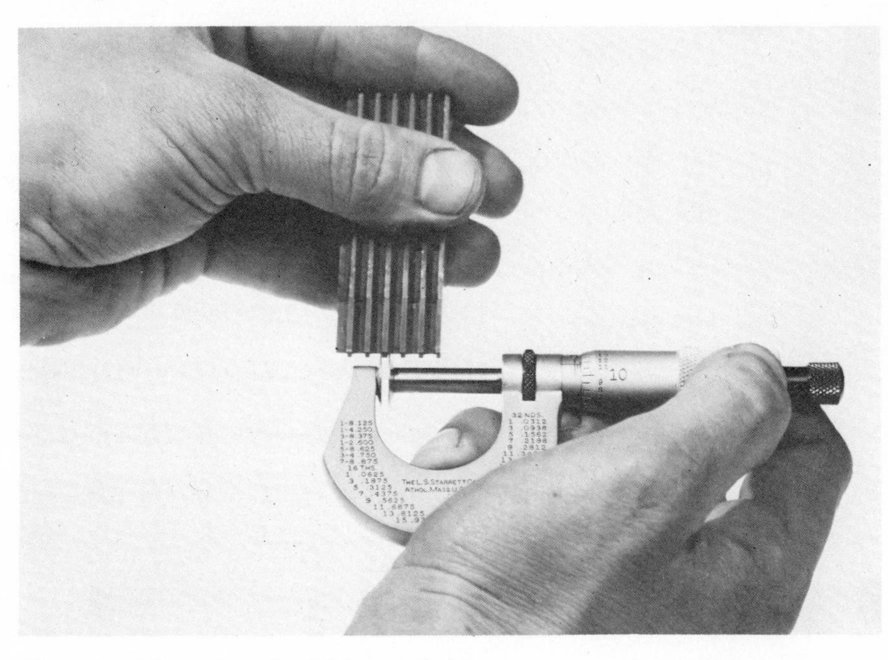

Fig. 16-3 Measuring the thickness of thin, closely-spaced sections with a disc-type micrometer. (Courtesy The L.S. Starrett Company)

A thread with 40 threads to the inch will advance $\frac{1}{40}$ in in one complete turn. Since $\frac{1}{40} = 0.025$, one complete turn of the thimble will move the thread 0.025 in. The sleeve is marked with 40 lines to the inch with each fourth division numbered. Also, the circumference of the thimble is divided into 25 equal parts. When the spindle touches the anvil, the zero on the thimble and its beveled edge line up with the line marked zero on the sleeve, as shown in Fig. 16-4*a*. When the thimble is turned counterclockwise through one full turn, the spindle will recede 0.025 in and expose the first small line on the sleeve. This gives a reading of 0.025 in,

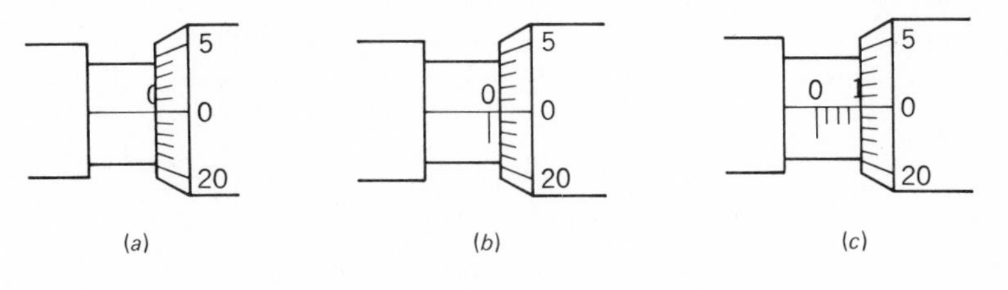

Fig. 16-4 (a) *0.000 in,* (b) *0.025 in,* (c) *0.100 in.*

as shown in Fig. 16-4*b*. A second turn will expose a second line, which will indicate a reading of 2 × 0.025 or 0.050 in. Thus, each small line on the sleeve indicates 0.025 in. *Four* complete turns will expose the fourth line, which is the first *numbered* line, as shown in Fig. 16-4*c*. This reading is 4 × 0.025 or 0.100 in. It is read as 100 thousandths. Four more turns will expose the second numbered line or 0.200 in. Thus, the numbered lines represent 0.100 in, 0.200 in, 0.300 in, etc.

Example 16-1 Read the settings on the following micrometers:

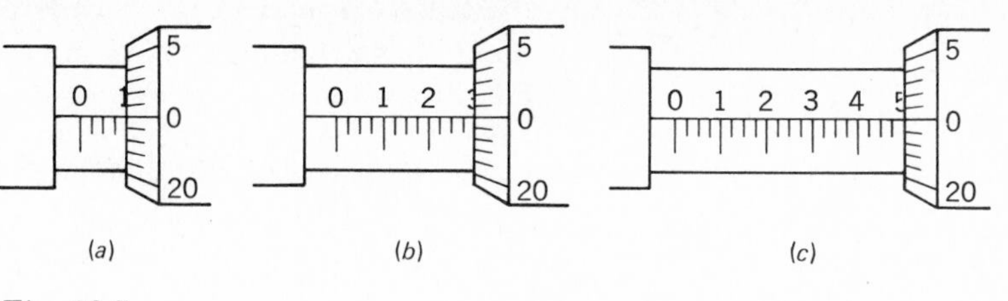

Fig. 16-5

Solution

a. The first numbered line is exposed and the zero on the thimble lines up with the gage line. The reading is 0.100 in. *Ans.*

b. The third numbered line is exposed and the zero on the thimble lines up with the gage line. The reading is 0.300 in. *Ans.*

c. The fifth numbered line is exposed and the zero on the thimble lines up with the gage line. The reading is 0.500 in. *Ans.*

The beveled edge of the thimble does not always fall on a numbered line. It may fall on one of the small lines. If we remember that each small line represents 0.025 in, these measurements will also be easy to read.

Example 16-2 Read the settings on the following micrometers:

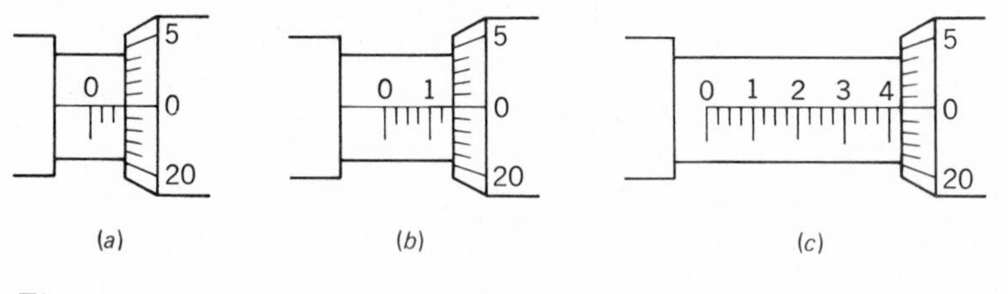

Fig. 16-6

Solution

a. We have not yet reached the first numbered line. Only 3 small lines have been exposed and the zero on the thimble lines up with the gage line.

Three small lines $= 3 \times 0.025 = 0.075$.

The reading is 0.075 in. *Ans.*

b. We have passed the first numbered line and the zeros line up.

Numbered lines exposed = 1	The value = 0.100 in	
Small lines exposed = 2	The value = 0.050 in	
The reading = total value =	0.150 in	*Ans.*

c. We have passed 4 numbered lines and the zeros line up.

Numbered lines exposed = 4	The value = 0.400 in	
Small lines exposed = 1	The value = 0.025 in	
The reading = total value =	0.425 in	*Ans.*

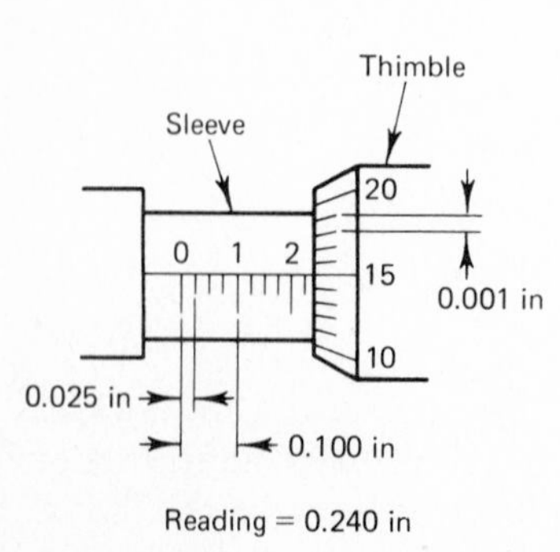

Fig. 16-7 Summary of the values of the markings on a micrometer.

So far we have only had readings in which the beveled edge fell on a line and in which the zero on the thimble lined up with the gage line. What happens when the beveled edge falls between the lines and a number other than zero on the thimble lines up with the gage line?

You will notice that the beveled edge of the thimble is divided into 25 equal parts. Since one complete turn of the thimble moves the spindle 0.025 in, a rotation of the thimble through only one of its divisions represents only $\frac{1}{25} \times 0.025$ or 0.001 in. The value of these thimble divisions must be added to the reading obtained from the lines exposed on the barrel. If a 9 lines up, we add 9×0.001 or 0.009 in. If a 22 lines up, we add 0.022 in, etc.

The values of the markings may now be summarized as shown in Fig. 16-7.

Example 16-3 Find the micrometer reading shown in Fig. 16-8.

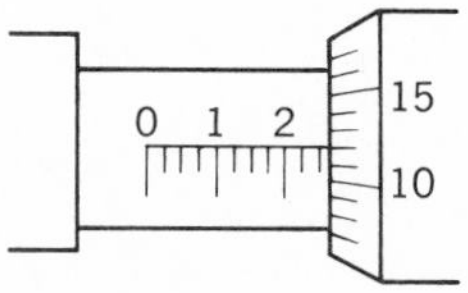

Fig. 16-8

Solution

Numbered lines exposed = 2	Value = 0.200
Small lines exposed = 2	Value = 0.050
Thimble reading = 12 (at 0.001 in each)	Value = <u>0.012</u>
Micrometer reading =	0.262 in *Ans.*

Example 16-4 Find the micrometer reading shown in Fig. 16-9.

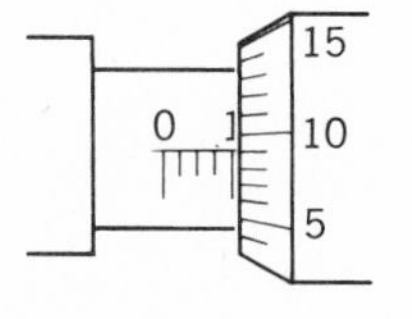

Fig. 16-9

Solution

Numbered lines exposed = 1	Value = 0.100
Small lines exposed = 0	Value = 0.000
Thimble reading = 9 (at 0.001 in each)	Value = <u>0.009</u>
Micrometer reading =	0.109 in *Ans.*

We must be particularly careful in our readings whenever the thimble marking is close to zero. When the micrometer is opened, the thimble is turned in a counterclockwise direction and the numbers on the thimble increase as they pass the gage line. In Fig. 16-10, we must decide whether we have passed the 0.600 mark or are just coming up to it. Just because you see the 6 doesn't mean that you have *passed* it. As this thimble turns to the left, we pass the 20 mark and approach the zero mark, which would indicate the full 0.600. The reading on the thimble is 23, meaning that we have not yet reached the 6 mark but are just coming up to it. The last numbered line must therefore be the 5.

Example 16-5 Find the micrometer reading shown in Fig. 16-10.

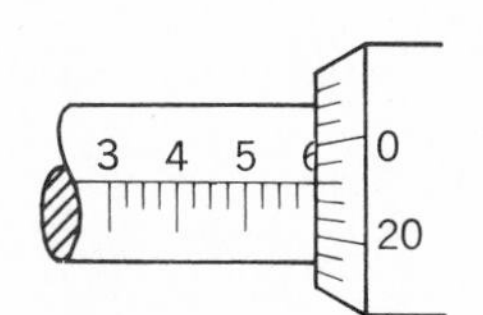

Fig. 16-10

Solution

Numbered lines exposed = 5	Value = 0.500
Small lines exposed = 3	Value = 0.075
Thimble reading = 23 (at 0.001 each)	Value = <u>0.023</u>
Micrometer reading =	0.598 in *Ans.*

Example 16-6 Find the micrometer reading shown in Fig. 16-11.

Fig. 16-11

Solution Do not be too hasty! Did we pass 2 small lines or did we pass 3 small lines? The fact that the thimble reading is on the 1 would indicate that we *have* passed 3 lines, but that the third line is hidden under the thimble.

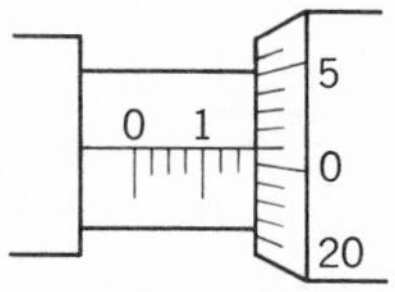

Numbered lines exposed = 1	Value = 0.100
Small lines exposed = 3	Value = 0.075
Thimble reading = 1 (at 0.001 each)	Value = <u>0.001</u>
Micrometer reading =	0.176 in *Ans.*

Problems

Find the reading for each of the settings shown in Fig. 16-12.

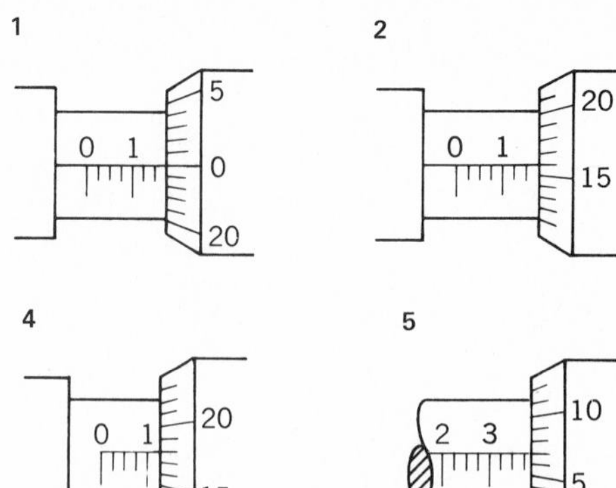

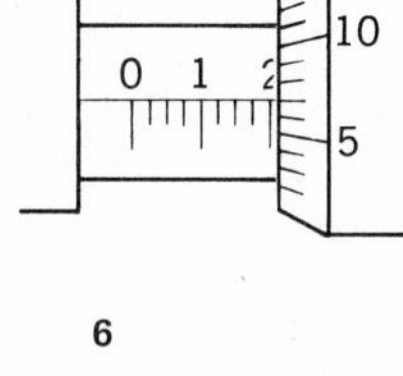

1 2 3

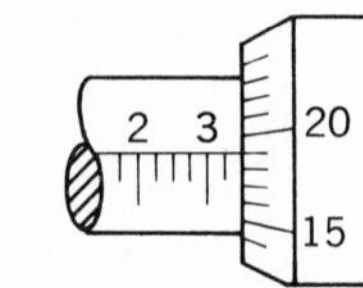

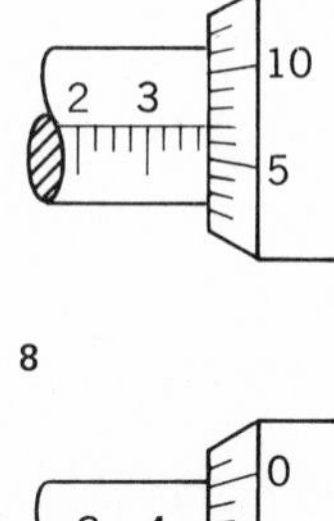

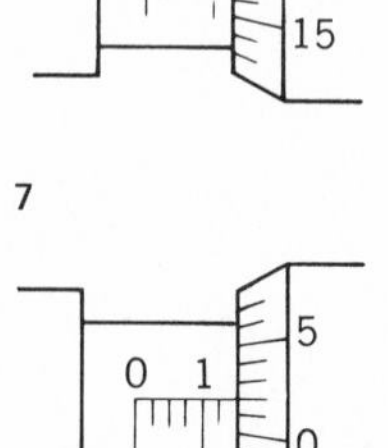

4 5 6

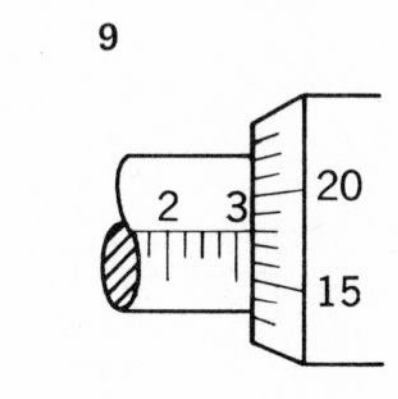

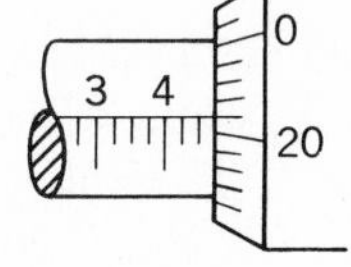

7 8 9

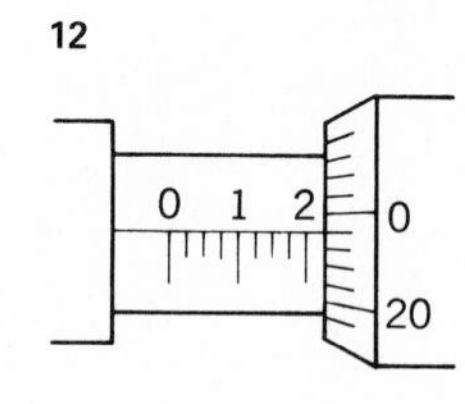

10 11 12

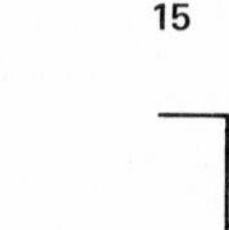

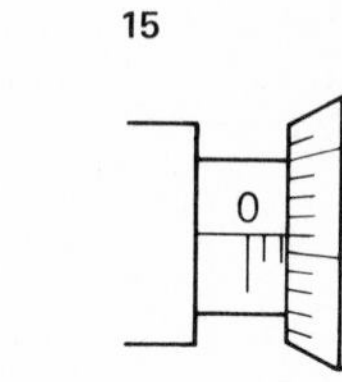

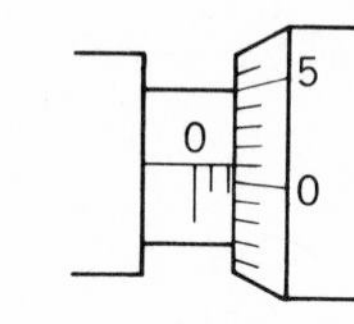

13 14 15

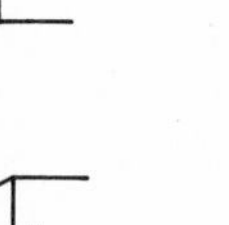

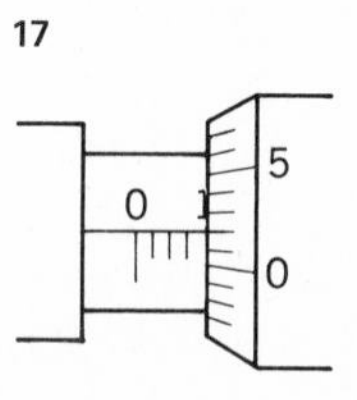

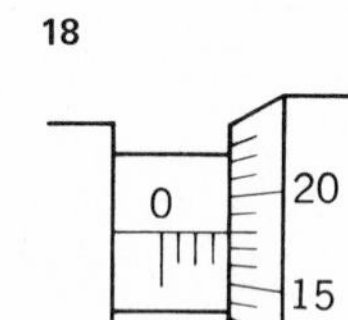

16 17 18

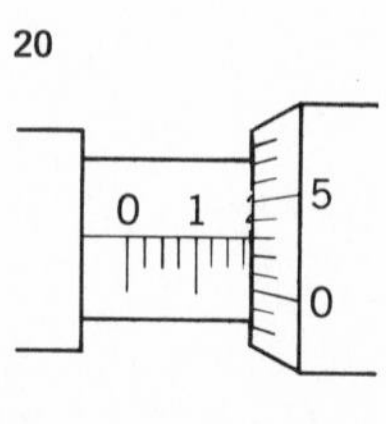

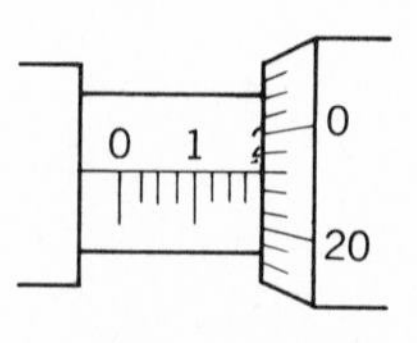

19 20 21

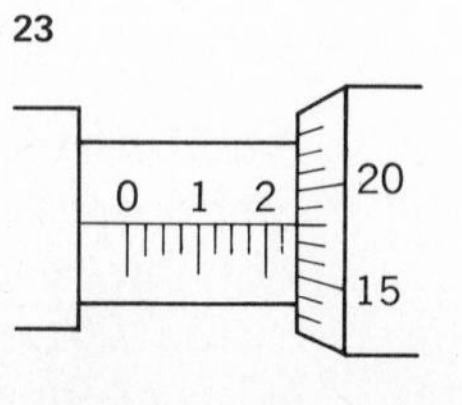

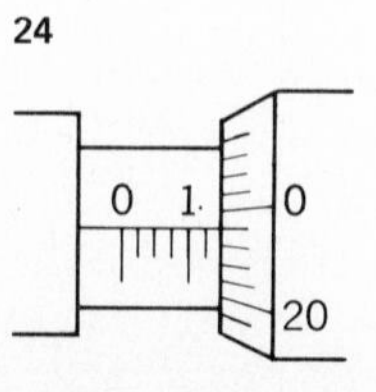

22 23 24

Fig. 16-12

JOB 17 The Vernier Micrometer

Modern industry frequently requires that parts be made to an accuracy greater than can be measured with an ordinary micrometer. A simple micrometer will measure to the thousandth of an inch, in which case the graduations on the thimble line up exactly on the gage line, as shown in Fig. 17-1*a*. When the gage line falls between two numbers on the thimble (Fig. 17-1*b*), this means that the measurement is a little bit *more* than it would be if it had fallen exactly on a thimble number. This extra amount, measured in ten-thousandths of an inch, can be measured by using a *vernier*, invented by Pierre Vernier in 1631.

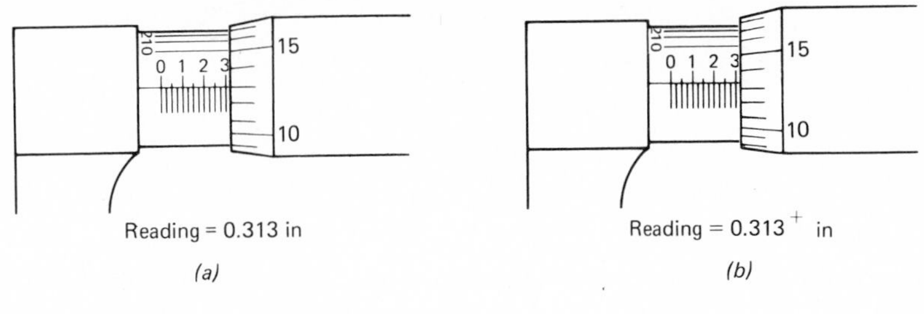

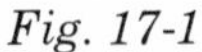

Fig. 17-1

THE VERNIER PRINCIPLE

Let us take 9 divisions on the thimble (a total of 0.009 in) and roll this distance out flat, as in Fig. 17-2. Each division = 0.001 in. Now let us take

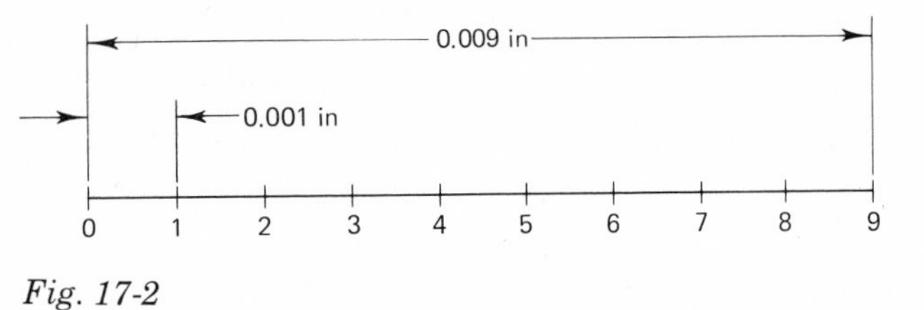

Fig. 17-2

this 0.009-in length and divide it into 10 equal parts, as shown in Fig. 17-3. Each one of these parts is equal to $\frac{1}{10}$ of 0.009 in, or 0.0009 in. This is the vernier scale.

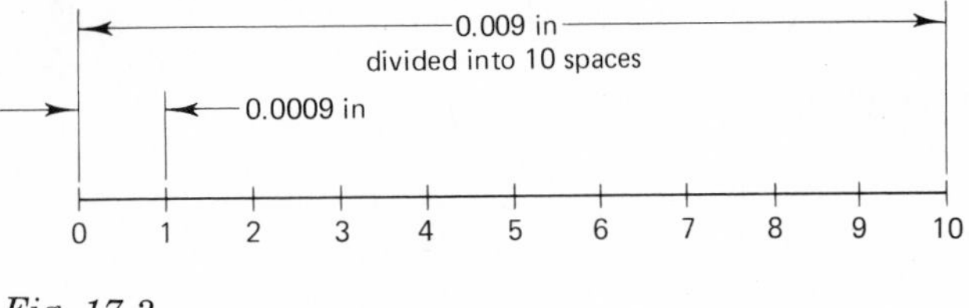

Fig. 17-3

Thimble spaces = 0.001 in = 0.0010 in
Vernier spaces = 0.0009 in
Difference in size of space = 0.0001 in

Place the vernier scale on top of the thimble scale, as shown in Fig. 17-4. Match up the zeros on the left and the 10 and 9 points on the right. This is the position of the vernier when the micrometer is measuring exact thousandths. The difference between matching numbers on the thimble and on the vernier at 1 is 0.0001 in, at 2 is 0.0002 in, at 3 is 0.0003 in, etc.

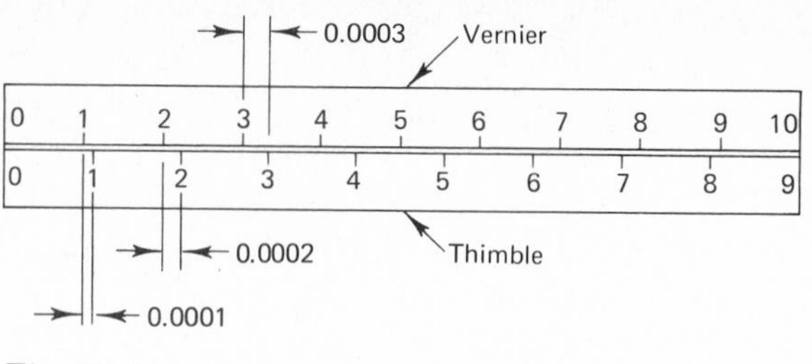

Fig. 17-4

This means that when any pair of numbers line up, or coincide, the thimble has moved past the zero setting that number of ten-thousandths. For example,

NUMBERS MATCHED	ADDITION TO MICROMETER READING
1	0.0001 in
2	0.0002 in
3	0.0003 in
9	0.0009 in

When the 10 on the vernier matches up with the 9 on the thimble, we are back to an exact number of thousandths.

READING A VERNIER MICROMETER

The vernier scale is etched on the sleeve of the micrometer as shown in Fig. 17-5*a*. In this figure, and in the enlargement shown in Fig. 17-5*b*, the zero on the thimble lines up with the gage line and only the zeros of the vernier match a thimble number. The reading is exactly 0.250 in.

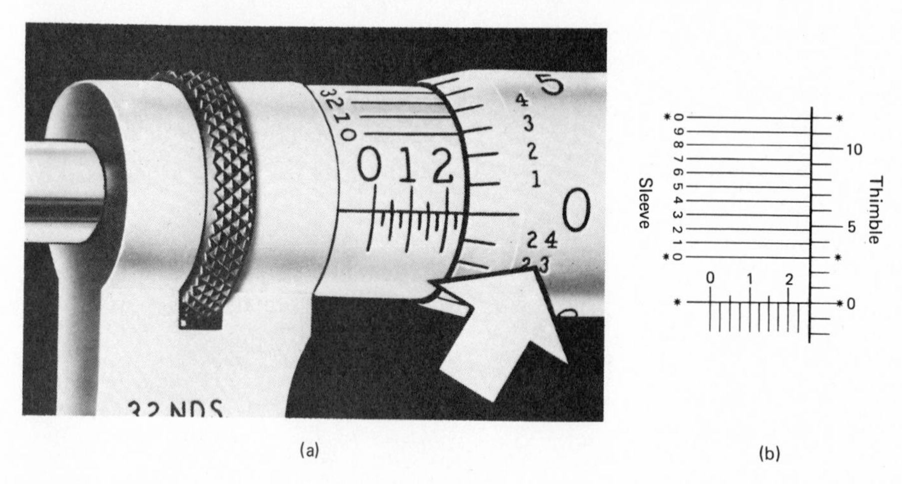

Fig. 17-5 The reading is exactly 0.250 in. (Courtesy The L.S. Starrett Company)

Example 17-1 Now suppose the thimble had moved a little bit past the 0.250-in mark, as shown in Fig. 17-6*a*. What is the reading?

Solution

1. Get the normal micrometer reading. Be careful! The number of *thousandths* on the thimble line is *not* the line closest to the gage line, but is the line just *under* the gage line. This is true because the thimble is always rotated counterclockwise to increase the reading.

   The reading in Fig. 17-6*a* indicates a measurement which is a little more than 0.250 in but not as much as 0.251 in. The vernier in Fig. 17-6*b* tells us how much extra to add to the reading of 0.250 in.

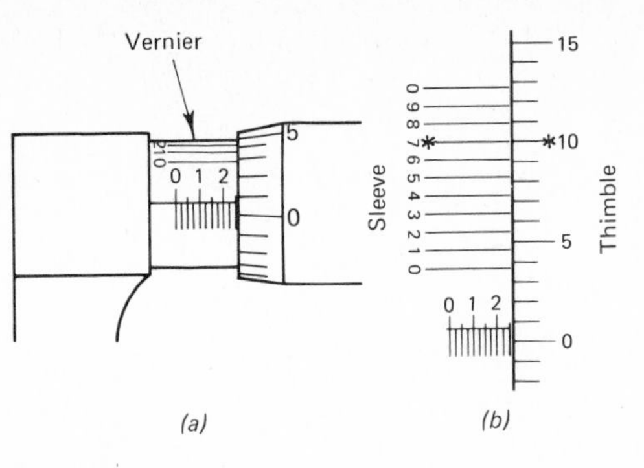

Fig. 17-6

2. Find the line on the vernier (located on the sleeve) that coincides (lines up) with a number on the thimble. Since the 7 lines up, it means that we must add 0.0007 in to the 0.2500 in mike reading.

$$\begin{array}{r} 0.2500 \\ +\ 0.0007 \\ \hline \end{array}$$

Reading = 0.2507 in *Ans.*

Find the micrometer reading shown in Fig. 17-7.

Solution

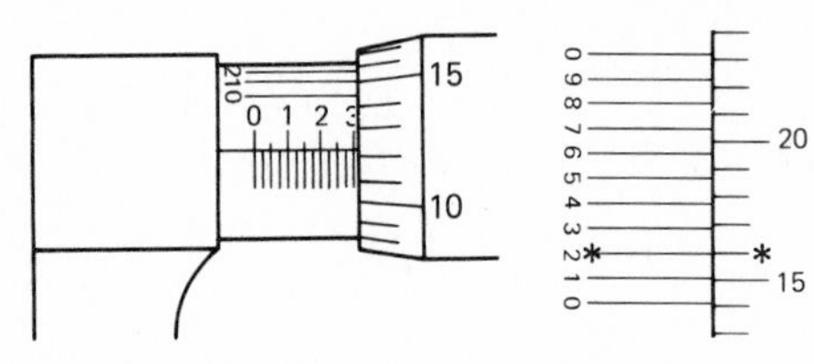

Fig. 17-7

1. Find the basic reading. Use the thousandth reading which is just (under)(above) the gage line.

   The basic reading = ______ in

2. Find the line on the vernier that coincides with a number on the thimble. Since the 2 coincides, it means that we must add ______ to the basic reading of ______ in.

$$\begin{array}{r} 0.3120 \\ +\ 0.0002 \\ \hline \end{array}$$

Reading = ______ in *Ans.*

under

0.312

0.0002 in

0.3120

0.3122

Problems

Find the reading for each of the settings shown in Fig. 17-8.

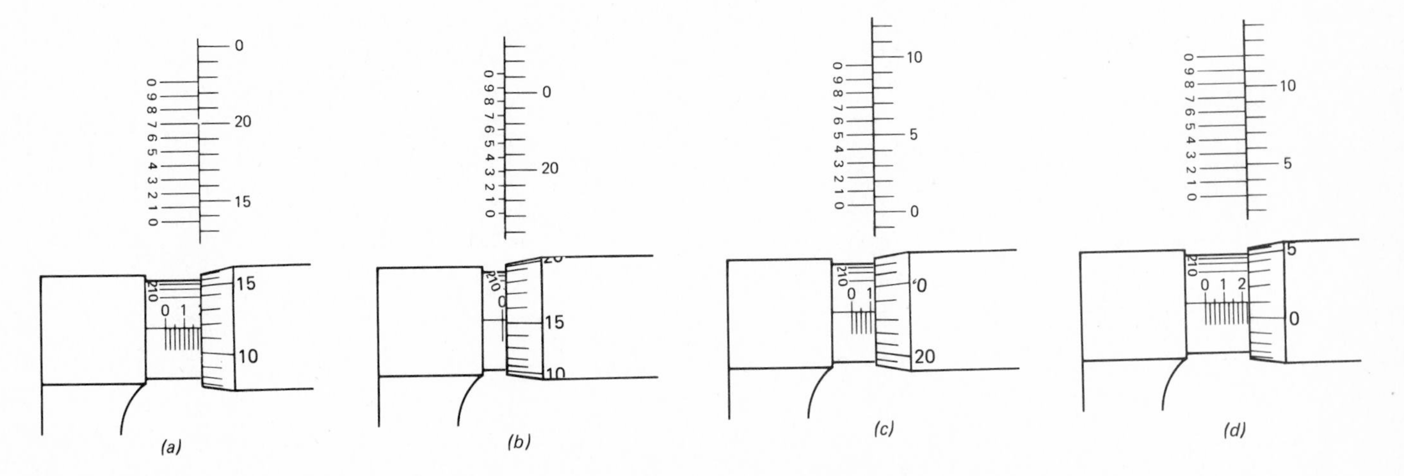

Fig. 17-8

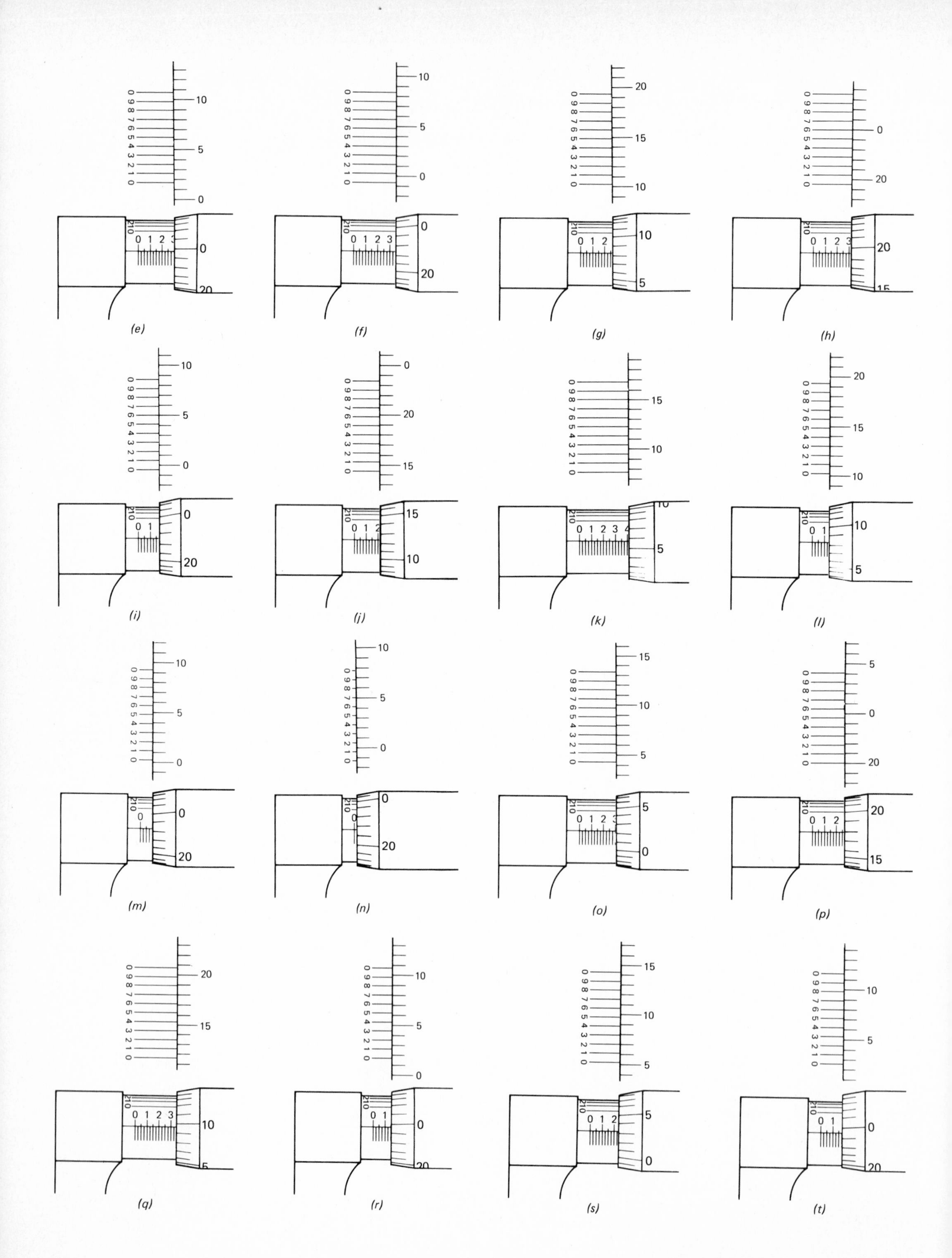
(e)
(f)
(g)
(h)
(i)
(j)
(k)
(l)
(m)
(n)
(o)
(p)
(q)
(r)
(s)
(t)

JOB 18 The Vernier Caliper

The vernier caliper was invented by Joseph R. Brown in 1851. The vernier caliper shown in Fig. 18-1 has two jaws, a fixed jaw containing the main scale and a vernier jaw which is free to move along the main scale.

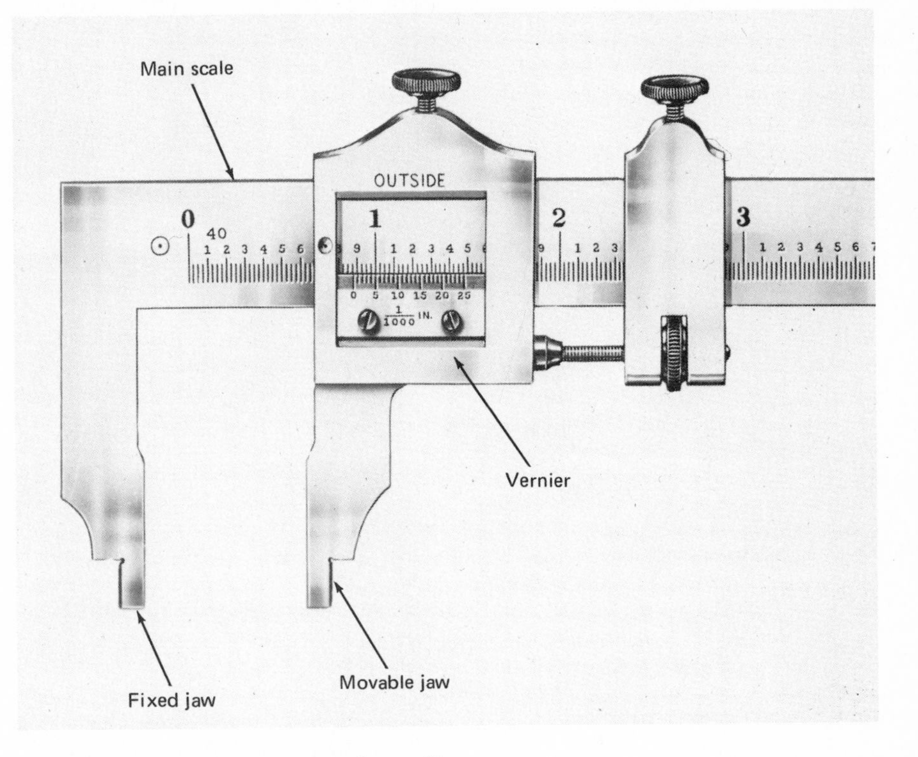

Fig. 18-1 A 25-division vernier caliper. (Courtesy The L.S. Starrett Company)

As shown in Fig. 18-2, this type of vernier may be used for both inside and outside measurements. Outside measurements are read from left to right on the front of the caliper, marked "outside" (Fig. 18-2*a*). Inside measurements are read from right to left on the back of the caliper, marked "inside" (Fig. 18-2*b*).

Each inch of the main scale of the vernier caliper shown in Fig. 18-1 is divided into 40 parts, so that each division equals $1 \div 40 = 0.025$ in.

Every fourth division is marked 1, 2, 3, etc., representing 0.100 in, 0.200 in, 0.300 in, etc. The vernier has 25 divisions which are numbered every fifth division. The total length of the vernier equals 24 divisions on the main scale, or $24 \times 0.025 = 0.600$ in. Therefore, one division on the vernier equals $0.600 \div 25 = 0.024$ in. Thus, the difference between a division on the vernier and a division on the main scale is $0.025 \text{ in} - 0.024 \text{ in} = 0.001$ in.

Set the tool so that the 0 line on the vernier lines up with the 0 line on the main scale, as shown in Fig. 18-3. In this closed position of the caliper, the first line on the vernier will differ from the first line on the main scale by 0.001 in, the second lines will differ by 0.002 in, etc. The difference will continue to increase by 0.001 in for each division until the line 25 on the vernier lines up with the line 6 on the main scale.

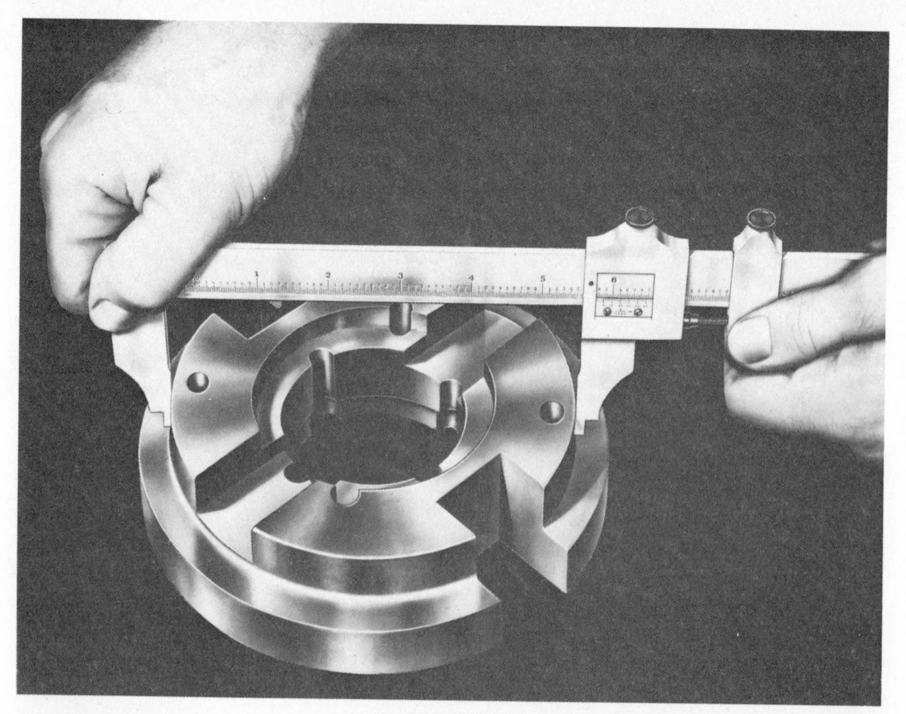

(a)

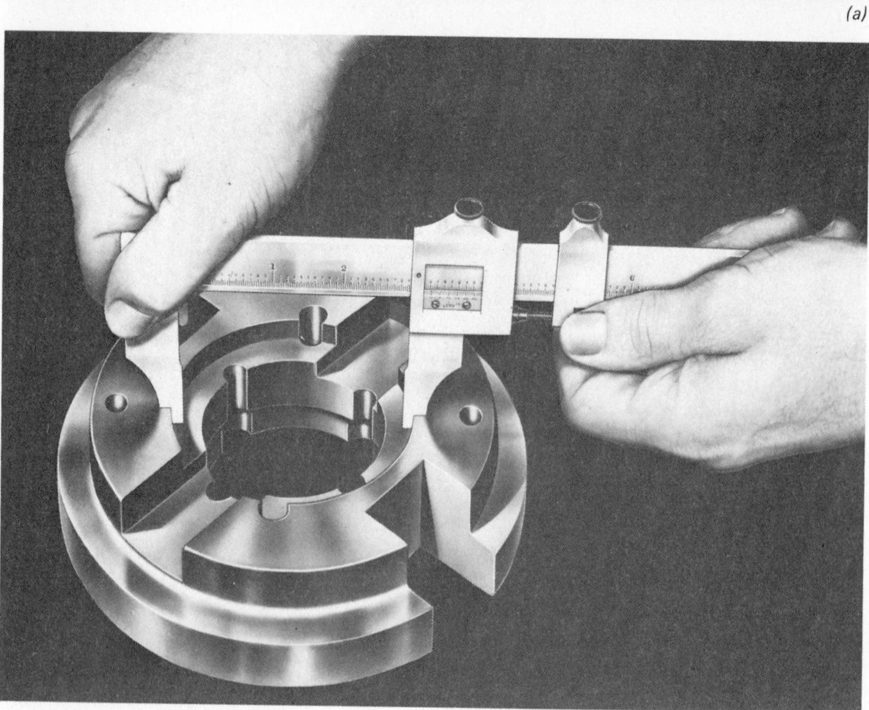

(b)

Fig. 18-2 Using a caliper for (a) *outside, and* (b) *inside measurements.*

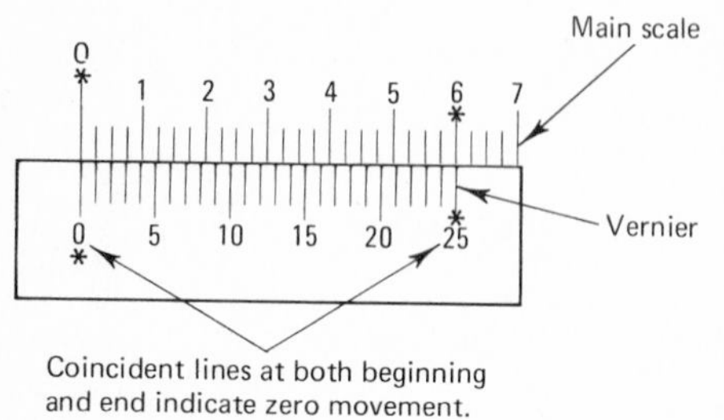

Main scale
Vernier
Coincident lines at both beginning and end indicate zero movement.

Fig. 18-3

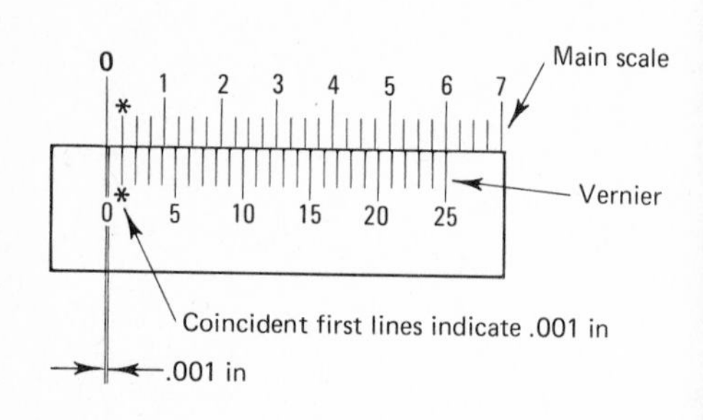

Main scale
Vernier
Coincident first lines indicate .001 in

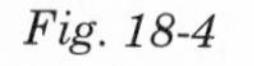

.001 in

Fig. 18-4

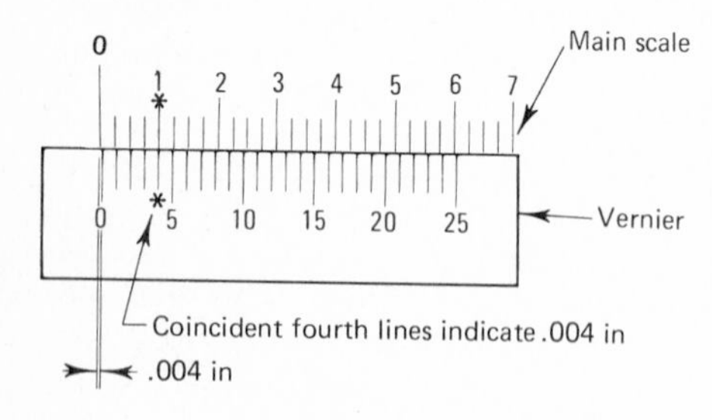

Main scale
Vernier
Coincident fourth lines indicate .004 in
.004 in

Fig. 18-5

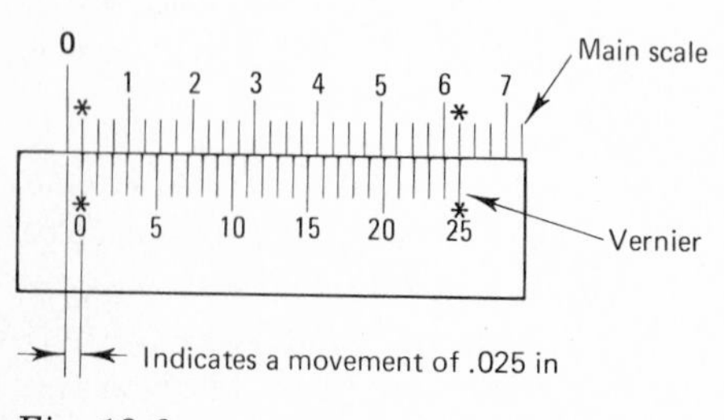

Main scale
Vernier
Indicates a movement of .025 in

Fig. 18-6

If the movable leg of the caliper, to which the vernier is attached, is moved to the right until the first line on the main scale and the vernier coincide, the caliper will open 0.001 in, as shown in Fig. 18-4.

When the caliper is opened so that the fourth line on the vernier coincides with a line on the main scale (Fig. 18-5), it indicates a movement of 0.004 in.

When the jaws are opened a full 0.025 in, the zero line on the vernier will coincide with the first line on the main scale, as shown in Fig. 18-6.

The values of the divisions are summarized in Fig. 18-7.

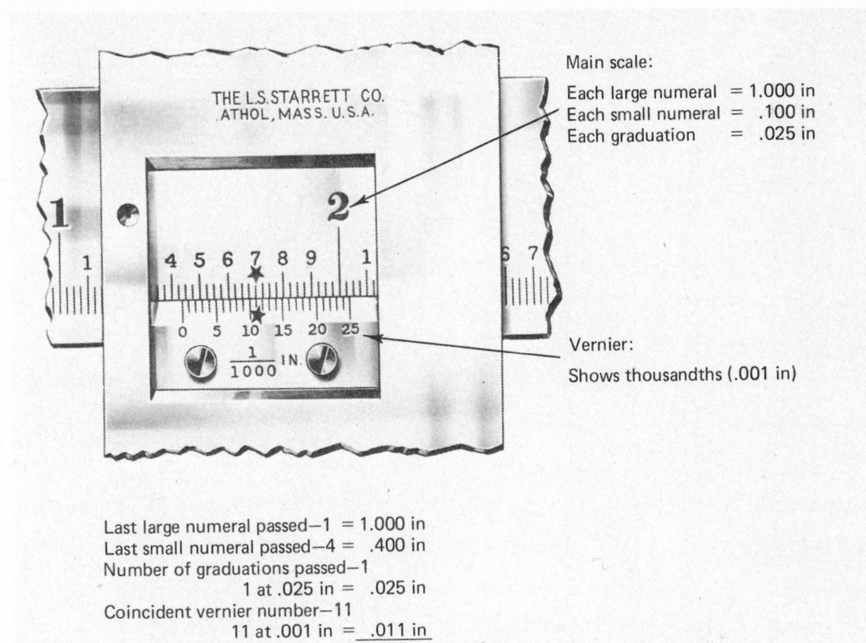

Fig. 18-7 Reading a 25-division vernier caliper. (Courtesy The L.S. Starrett Company)

Example 18-1 Read the vernier setting shown in Fig. 18-8.

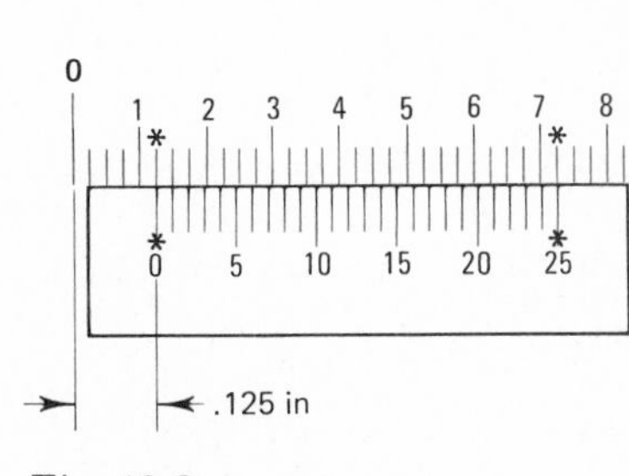

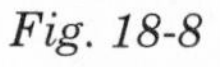

Fig. 18-8

Solution

1. Find the last large numeral passed (0).	0.000 in	
2. Find the last small numeral passed (1).	0.100 in	
3. Find the number of graduations passed (1). 1 at 0.025 in =	0.025 in	
4. Find the coincident vernier number (0).	0.000 in	
5. Add.	0.125 in	*Ans.*

Example 18-2 Read the vernier setting shown in Fig. 18-9.

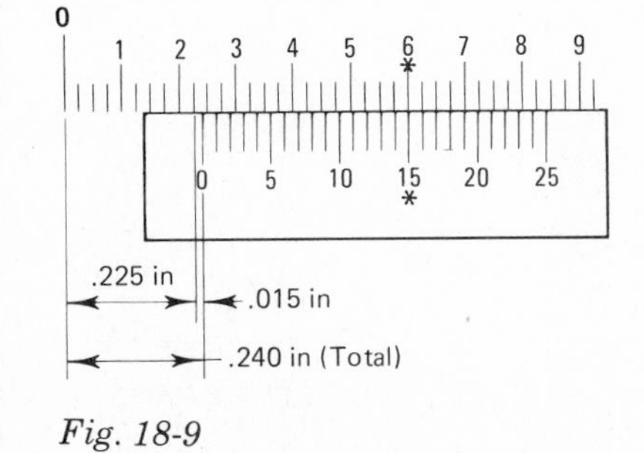

Fig. 18-9

Solution

1. Find the last large numeral passed (0).	0.000 in	
2. Find the last small numeral passed (2).	0.200 in	
3. Find the number of graduations passed (1). 1 at 0.025 in =	0.025 in	
4. Find the coincident vernier number (15). 15 at 0.001 in =	0.015 in	
5. Add.	0.240 in	*Ans.*

Example 18-3 Read the vernier setting shown in Fig. 18-10.

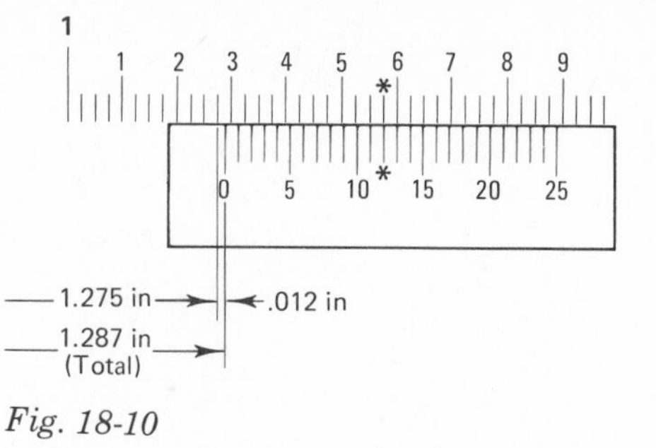

Fig. 18-10

Solution

1. Find the last large numeral passed (1). 1.000 in
2. Find the last small numeral passed (2). 0.200 in
3. Find the number of graduations passed (3).
   3 at 0.025 in = 0.075 in
4. Find the coincident vernier number (12).
   12 at 0.001 in = 0.012 in
5. Add. 1.287 in *Ans.*

Self-Test 18-4 Read the vernier settings shown in Fig. 18-11.

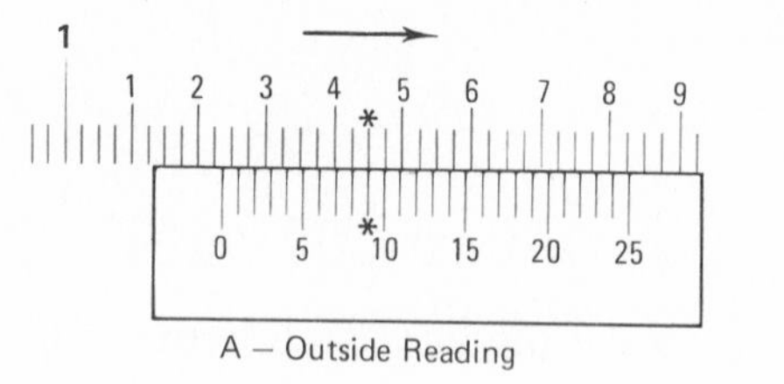

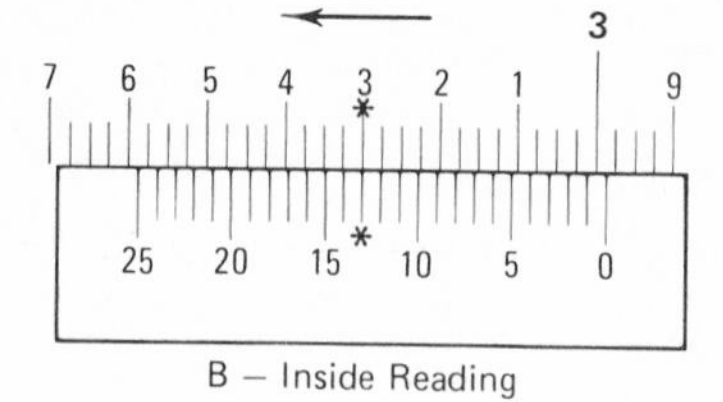

Fig. 18-11

Solution

a. Outside reading. (Read from left to right)
   1. Find the last large numeral passed (1) = ______ 1.000 in
   2. Find the last small numeral passed (2). = ______ 0.200 in
   3. Find the number of graduations passed (1).
      1 at 0.025 in = ______ 0.025 in
   4. Find the coincident vernier number (9).
      9 at 0.001 in = ______ 0.009 in
   5. Add. The answer is ______. *Ans.* 1.234 in

b. Inside reading. (Read from right to left)
   1. Find the last large numeral passed (2). = ______ 2.000 in
   2. Find the last small numeral passed (9). = ______ 0.900 in

3. Find the number of graduations passed (3).

   3 at 0.025 in = ______ **0.075 in**

4. Find the coincident vernier number (13).

   13 at 0.001 in = ______ **0.013 in**

5. Add. The answer is ______. *Ans.* **2.988 in**

Problems

Find the reading of each caliper setting shown in Fig. 18-12. Be careful! Both outside and inside measurements are shown.

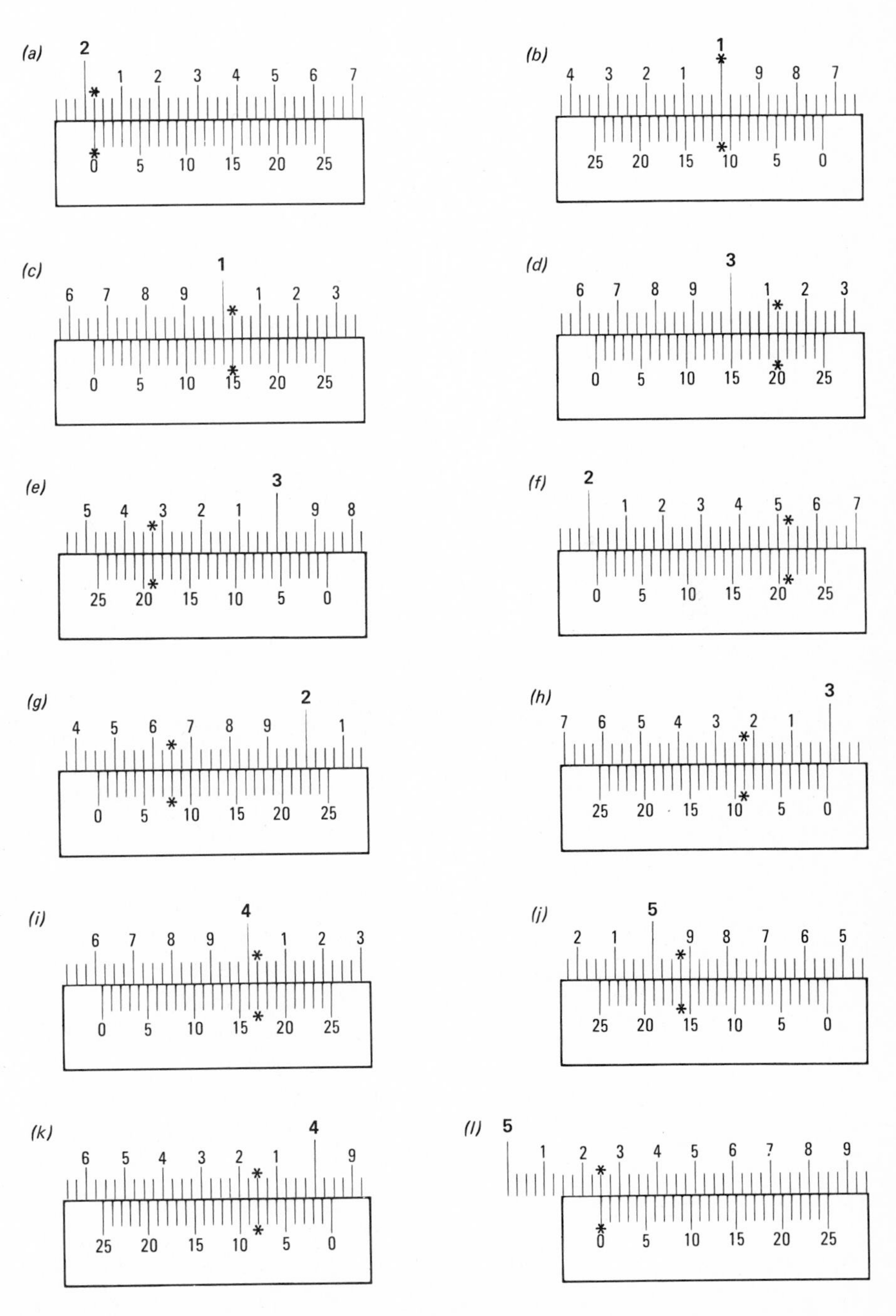

Fig. 18-12

THE 50-DIVISION VERNIER CALIPER

This type of vernier caliper (Fig. 18-13) has two main scales, the top scale for inside measurements and the bottom scale for outside measurements. These two scales differ by 0.300 in, which is the width of the nibs when the caliper is completely closed.

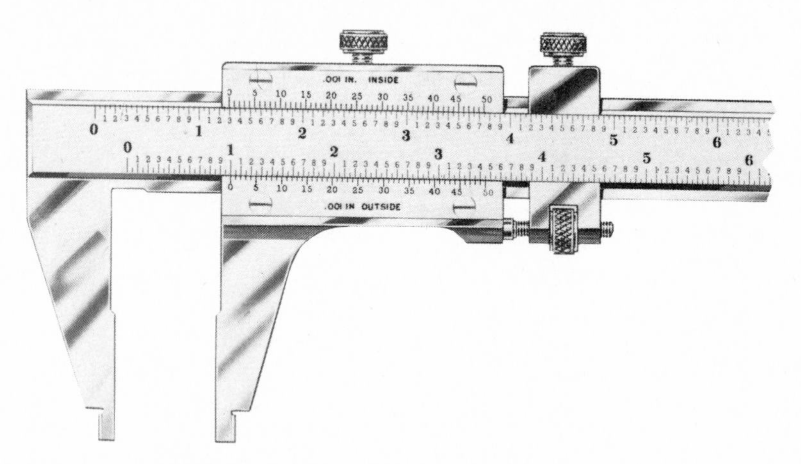

Fig. 18-13 The 50-division vernier caliper. (Courtesy The L.S. Starrett Company)

Figure 18-14 shows that each inch of the main scale is divided into twenty parts, so that each division = 1 ÷ 20 = 0.050 in. Every second division is marked 1, 2, 3, etc., representing 0.100 in, 0.200 in, 0.300 in, and so on.

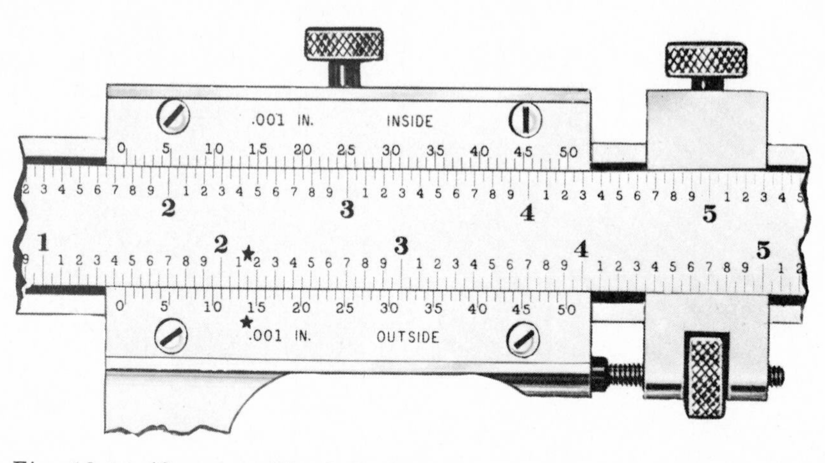

Fig. 18-14 (Courtesy The L.S. Starrett Company)

A distance equal to 49 main divisions is marked off on the vernier scale. This distance equals 49 × 0.050 = 2.45 in. This vernier distance is now divided into 50 parts, so that each part = 2.45 ÷ 50 or 0.049 in. The difference between the width of one of the 50 spaces on the vernier and one of the 49 spaces on the main scale = 0.050 in − 0.049 in = 0.001 in.

Set the tool for an outside measurement so that the 0 line on the vernier coincides with the 0 line on the bottom main scale, as shown in Fig. 18-15. In this closed position of the caliper, the first line on the vernier will differ from the first line on the main scale by 0.001 in, the second lines on each scale will differ by 0.002 in, etc. The difference will continue to increase by 0.001 in for each division until the fiftieth line on the vernier coincides with the forty-ninth line on the main scale.

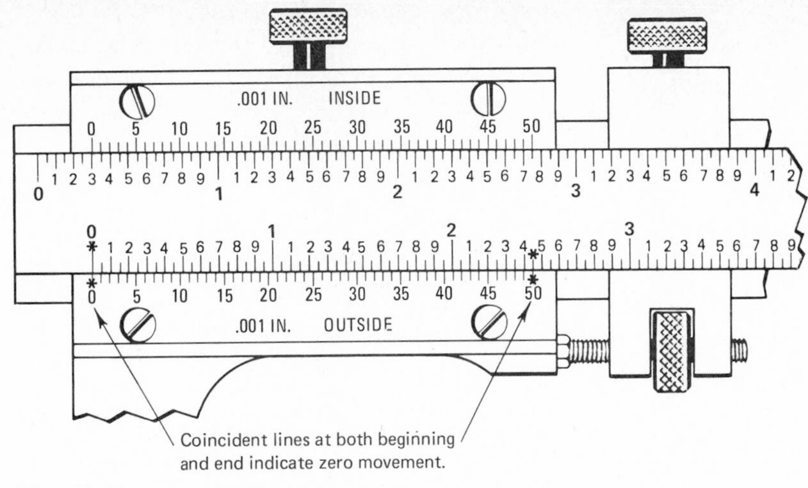

Fig. 18-15

Suppose the caliper is opened so that, for example, the 20 on the vernier coincides with a line on the main scale, as shown in Fig. 18-16. This would indicate a reading of 0.020 in.

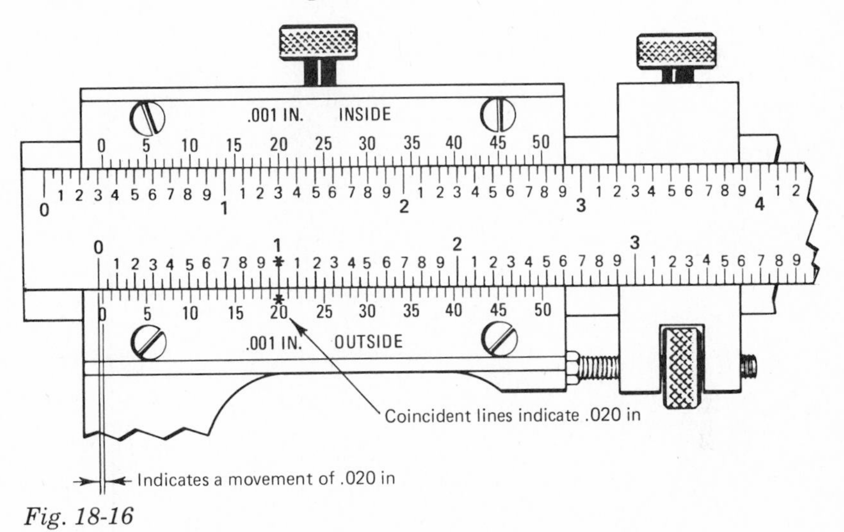

Fig. 18-16

When the caliper is opened a full 0.050 in, the zero line on the vernier will coincide with the first graduation on the main scale (Fig. 18-17).

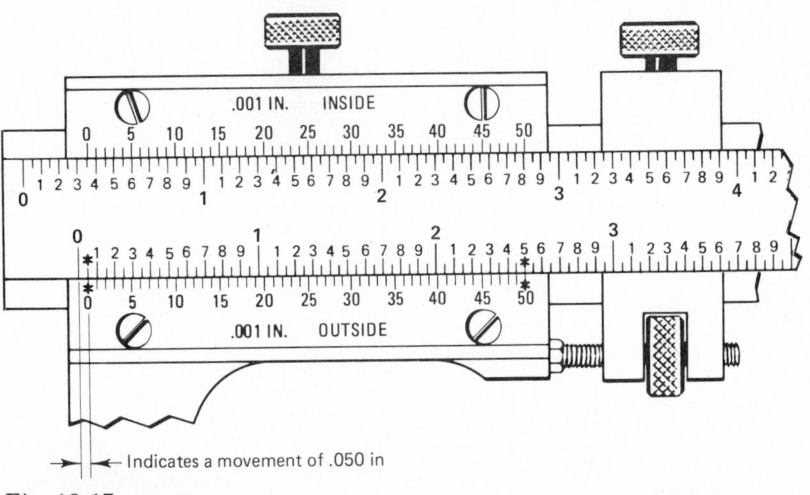

Fig. 18-17

The values of the divisions are summarized in Fig. 18-18.

Example 18-5 Read the vernier caliper setting shown in Fig. 18-18 for the outside measurement (bottom scale).

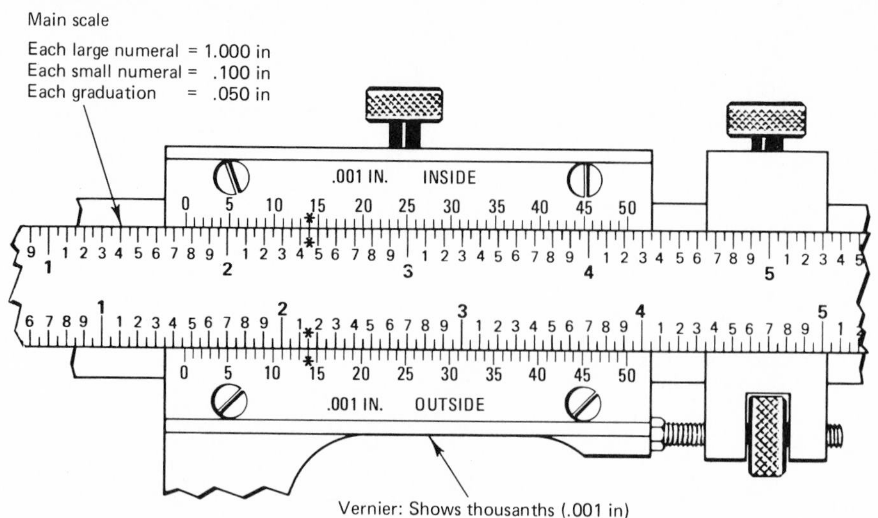

Fig. 18-18 A 50-division vernier caliper.

Solution

1. Find the last large numeral on the main scale that the 0 on the vernier has passed (1). — 1.000 in
2. Find the last small numeral passed (4). — 0.400 in
3. Find the number of main scale graduations passed (1). — 0.050 in
4. Find the coincident vernier number (14).
   14 × 0.001 in = — 0.014 in
5. Add — 1.464 in *Ans.*

Self-Test 18-6 Read the vernier caliper setting shown in Fig. 18-18 for the inside measurement (top scale).

Solution

1. Find the last large numeral on the main scale that the 0 on the vernier has passed (1). = ______ — 1.000 in
2. Find the last small numeral passed (7). = ______ — 0.700 in
3. Find the number of main scale graduations passed (1). = ______ — 0.050 in
4. Find the coincident vernier number (14).
   14 × 0.001 in = ______ — 0.014 in
5. Add. The answer is ______. *Ans.* — 1.764 in

Note: as the last two examples show, the inside and outside measurements will always differ by 0.300 in, which is the width of the nibs.

Problems

Find the inside and outside measurements for each of the caliper settings shown in Fig. 18-19.

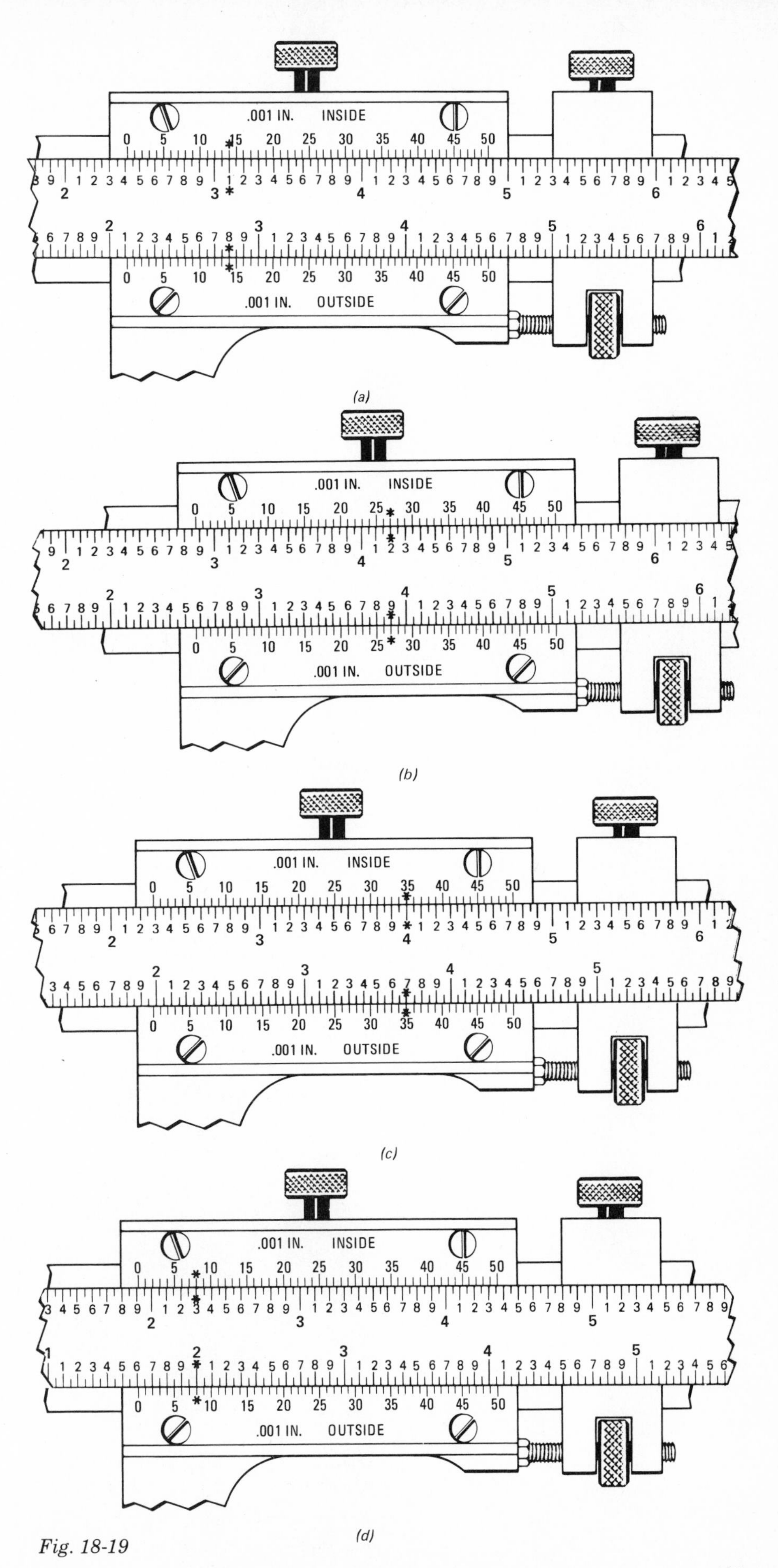

(a)

(b)

(c)

(d)

Fig. 18-19

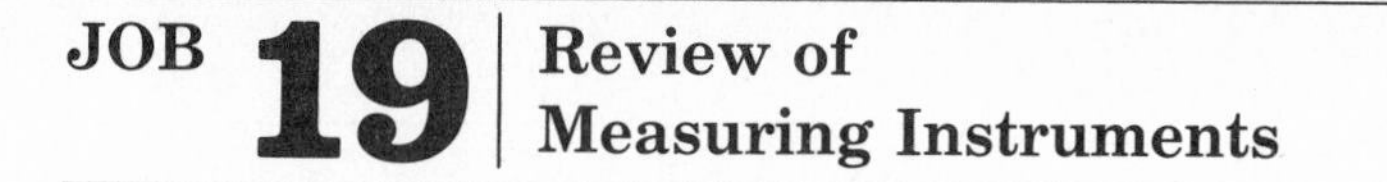

JOB 19 Review of Measuring Instruments

1. Find the distance from the left edge of the scale to the points indicated in both centimeters and millimeters (Fig. 19-1).

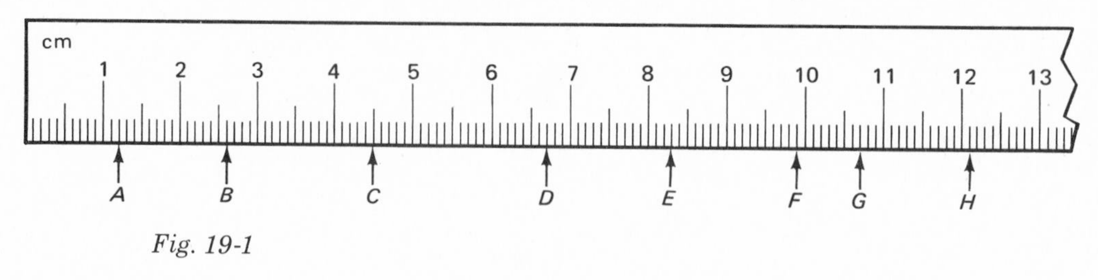

Fig. 19-1

Measure (in mm) the distance L in the wrench shown in Fig. 19-2.

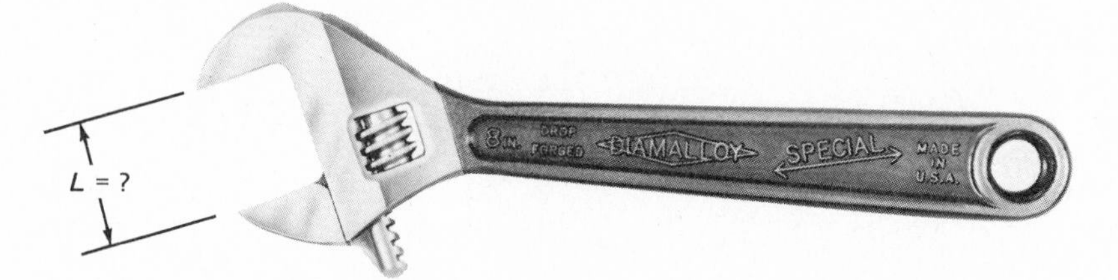

Fig. 19-2 A monkey wrench. (Courtesy The Diamond Tool and Horseshoe Company)

3. Measure the distances A, B, and C on the picture of the ball check valve shown in Fig. 19-3. Give your answers in both millimeters and centimeters.

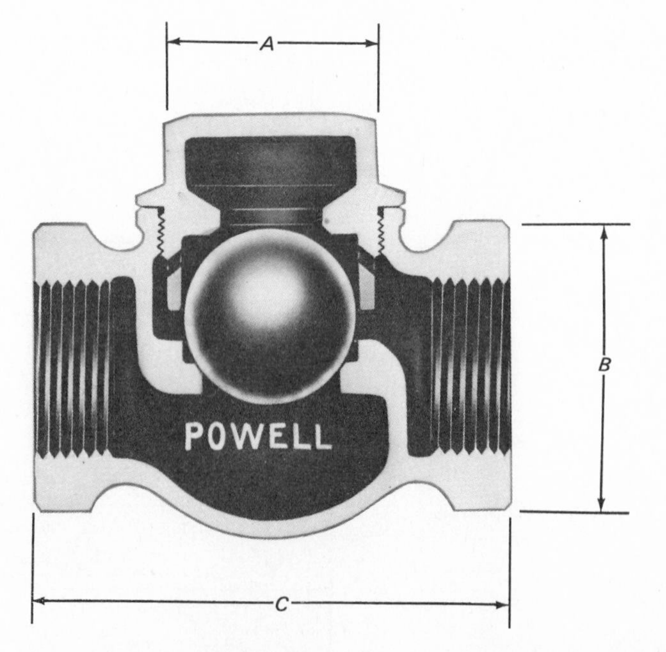

Fig. 19-3 Section of a ball check valı (Courtesy The William Powell Company)

4. Read the micrometer settings shown in Fig. 19-4.

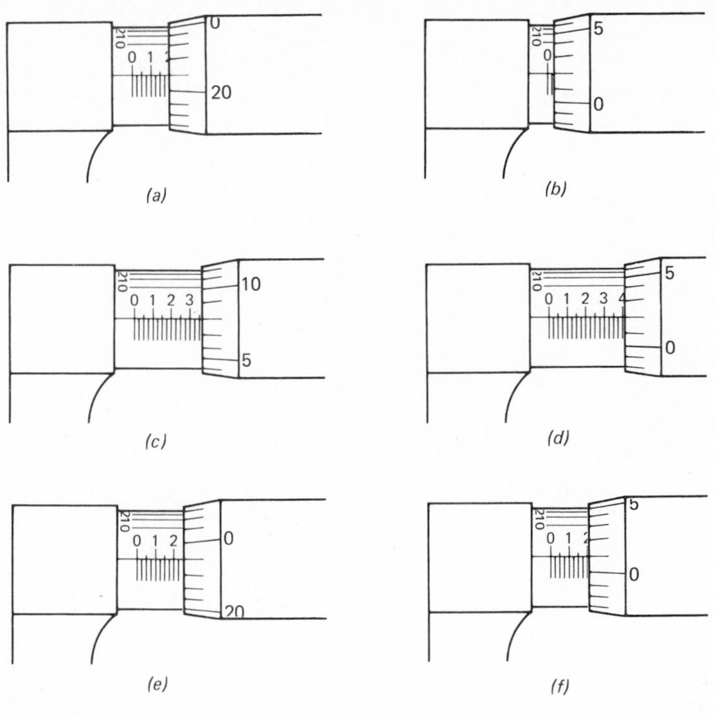

Fig. 19-4

5. Read the vernier micrometer settings shown in Fig. 19-5.

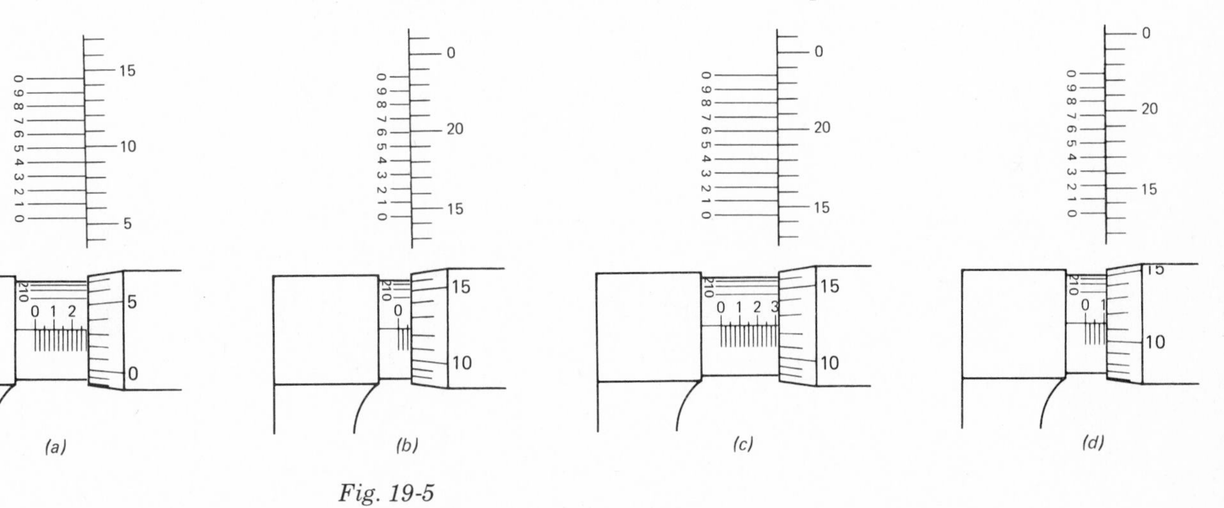

Fig. 19-5

6. Read the vernier caliper setting shown in Fig. 19-6.

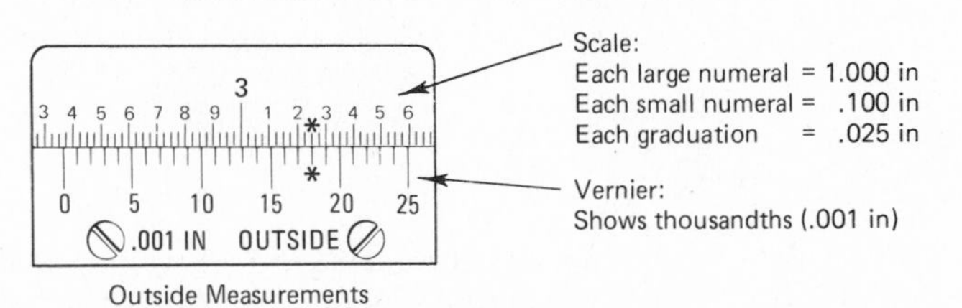

Fig. 19-6 (Courtesy Brown & Sharpe Manufacturing Company)

7. Read the inside vernier caliper settings shown in Fig. 19-7.

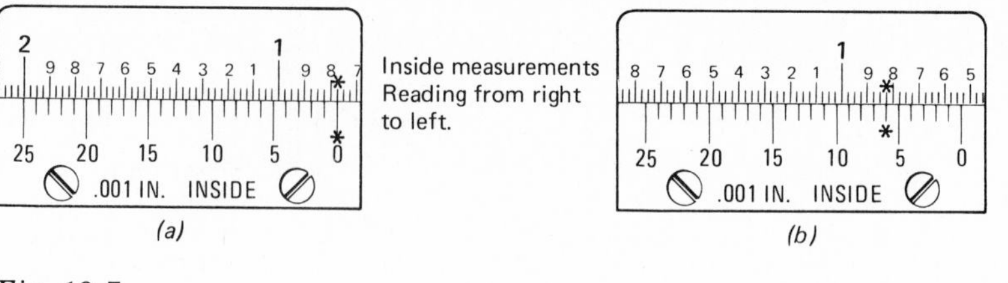

Fig. 19-7

8. Read the setting of the vernier depth gage shown in Fig. 19-8.

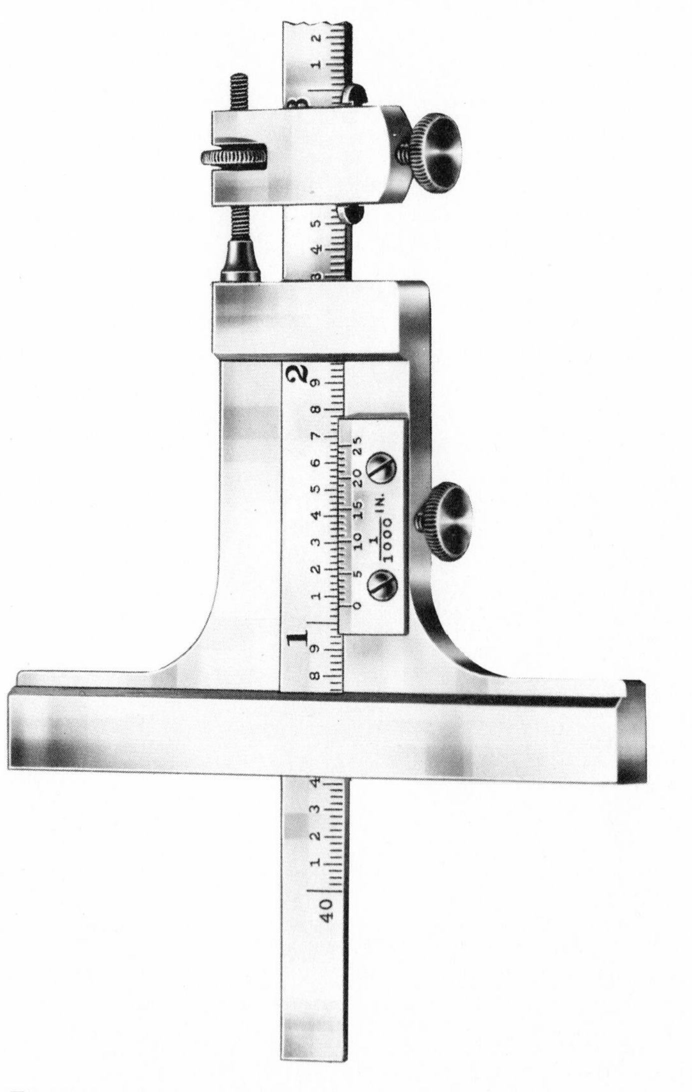

Fig. 19-8 A vernier depth gage. (Courtesy The L.S. Starrett Company)

9. Read the inside and outside measurements on the 50-division vernier calipers shown in Fig. 19-9.

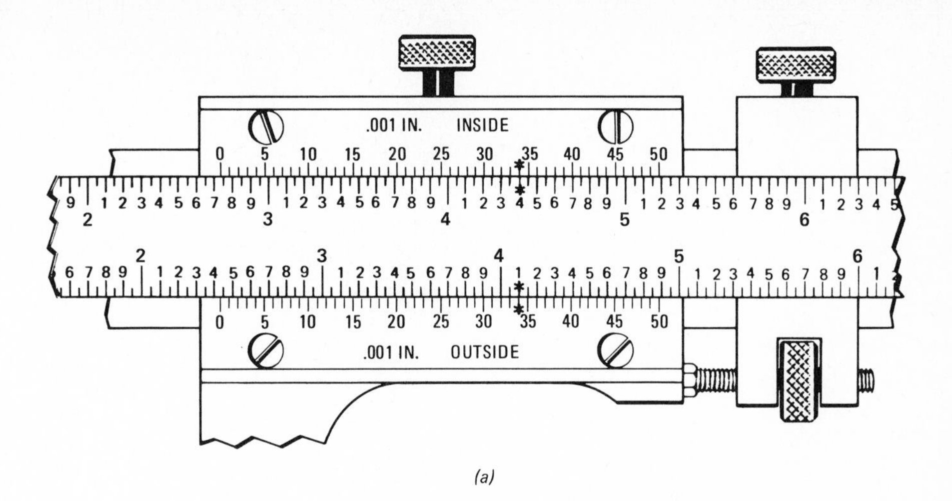

(a)

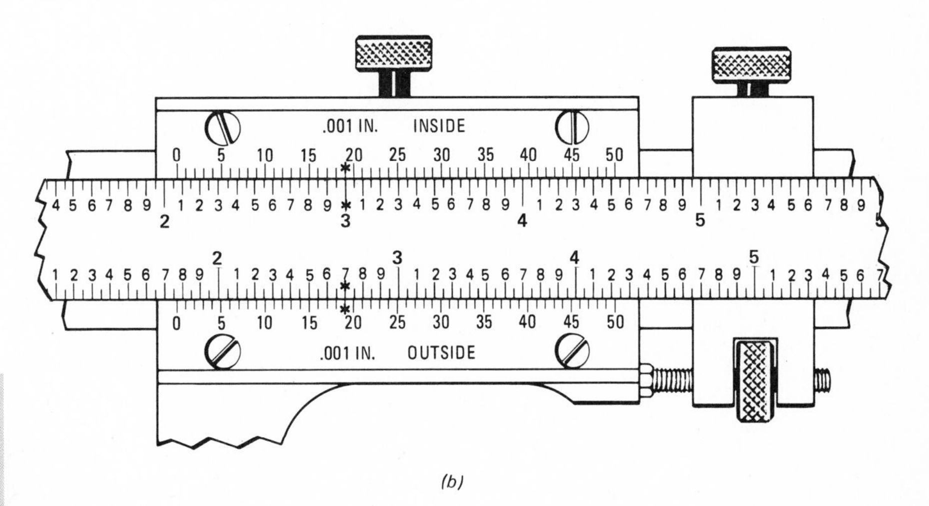

(b)

Fig. 19-9

(See Answer Key for Test 5—Measuring Instruments)

Fig. 20-1 A metric converter/calculator. (Courtesy Continental West, Inc.)

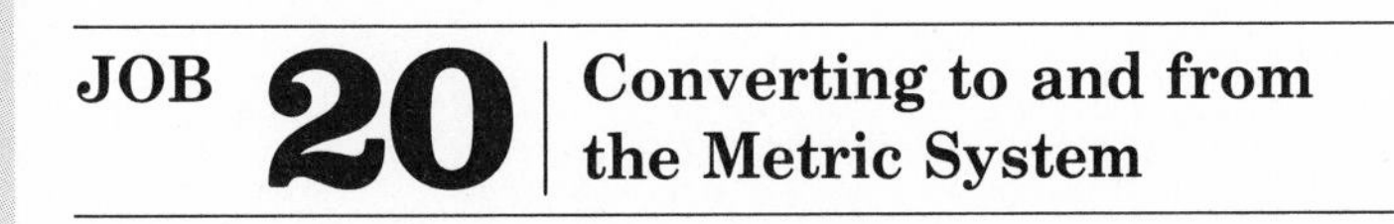

JOB 20 Converting to and from the Metric System

During the change from the English system to the metric system, we will have to be able to convert from one system to the other. Special hand calculators such as the one pictured in Fig. 20-1 are already widely used. However, simple calculations do not require such sophisticated and expensive tools.

As shown in Fig. 20-2, there are 10 mm in 1 cm, and 1 in equals a little more than 25 mm.

1 cm = 10 mm
1 cm = 0.394 in
1 in = 2.54 cm = 25.4 mm
1 ft = 0.305 m
1 m = 3.28 ft

Fig. 20-2 Comparing millimeter and inch scales. (Courtesy The L.S. Starrett Company)

Example 20-1 Measure the line segment in Fig. 20-3 in inches, and convert the measurement to centimeters and millimeters.

Solution The line measures 2 in long.

Change the 2 in to centimeters and millimeters.

A ———————————— B

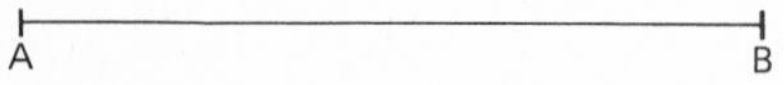

Fig. 20-3

Since 1 in = 2.54 cm

2 in = 2 × 2.54 = 5.08 cm = 50.8 mm *Ans.*

Check this answer with your metric ruler.

Example 20-2 Find the outside diameter of the cap screw shown in Fig. 20-4 in millimeters.

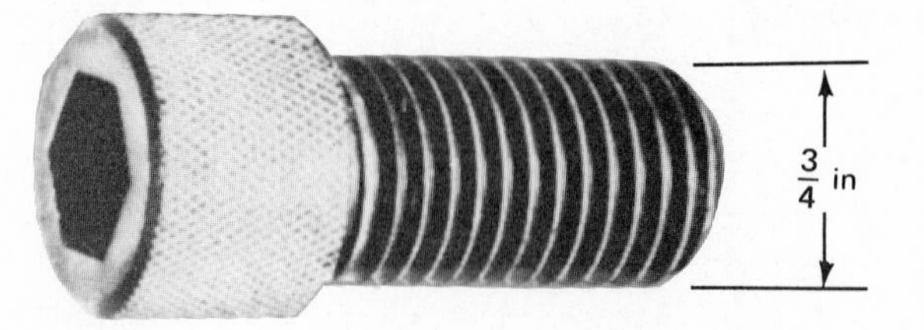

Fig. 20-4

Solution Since $\frac{3}{4}$ in = 0.75 in,

0.75 × 2.54 = 1.91 cm = 19.1 mm *Ans.*

Example 20-3 The tolerance on the top compression ring gap for a foreign car is given as 0.033 to 0.058 cm. Express these measurements in inches.

Solution Since 1 cm = 0.394 in,

0.033 cm = 0.033 × 0.394 = 0.013 in *Ans.*

0.058 cm = 0.058 × 0.394 = 0.023 in *Ans.*

Example 20-4 The grindstone shown in Fig. 20-5 has a cutting speed of 400 ft/min. Express this speed as meters per minute.

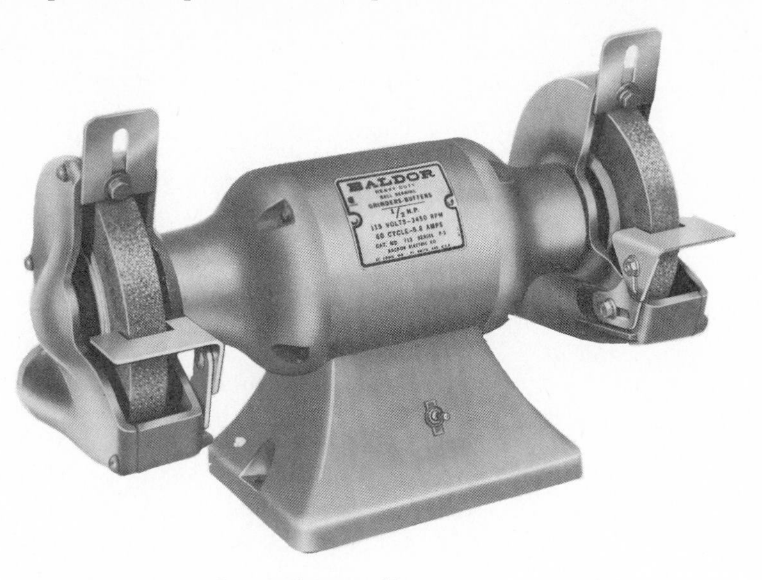

Fig. 20-5 A bench grindstone. (Courtesy The Baldor Electric Company)

Solution As noted in Fig. 20-2,

1 ft = 0.305 m

1 m = 3.28 ft

Therefore, 400 ft = 400 × 0.305 = 122 m/min *Ans.*

Example 20-5 The electrolytic tinning line shown in Fig. 20-6 can handle 20-ton coils of steel strip at speeds of up to 460 m/min. Express this speed in feet per minute.

Fig. 20-6 An electrolytic tinning line. (Courtesy U.S. Steel Company)

Solution Since 1 m = 3.28 ft,

460 m = 460 × 3.28 = 1,509 ft/min *Ans.*

The 1974 8-350 Camaro engine uses a spark plug with a gap of 0.035 in. Change this measurement to centimeters and millimeters.

Solution

Since 1 in = ______ cm, to change 0.035 in to centimeters, (×)(÷) 0.035 in by ______.	2.54 ×, 2.54
Thus, 0.035 in = 0.035 × 2.54 = ______ cm *Ans.*	0.089
To change 0.089 cm to millimeters, (×)(÷) 0.089 by 10.	×
Thus, 0.089 cm = 0.089 × 10 = ______ mm *Ans.*	0.89

Problems

1. (*G*) Draw a 4-in-long line.
   a. Calculate the length in centimeters and millimeters.
   b. Check your answers with a metric ruler.
2. (*G*) Repeat Prob. 1 for a $2\frac{1}{2}$-in-long line.
3. (*M*) Before turning the $3\frac{1}{2}$-in-diameter shaft shown in Fig. 20-7, a center hole must be drilled into which the tailstock of the lathe will fit. Express the diameter of the shaft in centimeters.

Fig. 20-7 Center drilling using a steady rest. (Courtesy LeBlond Inc.)

4. (*G*) How high is the 10-m diving board in feet?
5. (*G*) What is the metric equivalent of the 880-yd run?
6. (*G*) Which measurement is larger? (Change inches to metric measurements and compare.)
   a. 4 cm or 1 in
   b. 90 cm or 12 in
   c. $\frac{1}{2}$ in or 1 cm
   d. 2 ft or 35 cm
   e. $1\frac{1}{2}$ in or 4 cm
   f. 62 mm or 2.5 in
7. (*G*) How wide is 35-mm film in inches?

8. (*A*) At an altitude of 3,000 ft, a vacuum gage test on an engine gave a reading of 14 in of mercury. Express this reading in centimeters and millimeters.
9. (*A*) A foreign car requires a 0.41-mm distributor point gap. What size feeler gage (to the nearest thousandth of an inch) would you use?
10. (*M, A*) The belt in Fig. 20-8 connects pulleys of 7 in and 4 in in diameter. Express these diameters in centimeters.

Fig. 20-8 The pulley system of a drill press.
(Courtesy Rockwell International)

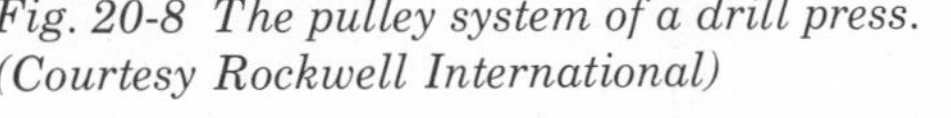

11. (*M*) Convert the given measurements in the spiral flute reamer shown in Fig. 20-9 to centimeters.

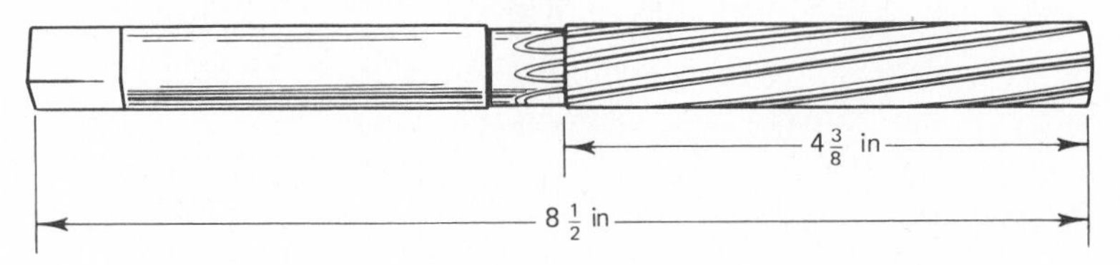

Fig. 20-9 A spiral flute reamer.

12. (*M*) Change the measurements of the dovetail cutter shown in Fig. 20-10 to centimeters and millimeters.
13. (*F*) The International Harvester 241 "Bigroll Baler" shown in Fig. 20-11 can harvest bales of hay 5 ft wide and up to 5.9 ft in diameter. Express these measurements in meters.
14. (*M*) Measure dimensions $A, B, C, D,$ and E on the toolmaker's parallel clamps shown in Fig. 20-12 in millimeters. Convert these measurements to inches.
15. (*G*) Change the following neck sizes of American shirts to their metric equivalents: (a) 15 in, (b) 16 in, (c) $16\frac{1}{2}$ in, and (d) $17\frac{1}{2}$ in.
16. (*C*) The tire on the PAY loader shown in Fig. 20-13 measures $9\frac{1}{2}$ ft in diameter. Convert this measurement to meters.

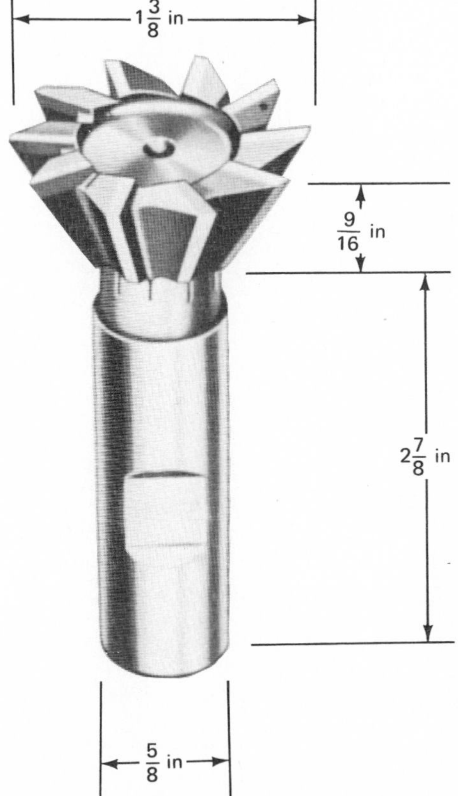

Fig. 20-10 A dovetail cutter.

Fig. 20-11 The International 241 Bigroll Baler: (Courtesy The International Harvester Company)

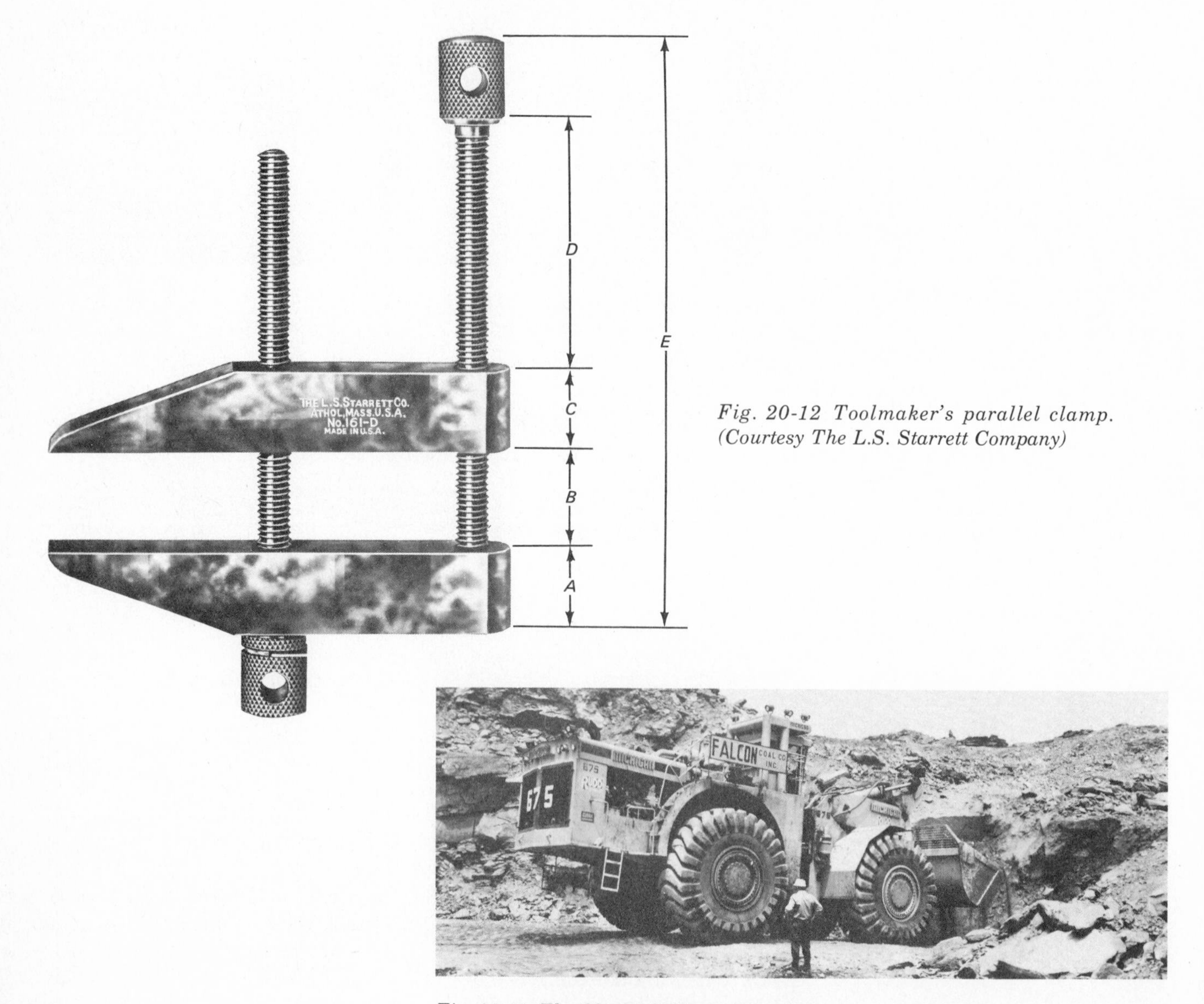

Fig. 20-12 Toolmaker's parallel clamp. (Courtesy The L.S. Starrett Company)

Fig. 20-13 The Clark Michigan 675 uncovers new seams of coal. (Courtesy The Clark Equipment Company)

17. (*A, M, C*) What size American wrenches (to the nearest sixty-fourth of an inch) would best fit the following metric bolts? (a) 9 mm, (b) 11 mm, (c) 13 mm, (d) 15 mm, and (e) 19 mm.
18. (*M*) The part shown in Fig. 20-14 moves past the ball-end milling cutter at a speed of 0.015 in/rev of the cutter. If the cutter rotates at 200 rpm, find (a) the movement of the part (the feed) expressed as inches per minute, and (b) the feed expressed as centimeters per minute.

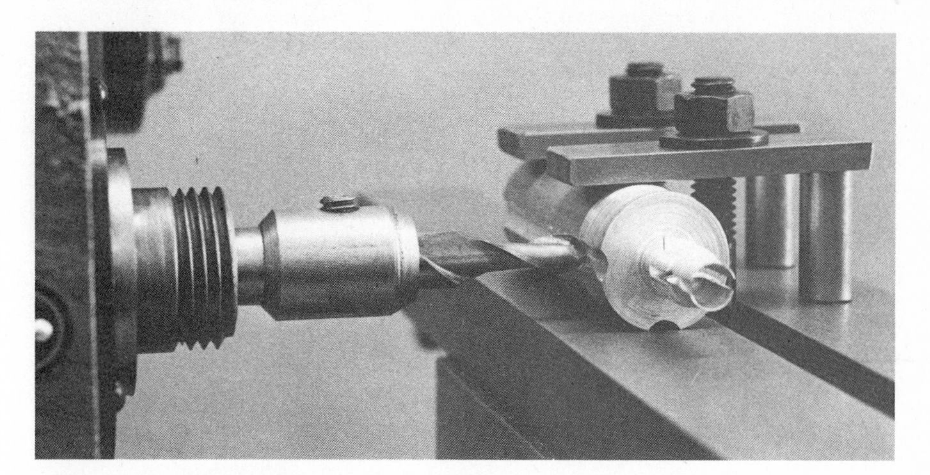

Fig. 20-14 Milling a groove with a ball-end mill on a horizontal milling machine. (Courtesy Hamilton Associates)

19. (*F*) The pivot point of the bucket on the International 2350 Mount-O-Matic Farm Loader shown in Fig. 20-15 is 12 ft above the ground. This high lift enables farmers to clear high-sided trucks and wagons. Express the distance 12 ft as meters.

Fig. 20-15 The International 2350 Mount-O-Matic Farm Loader. (Courtesy The International Harvester Company)

20. (*A*) The ring side clearance on the 1974 Buick special for the bottom compression ring is listed as 0.0012 to 0.0032 in. Express these measurements as millimeters.
21. (*G*) The athletes competing in the Olympic games run 100 m instead of 100 yd. How much longer is 100 m than 100 yd?
22. (*M, A*) The pitch diameter of the thread shown in Fig. 20-16 is "miked" at 0.5746 in. Express this measurement as millimeters.

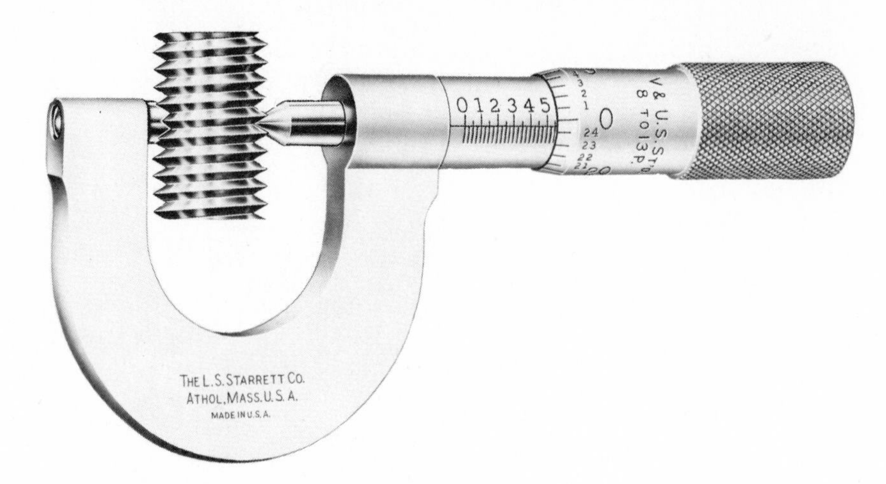

Fig. 20-16 This special micrometer reads the pitch diameter of the thread directly. (Courtesy The L.S. Starrett Company)

23. (*G*) a. Read the indicated measurements in Fig. 20-17 in centimeters and millimeters.
    b. Convert each measurement to inches.

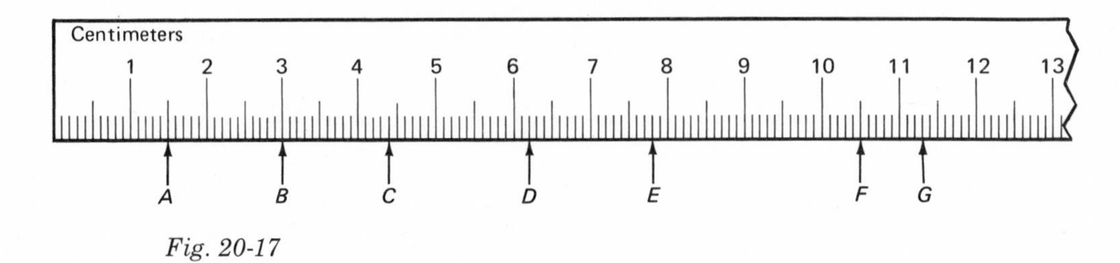

Fig. 20-17

24. (*C*) What would be the metric measurements for a board $\frac{7}{8}$ in thick by $11\frac{1}{2}$ ft long?
25. (*M*) a. Measure the tap shown in Fig. 20-18 in inches.
    b. Convert this measurement to centimeters and millimeters.

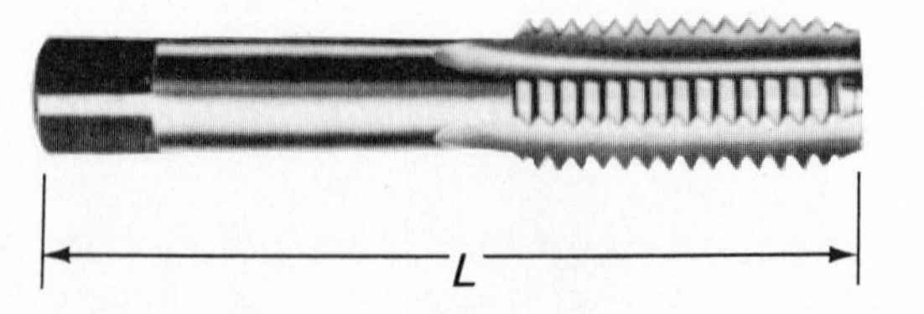

Fig. 20-18

CONVERTING LARGE UNITS OF MEASUREMENT

Refer to Fig. 3-1 on page 17, in which distances were described in terms of miles and kilometers. You will recall that the prefix "kilo" means 1,000 times as much, and that therefore the word kilometer (km) means 1,000 meters.

$$1{,}000 \text{ m} = 1 \text{ km}$$
$$1 \text{ km} = 0.62 \text{ mi}$$
$$1 \text{ mi} = 1.61 \text{ km}$$

Example 20-7 Check the equivalent distances shown in Fig. 3-1.

Solution

$10 \text{ km} = 10 \times 0.62 = 6.2 \text{ mi}$ *Check*

$4.8 \text{ km} = 4.8 \times 0.62 = 2.976 = 3 \text{ mi}$ *Check*

or

$6.2 \text{ mi} = 6.2 \times 1.61 = 9.98 = 10 \text{ km}$ *Check*

$3 \text{ mi} = 3 \times 1.61 = 4.83 \text{ km}$ *Check*

Example 20-8 What is the metric equivalent of the 440-yd run ($\frac{1}{4}$ mi)?

Solution

$\frac{1}{4} \text{ mi} = \frac{1}{4} \times 1.61 = 0.4 \text{ km}$

$0.4 \text{ km} = 0.4 \times 1{,}000 = 400 \text{ m}$ *Ans.*

Problems

Change the following airline distances between the given cities into kilometers.

1. New York to London, 3,500 mi
2. Madrid to London, 800 mi
3. Paris to Moscow, 1,550 mi
4. San Francisco to Tokyo, 5,150 mi
5. New York to Rome, 4,300 mi

Change the following airline distances between the given cities into miles.

6. London to Berlin, 960 km
7. Tokyo to Honolulu, 6,080 km
8. Madrid to Stockholm, 2,640 km
9. New York to Berlin, 6,400 km
10. Paris to Stockholm, 1,600 km
11. The circumference of the earth at the equator is about 25,000 mi. Express this distance in kilometers.
12. The speed limit in the following countries is given as kilometers per hour. Express the speeds as miles per hour. (a) France, 100 km/hr; (b) Great Britain, 65 km/hr; (c) Spain, 80 km/hr; (d) Germany, 130 km/hr.
13. State the ratio of 20 km to 4 mi.
14. An athlete ran 1,500 m in 3 min 45 sec. At this rate, what would be his time (in minutes) for the mile run?

CONVERTING UNITS OF VOLUME

A cube measuring 1 cm on a side contains a volume of 1 cubic centimeter (cc). As shown in Fig. 20-19, 1,000 cc is called 1 liter (l).

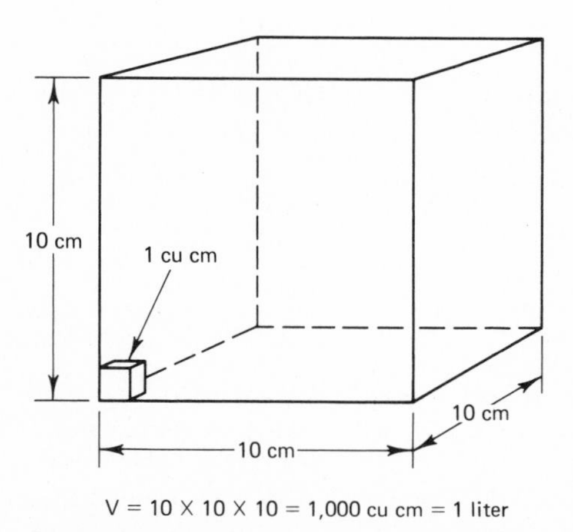

Fig. 20-19

USCS Units	Metric Units of Volume		
1 cu in	= 16.4 cc	1 cc	= 0.061 cu in
1 cu ft	= 0.0283 cu m	1 cu m	= 35.32 cu ft
1 cu ft	= 28.32 l	1 l	= 0.0353 cu ft
1 qt	= 0.95 l	1 l	= 1.06 qt
1 gal	= 3.8 l	1 l	= 0.264 gal

Example 20-9 The 1975 Plymouth V-8 engine has a 400 cu in displacement (cid). Express this displacement as (a) cubic centimeters, and (b) liters.

Solution (a) Since 1 cu in = 16.4 cc,

$$400 \text{ cu in} = 400 \times 16.4 = 6{,}560 \text{ cc} \qquad \textit{Ans.}$$

(b) Since 1 l = 1,000 cc,

$$6{,}560 \text{ cc} = \frac{6{,}560}{1{,}000} = 6.56 \text{ l} \qquad \textit{Ans.}$$

Example 20-10 A grain box has a volume of 30 cu m. Convert this measurement to cu ft.

Solution Since 1 cu m = 35.32 cu ft,

$$30 \text{ cu m} = 30 \times 35.32 = 1{,}060 \text{ cu ft} \qquad \textit{Ans.}$$

Example 20-11 How many liters are contained in a 50-gal oil drum?

Solution Since 1 gal = 3.8 l,

$$50 \text{ gal} = 50 \times 3.8 = 190 \text{ l} \qquad \textit{Ans.}$$

Example 20-12 In France, antifreeze costs $3.18 for a 2-l container. Find the cost per quart.

Solution $2 \text{ l} = 2 \times 1.06 \text{ qt} = 2.12 \text{ qt}$

$$\$3.18 \text{ for } 2.12 \text{ qt} = \frac{3.8}{2.12} = \$1.50 \text{ per qt} \qquad \textit{Ans.}$$

CONVERTING UNITS OF WEIGHT

The basic unit of weight in the metric system is the *gram* (g). An ordinary paper clip weighs about 1 g. Since the prefix "kilo" means 1,000, 1 kilogram (kg) = 1,000 g.

Value of the kilogram

1 kg = 2.2 lb
1 lb = 0.454 kg

Example 20-13 A sign on a bridge said "Maximum weight—4,545 kg." Express the maximum weight to the nearest ton.

Solution 4,545 kg = 4,545 × 2.2 = 9,999 lb, or 10,000 lb (approx)

Since 1 ton = 2,000 lb,

$$10{,}000 \text{ lb} = \frac{10{,}000}{2{,}000} = 5 \text{ tons} \qquad \textit{Ans.}$$

Example 20-14 A certain steel pipe weighs 2 lb/ft of length. Find the weight (in kilograms) of a standard 20-ft length.

Solution 20 ft weigh 20 × 2 = 40 lb.

$$40 \text{ lb} = 40 \times 0.454 \text{ kg} = 18.2 \text{ kg} \qquad \textit{Ans.}$$

Problems

1. (*A*) How many liters of gasoline are needed to fill a 20-gal tank?
2. (*G*) How many liters are contained in a 2-qt container of milk?
3. (*A, G*) A woman bought 250 l of gasoline on an automobile trip in Europe. How many gallons was this?
4. (*B, G*) If milk costs $0.53/l in London, find the cost per quart.
5. (*A*) How many miles per gallon will a car travel if it can travel 7 mi on 1 l of gasoline?
6. (*A*) The aluminum motor block on a foreign car weighed 120 kg. Express this weight in pounds.
7. (*M*) If 1 ft of 3-in-diameter solid steel round stock weighs 24 lb, how many kilograms does it weigh?
8. (*M, C*) If 100 sq ft of No. 20 gage sheet iron weighs 128 lb, how many kilograms does it weigh?
9. (*A*) A car can travel 14 mi on 1 gal of gas. How many liters would it use to travel 210 mi?
10. (*F, G*) How many 5-gal cans can be filled from a kiloliter drum of insecticide?
11. (*A*) The 1970 Dodge Polara has a gas tank capacity of 24 gal. How many liters is this?
12. (*A, G*) A gasoline storage can measures 25 cm × 25 cm × 30 cm. Find (a) the volume in cubic centimeters, (b) the volume in liters, and (c) the volume in gallons. *Note:* Volume = $L \times W \times H$.
13. (*C, B*) An oil storage tank has a capacity of 760 l. Find the cost to fill it at $0.32/gal.
14. (*A*) A car has a 427-cu in displacement (CID) engine. Find the displacement of the engine in (a) cubic centimeters and (b) liters.
15. (*A*) A foreign car has a 2.3-l engine. Find the cubic inch displacement of the engine.
16. (*B, F*) If a bag of coffee weighs 50 kg, find the value of 2,500 bags at $0.38/lb.
17. (*B, G*) The baggage limit on overseas flights is 44 lb. Express this weight in kilograms.
18. (*F, G*) Custom officials figure the duty on wine imported into the United States on the number of gallons imported. What is the gallon equivalent of 14 l of wine?
19. (*A*) If gasoline weighs 0.67 kg/l, express the weight of gasoline in pounds per gallon.
20. (*A*) The mileage on a foreign car was described as 8 km/l. Express this as miles per gallon.
21. (*G*) A "seatainer" for overseas transport has a volume of 3,000 cu ft. Convert this volume to cubic meters.
22. (*F, G*) How many cubic feet of space will be left after 2,000 l of milk is poured into a tank with a volume of 75 cu ft?
23. (*A, G*) A woman touring Europe drove 1,500 km on 235 l of gasoline. Find the average number of miles per gallon.
24. (*E, G*) A certain type of telephone cable weighs 4.5 kg/m. Express this weight as lb/ft.
25. (*G*) Water weighs 1,000 kg/cu m. Express this density as lb/cu ft.

CONVERTING UNITS OF AREA

The word *area* refers to the amount of *surface* bounded by some closed figure. The unit of measurement that is used to measure an area is *another* area of standard measurement. A *square inch* (sq in) is the area within a square 1 in on each side. If each 1-in side of the square is described as 2.54 cm, then the area of the square would be 2.54 cm × 2.54 cm, or 6.45 sq cm.

USCS Units	Metric Units
1 sq in	= 6.45 sq cm
1 sq ft	= 0.093 sq m
1 sq cm	= 0.155 sq in.
1 sq m	= 10.76 sq ft

Example 20-15 A piston has a cross-sectional area of 15.9 sq in. Find the area in sq cm.

Solution Since 1 sq in = 6.45 sq cm

15.9 sq in = 15.9 × 6.45 = 102.6 sq cm *Ans.*

Example 20-16 A cement floor with an area of 50 sq m is to receive one coat of floor paint. Find the cost of the paint if it spreads at the rate of 250 sq ft/gal and costs $5.40/gal.

Solution

1. Change 50 sq m to sq ft.
   Since 1 sq m = 10.76 sq ft,

   50 sq m = 50 × 10.76 = 538 sq ft

2. Find the number of gallons of paint required.

   538 sq ft ÷ 250 sq ft/gal = 2.152 gal

3. Find the cost.

   2.152 gal × $5.40/gal = $11.62 *Ans.*

CONVERTING UNITS OF PRESSURE

1 lb/sq in = 0.0703 kg/sq cm

1 kg/sq cm = 14.22 lb/sq in

Example 20-17 A tire is inflated to a pressure of 28 lb/sq in. Express this pressure as kg/sq cm.

Solution Since 1 lb/sq in = 0.0703 kg/sq cm,

28 lb/sq in = 28 × 0.0703 = 1.97 kg/sq cm *Ans.*

Alternate Method: If you forget the conversion factors given above, you can convert by changing each of the measurements given into metric units and then dividing.

1. Change 28 lb into kg.
   Since 1 lb = 0.454 kg,

   28 lb = 28 × 0.454 = 12.7 kg.

2. Change 1 sq in into sq cm.

   1 sq in = 6.45 sq cm

3. Change 28 lb/sq in into kg/sq cm.

   $$\frac{28\text{ lb}}{\text{sq in}} = \frac{12.7\text{ kg}}{6.45\text{ sq cm}} = 1.97\text{ kg/sq cm} \qquad \textit{Ans.}$$

Problems

1. (*C, M, G*) An air-conditioning duct has a cross-sectional area of 0.84 sq m. Express this area as sq ft.
2. (*C, G*) A rivet has a cross-sectional area of 0.44 sq in. Express this area as sq cm.
3. (*E*) The maximum current permitted in a copper bus bar is 155 A/sq cm. Find the permitted current density in amperes per square inch.
4. (*F, C, G*) Find the cost to sod 1,200 sq ft of lawn at $3.75/sq m.
5. (*C, G*) The area of a boiler valve is 7 sq in. (a) Find the area in square centimeters. (b) Find the total force on the valve if the pressure is 9 kg/sq cm.
6. (*A*) A tire is inflated to a pressure of 2.2 kg/sq cm. Express this pressure in pounds per square inch.

JOB 21 Metric Measuring Instruments

The metric micrometer shown in Fig. 21-1 is graduated to read the measurement in millimeters and hundredths of a millimeter.

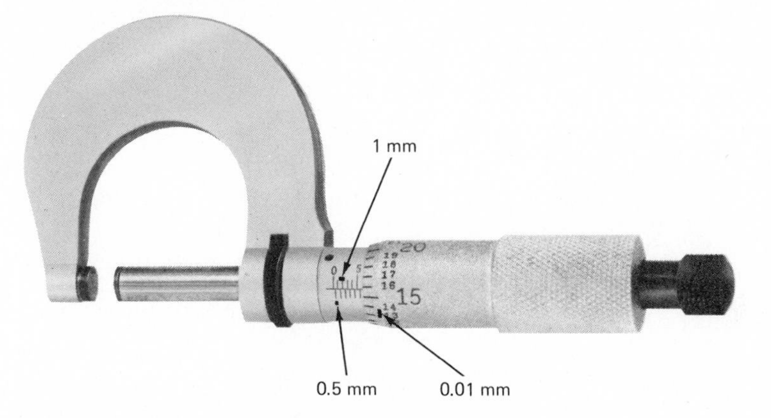

Fig. 21-1 Values of the graduations on a metric micrometer.

The sleeve above the gage line is graduated in millimeters from 0 to 25 mm. Each graduation below the gage line represents 0.5 mm. Since the pitch of the spindle screw is 0.5 mm, one complete revolution of the thimble will move the spindle exactly 0.5 mm and expose the first graduation *below* the gage line, as shown in Fig. 21-2*a*. Two revolutions will expose the first line *above* the gage line and represent 1 mm, as shown in Fig. 21-2*b*. Three revolutions will expose 1 line above and an additional line below the gage line and represent 1.5 mm, as shown in Fig. 21-2*c*. Four revolutions will expose 2 lines above the gage line and represent 2 mm, as shown in Fig. 21-2*d*.

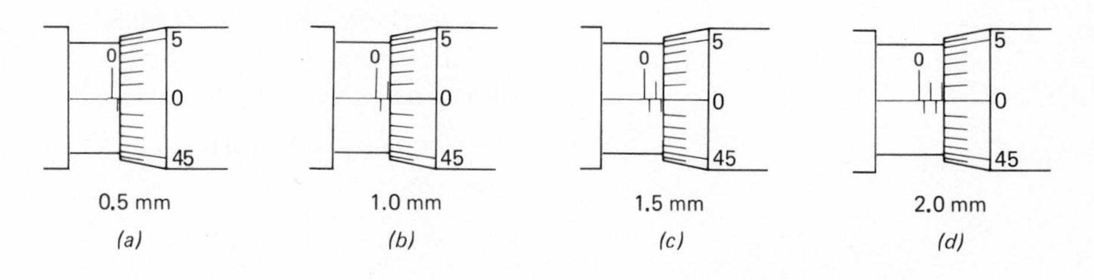

Fig. 21-2

The beveled edge of the thimble is divided into 50 parts, with every fifth line being numbered. Since 1 complete revolution of the thimble moves the spindle 0.5 mm, each graduation on the thimble represents $\frac{1}{50} \times 0.5$ mm or 0.01 mm. Two thimble graduations represent 0.02 mm, three thimble graduations represent 0.03 mm, etc.

READING A METRIC MICROMETER

1. Obtain the total reading in millimeters shown on the sleeve.
2. Find the number on the thimble coincident with the gage line. This number represents hundredths of a millimeter.
3. Add the values obtained in steps 1 and 2.

Example 21-1 Read the micrometer setting shown in Fig. 21-3.

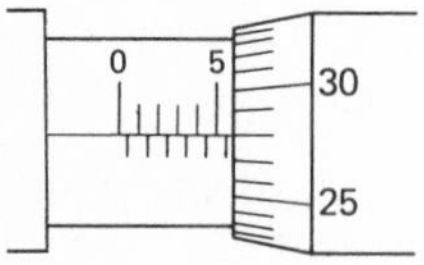

Solution

The 5-mm graduation is visible.	5 mm
An additional line below the gage line is visible.	0.5 mm
Line 28 on the thimble coincides with the gage line, each number representing 0.01 mm.	
$28 \times 0.01 =$	0.28 mm
Add	5.78 mm *Ans.*

Fig. 21-3

Example 21-2 Read the micrometer setting shown in Fig. 21-4.

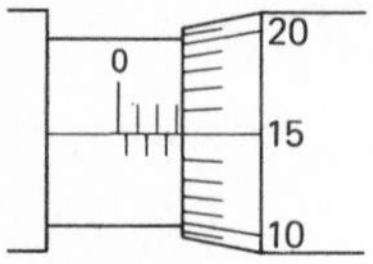

Solution

3 lines above the gage line are visible.	3 mm
No additional lines below the gage line are visible.	0.00 mm
Line 15 on the thimble coincides with the gage line, each number representing 0.01 mm.	
$15 \times 0.01 =$	0.15 mm
Add	3.15 mm *Ans.*

Fig. 21-4

Self-Test 21-3 Read the micrometer setting shown in Fig. 21-5.

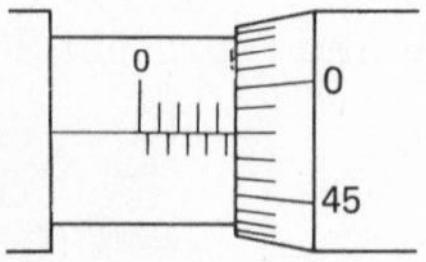

Solution

As with the ordinary micrometer, we must be careful when the thimble registers close to zero. Although we can see part of the 5, we (have)(have not) yet passed 5 millimeter marks.		have not
The 4-mm graduation is visible.	______ mm	4
An additional line below the gage line is visible.	______ mm	0.5
Line 48 coincides with the gage line.		
$48 \times 0.01 =$	______ mm	0.48
Add	______ mm *Ans.*	4.98

Fig. 21-5

Problems

1-10. Read the metric micrometer settings shown in Fig. 21-6.

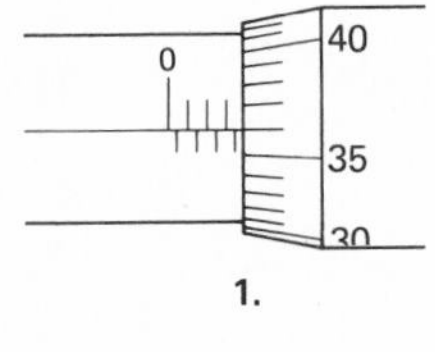

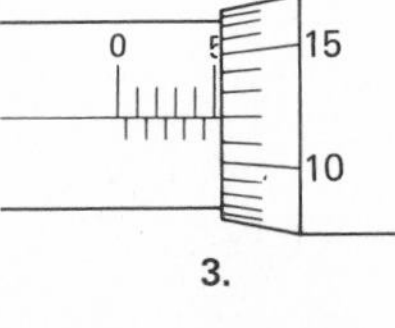

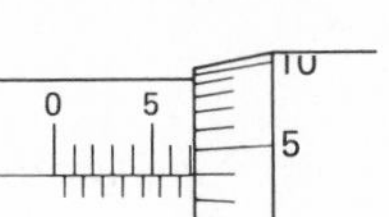

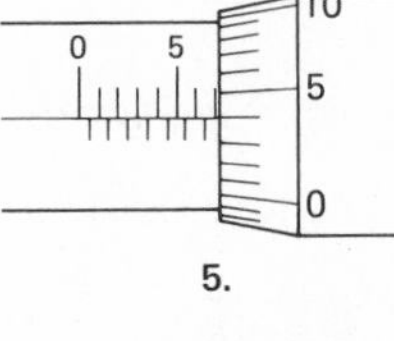

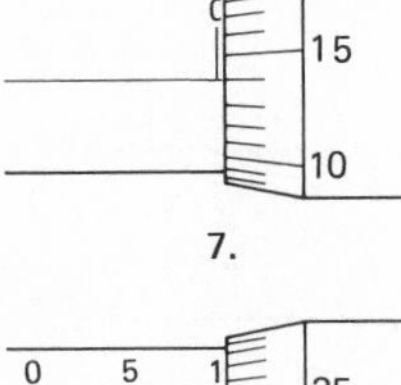

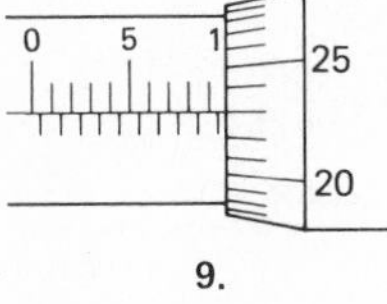

Fig. 21-6

THE METRIC VERNIER

Example 21-4 Read the metric vernier shown in Fig. 21-7.

Solution

1. Find the last large numeral passed (4).

   4 cm = 4 × 10 = 40.00 mm

2. Find the number of long graduations passed (1).

   1 × 1 mm = 1.00 mm

3. Find the number of short graduations passed (1).

   1 × 0.5 mm = 0.50 mm

4. Find the coincident vernier number (9).

   9 × **0.02** mm = 0.18 mm

5. Add. 41.68 mm *Ans.*

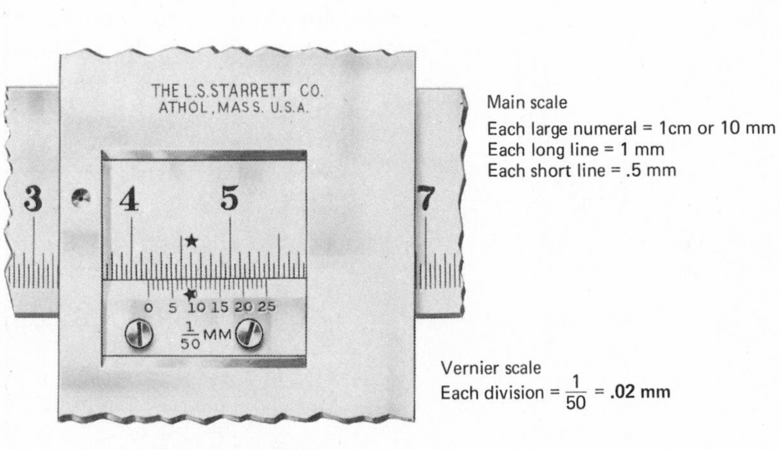

Fig. 21-7 A metric vernier. (Courtesy The L.S. Starrett Company)

Problems

Read the metric verniers shown in Fig. 21-8.

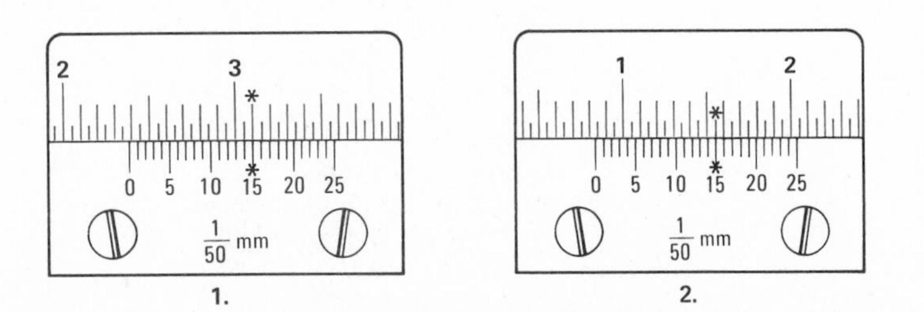

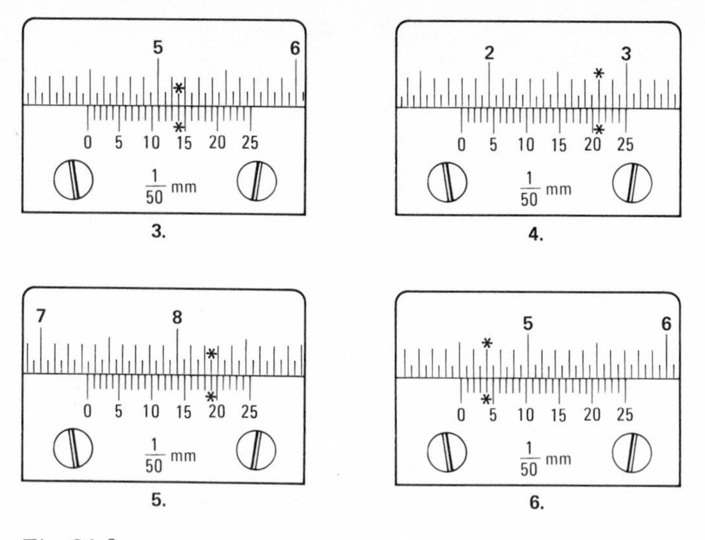

Fig. 21-8

COMBINATION VERNIERS

A combination vernier caliper which gives readings in both inches and millimeters is available, as shown in Fig. 21-9.

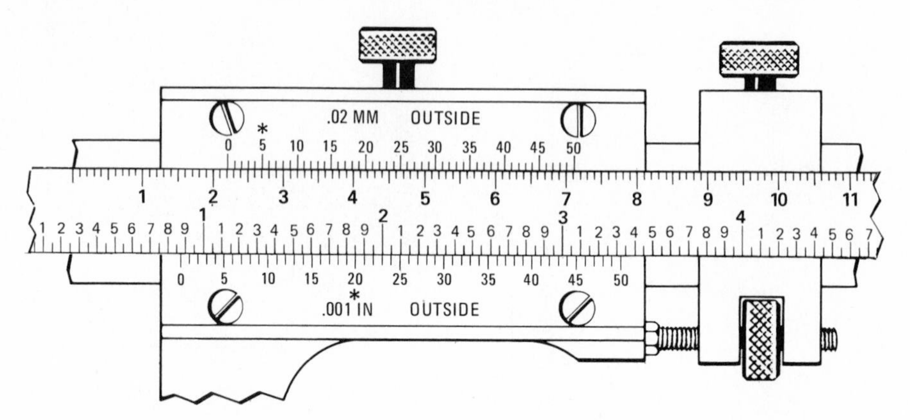

Fig. 21-9 How to read 50-division vernier with USCS/Metric graduations.

METRIC READING: Use the upper vernier and beam scales. Each beam graduation is 1 mm; each numbered beam graduation equals the number times 10 (numeral 1 = 10 mm, 2 = 20 mm, etc.). Vernier shows each .02 mm and has numbered graduations by 10's (thus, 5 × .02 = .1 mm, 10 × .02 = .2 mm, etc.).

Outside Measurements: Read outside measurements directly.

Example: 20 plus 2 beam graduations = 22 mm, plus vernier reading of line which coincides with beam line, 5 × .02 = .10 mm. Therefore, the total external reading is 22.10 mm.

Inside Measurements: Read outside measurement as described above and add 7.62 mm (width of nibs) to the reading.

Example: 22.10 mm + 7.62 mm = 29.72 mm

USCS READING: Use the lower vernier and beam scales. Each beam graduation is .050 in; each numbered beam graduation between inches is .100 in. Vernier shows each .001 in and has numbered graduations by .010 in.

Outside Measurements: Read outside measurements directly, as described above.

Example: .850 in + .020 in (vernier reading) = .870 in

Inside Measurements: Read outside measurement as described above and add .300 in (width of nibs) to the reading.

Example: .870 in + .300 in = 1.170 in

Problems

Read the combination verniers shown in Fig. 21-10 for the outside and inside measurements on both scales.

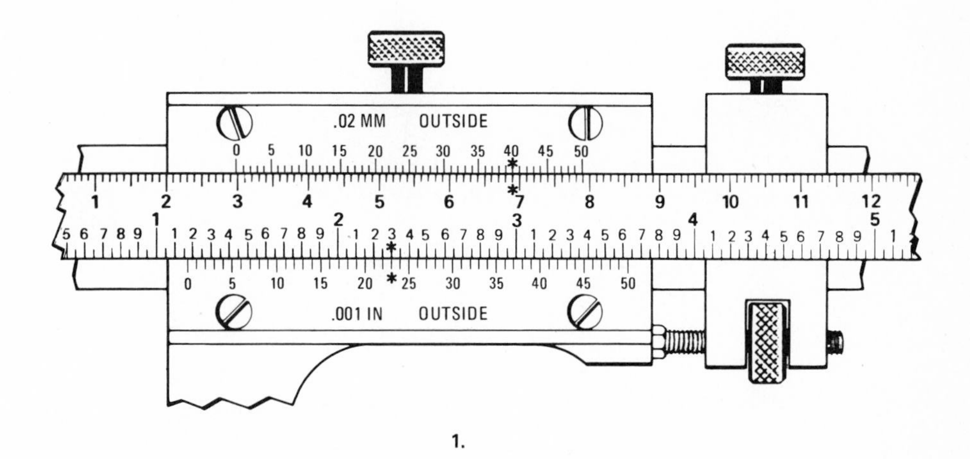

1.

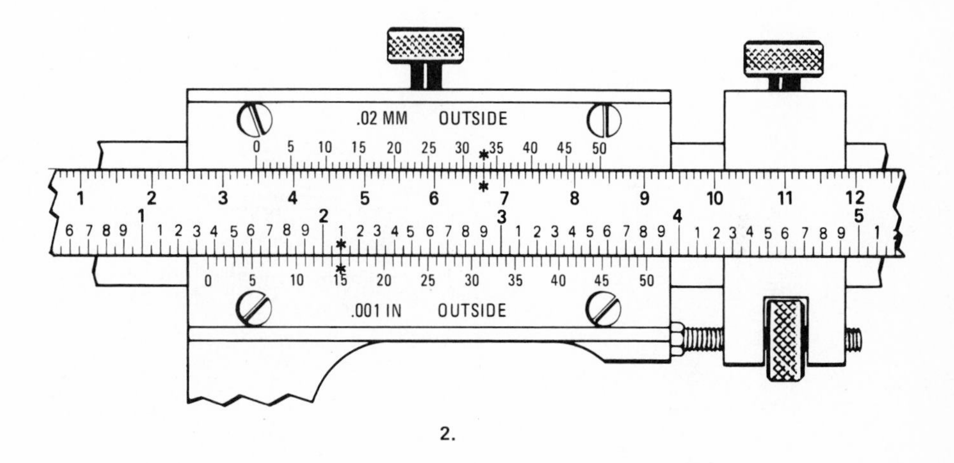

2.

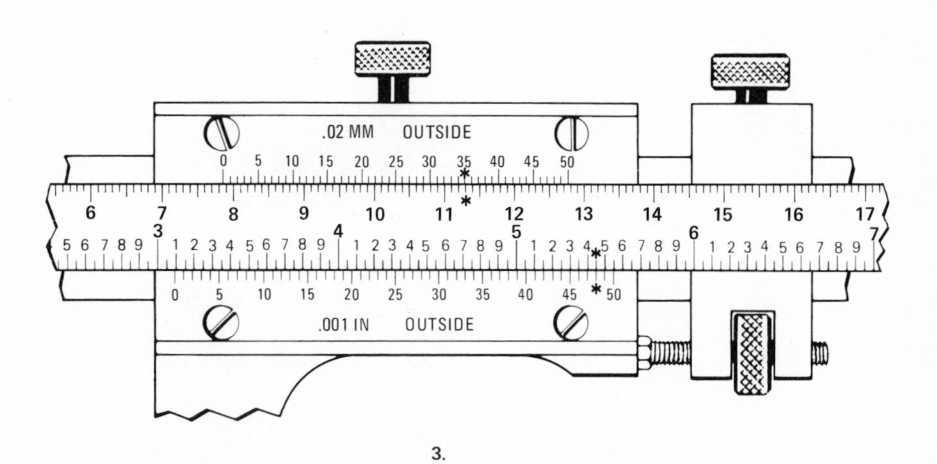

3.

Fig. 21-10

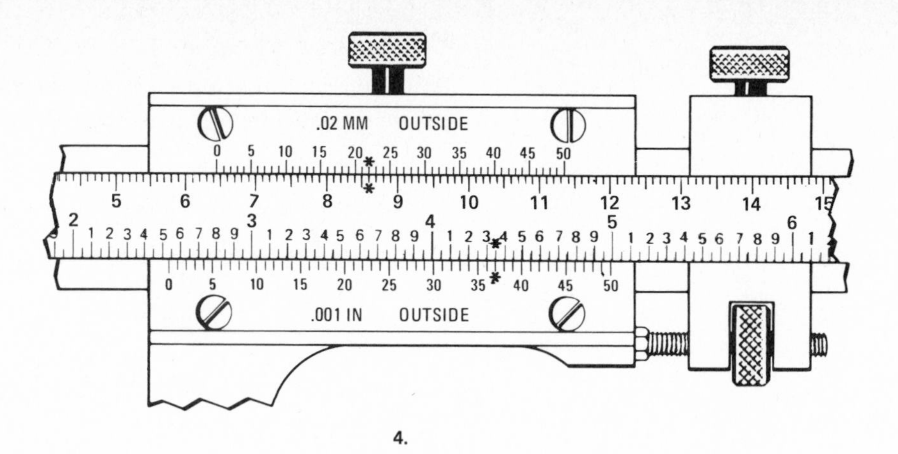

4.

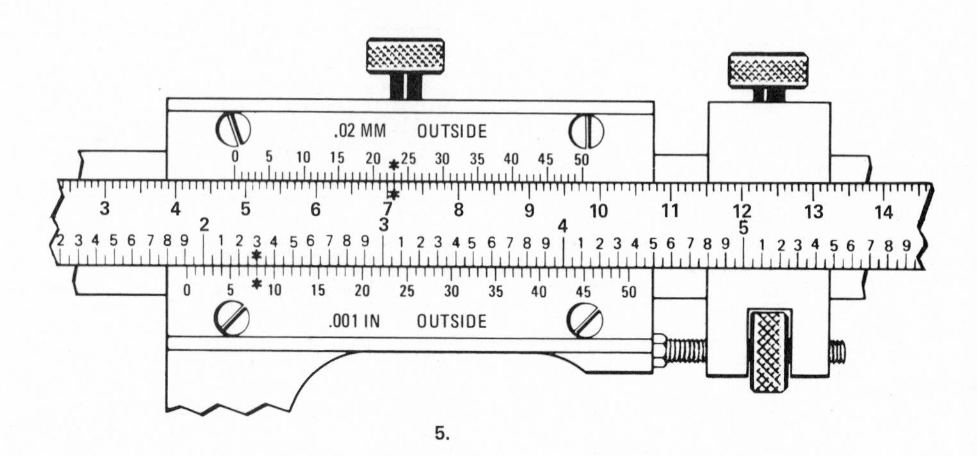

5.

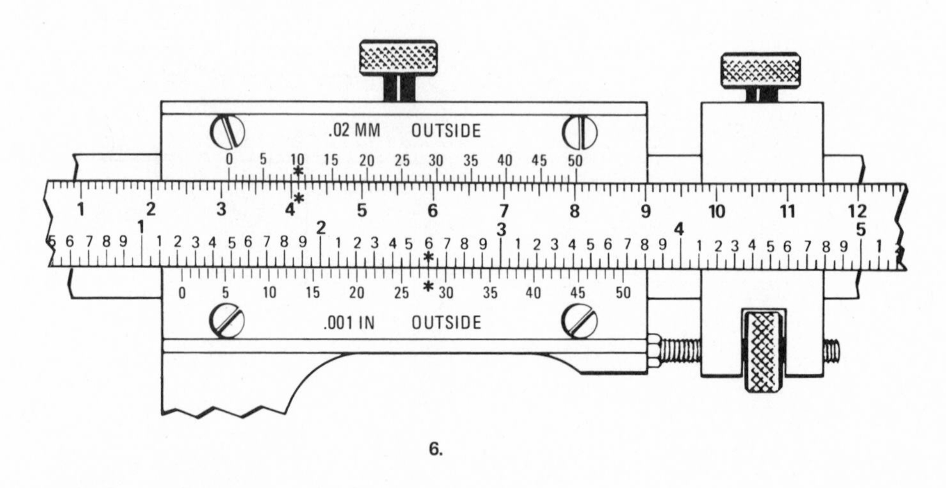

6.

JOB 22 Reading Meters and Indicators

The entire range of an instrument dial is divided into a number of equal parts called *main divisions*, the value of which varies from dial to dial. The main divisions are usually numbered, but some numbers may be omitted in order to make the dial easier to read. The main divisions are further indicated as either the heaviest or the longest lines on the dial.

Each of the main divisions is divided into a number of smaller parts, which we will call *spaces*. Just as the value of the main divisions of different instruments is not always the same, the value of the small spaces will also vary. There may be several scales on the same dial, as the meter may be used for different ranges. The *range* of a meter means the highest value on that particular scale.

READING METER DIALS

To read the value indicated,

1. Determine which of the several scales is to be read.

2. Determine the value of each main division and how many main divisions have been passed by the indicator.

3. Determine the value of each small space and how many small spaces past the last main division have been passed.

4. Add the small-space value to the main-division value to obtain the total reading.

VALUE OF THE MAIN DIVISIONS

To determine the value of the main division, merely subtract the value of *any* main division from the next *larger* one.

Example 22-1 Find the value of the main division on the DC and AC scales of the meter shown in Fig. 22-1.

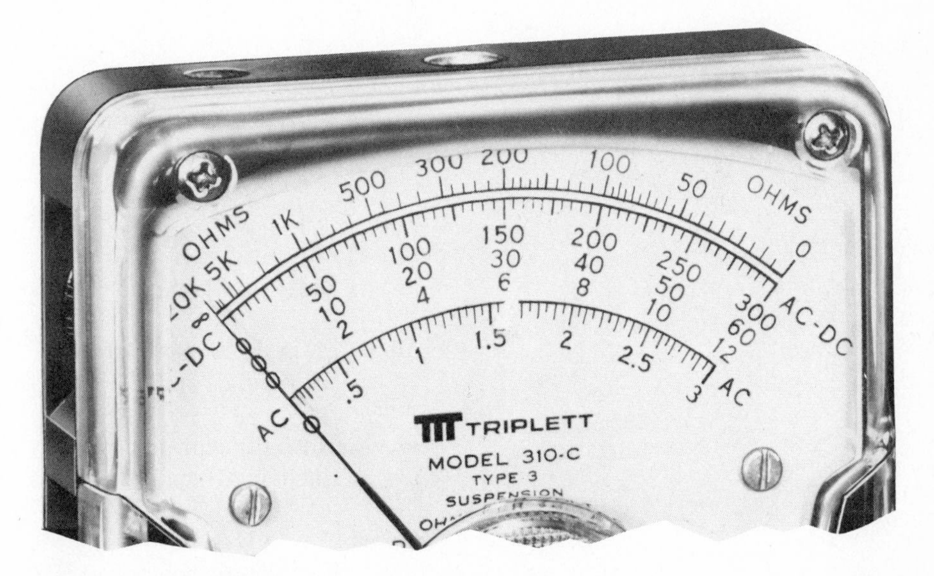

Fig. 22-1 Model 310-C Volt-Ohm-Milliammeter. (Courtesy The Triplett Company)

Solution Top DC scale: Main division $= 150 - 100 = 50$ *Ans.*

Middle DC scale: Main division $= 30 - 20 = 10$ *Ans.*

Bottom DC scale: Main division $= 6 - 4 = 2$ *Ans.*

AC scale: Main division $= 1.5 - 1 = 0.5$ *Ans.*

Note: scales measuring resistance (Ω) are generally not uniform, and we must be very careful to note the changes in value along the scale. For example, in the top Ω scale,
From 0 to 100 Ω, the main division = 100 − 50 = 50 Ω.

From 100 to 300 Ω, the main division = 300 − 200 = 100 Ω.

From 300 to 500 Ω, the main division = 500 − 300 = 200 Ω.

From 500 to 1,000 Ω (1 kΩ), the main division = 1,000 − 500 = 500 Ω.

VALUE OF THE SMALL SPACES

To determine the value of the small spaces,

1. Count the number of spaces between any two main divisions.

2. Each small space equals the value of the main division divided by the number of spaces.

Example 22-2 Find the value of each small space on the dial shown in Fig. 22-2.

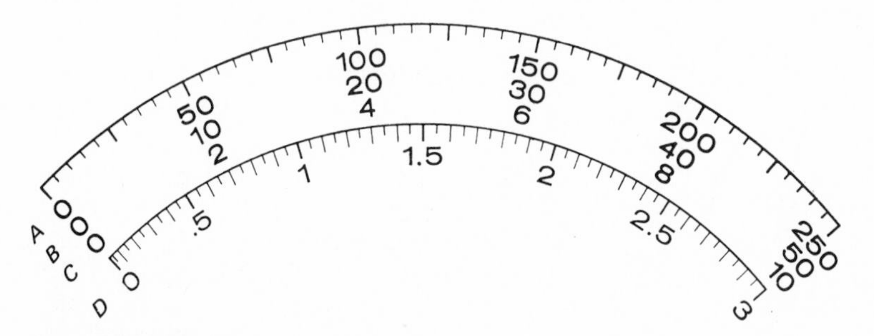

Fig. 22-2 Dial of the Model 630-PL V-O-M. (Courtesy The Triplett Company)

Solution For each scale shown, each main division contains 10 small spaces.

Scale *A*: Main division = 50
Each small space = 50/10 = 5 *Ans.*

Scale *B*: Main division = 10
Each small space = 10/10 = 1 *Ans.*

Scale *C*: Main division = 2
Each small space = 2/10 = 0.2 *Ans.*

Scale *D*: Main division = 0.5
Each small space = 0.5/10 = 0.05 *Ans.*

Example 22-3 Find the value of each small space on the scale shown in Fig. 22-3.

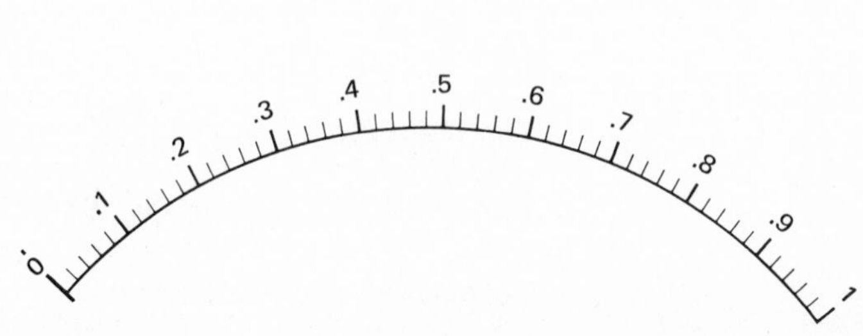

Fig. 22-3

Solution Main division = 0.1

Each main division = 5 small spaces

Each small space $= \dfrac{0.1}{5} = 0.02$ *Ans.*

FINDING THE METER READING

1. Find the value of the main division just before the indicator.

2. Find the number of spaces between this main division and the indicator.

3. Find the value of each small space.

4. Multiply the number of spaces found in step 2 by the value of each space found in step 3.

5. Add steps 1 and 4 to get the meter reading.

Example 22-4 Find the value indicated by the arrow for each scale of the dial shown in Fig. 22-4.

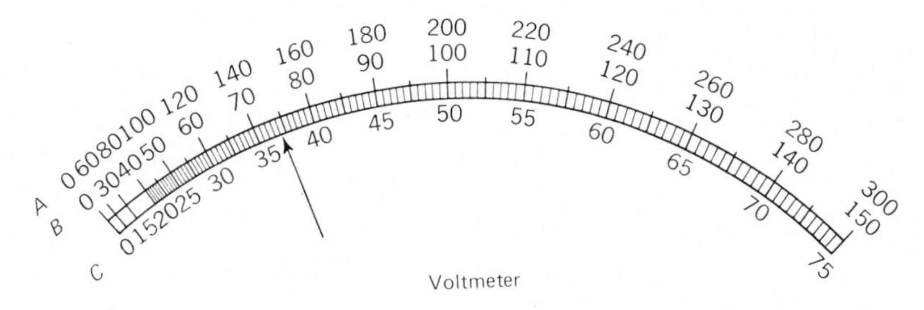

Fig. 22-4 (Courtesy The Weston Electrical Instrument Corporation)

Solution Scale *A:* The main division just before the arrow is 140 V.

Number of spaces covered by arrow = 4

Main division = 20 V, divided into 10 spaces

Each small space $= \dfrac{20}{10} = 2$ V

Value of spaces covered $= 4 \times 2 = 8$ V

Meter reading $= 140 + 8 = 148$ V *Ans.*

Scale *B:* The main division just before the arrow is 70 V.

Number of spaces covered by arrow = 4

Main division = 10 V, divided into 10 spaces

Each small space $= \dfrac{10}{10} = 1$ V

Value of spaces covered $= 4 \times 1 = 4$ V

Meter reading $= 70 + 4 = 74$ V *Ans.*

Scale *C:* The main division just before the arrow is 35 V.

Number of spaces covered by arrow = 4

Main division = 5 V, divided into 10 spaces

Each small space $= \dfrac{5}{10} = 0.5$ V

Value of spaces covered $= 4 \times 0.5 = 2$ V

Meter reading $= 35 + 2 = 37$ V *Ans.*

Example 22-5 Find the reading indicated by the arrow for each scale of the dial shown in Fig. 22-5.

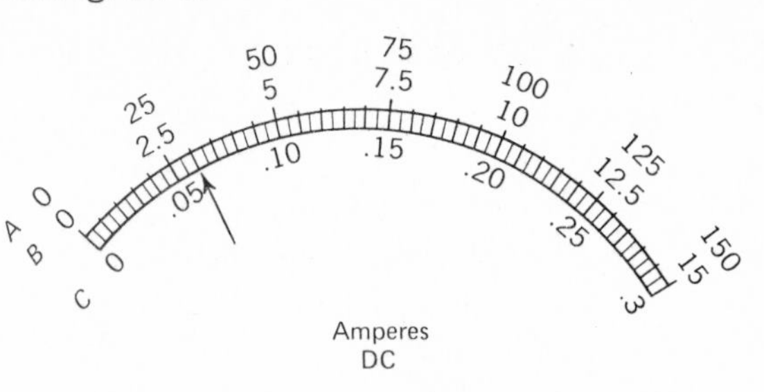

Fig. 22-5 (Courtesy The Weston Electrical Instrument Corporation)

Solution Scale A: The main division just before the arrow is 25 A.

$$\text{Number of spaces covered by arrow} = 2 \text{ spaces}$$
$$\text{Main division} = 25 \text{ A, divided into 10 spaces}$$
$$\text{Each small space} = \frac{25}{10} = 2.5 \text{ A}$$
$$\text{Value of spaces covered} = 2 \times 2.5 = 5 \text{ A}$$
$$\text{Meter reading} = 25 + 5 = 30 \text{ A} \qquad \textit{Ans.}$$

Scale B: The main division just before the arrow is 2.5 A.

$$\text{Number of spaces covered by arrow} = 2 \text{ spaces}$$
$$\text{Main division} = 2.5 \text{ A, divided into 10 spaces}$$
$$\text{Each small space} = \frac{2.5}{10} = 0.25 \text{ A}$$
$$\text{Value of spaces covered} = 2 \times 0.25 = 0.5 \text{ A}$$
$$\text{Meter reading} = 2.5 + 0.5 = 3.0 \text{ A} \qquad \textit{Ans.}$$

Scale C: The main division just before the arrow is 0.05 A.

$$\text{Number of spaces covered by arrow} = 2 \text{ spaces}$$
$$\text{Main division} = 0.05 \text{ A, divided into 10 spaces}$$
$$\text{Each small space} = \frac{0.05}{10} = 0.005 \text{ A}$$
$$\text{Value of spaces covered} = 2 \times 0.005 = 0.01 \text{ A}$$
$$\text{Meter reading} = 0.05 + 0.01 = 0.06 \text{ A} \qquad \textit{Ans.}$$

Self-Test 22-6 Find the reading indicated by the arrow for scales A, C, D, and E of the dial shown in Fig. 22-6.

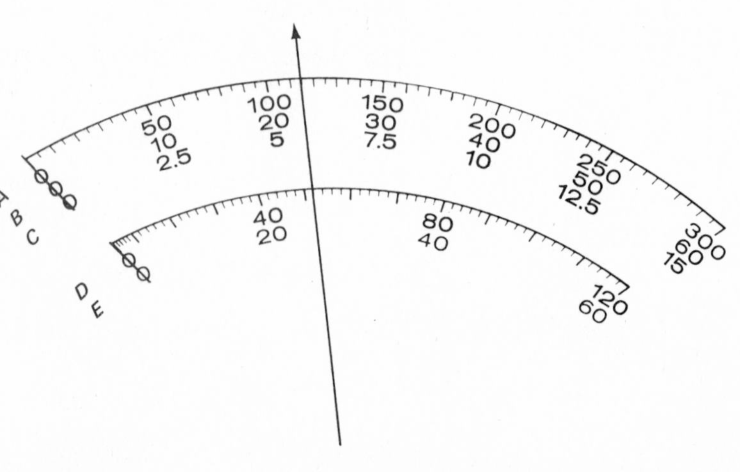

Fig. 22-6

Solution Scale *A:* The main division just before the arrow is _______. | 100

Number of spaces covered by arrow = _______ | 3

Main division = _______, divided into _______ spaces | 50, 10

Each small space = 50 ÷ 10 = _______ | 5

Value of spaces covered = 3 × 5 = _______ | 15

Meter reading = 100 + 15 = _______ *Ans.* | 115

Scale *C:* The main division just before the arrow is _______. | 5

Number of spaces covered by arrow = _______ | 3

Main division = _______, divided into _______ spaces | 2.5, 10

Each small space = 2.5 ÷ _______ = _______ | 10, 0.25

Value of spaces covered = 3 × _______ = _______ | 0.25, 0.75

Meter reading = 5 + 0.75 = _______ *Ans.* | 5.75

Scale *D:* The main division just before the arrow is _______. | 50

Number of spaces covered by arrow = _______ | 1

Main division = 10, divided into _______ spaces | 5

Each small space = 10 ÷ _______ = _______ | 5, 2

Value of spaces covered = 1 × _______ = _______ | 2, 2

Meter reading = 50 + _______ = _______ *Ans.* | 2, 52

Scale *E:* The main division just before the arrow is _______. | 25

Number of spaces covered by arrow = _______ | 1

Main division = 5, divided into _______ spaces | 5

Each small space = _______ ÷ _______ = _______ | 5, 5, 1

Value of spaces covered = 1 × _______ = _______ | 1, 1

Meter reading = 25 + _______ = _______ *Ans.* | 1, 26

Problems

Find the meter reading for each scale at the points indicated by the arrows in the following figures:

1.

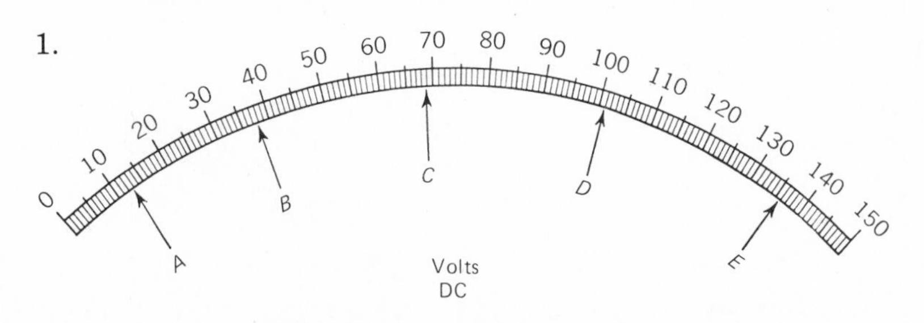

Fig. 22-7 (Courtesy The Weston Electrical Instrument Corporation)

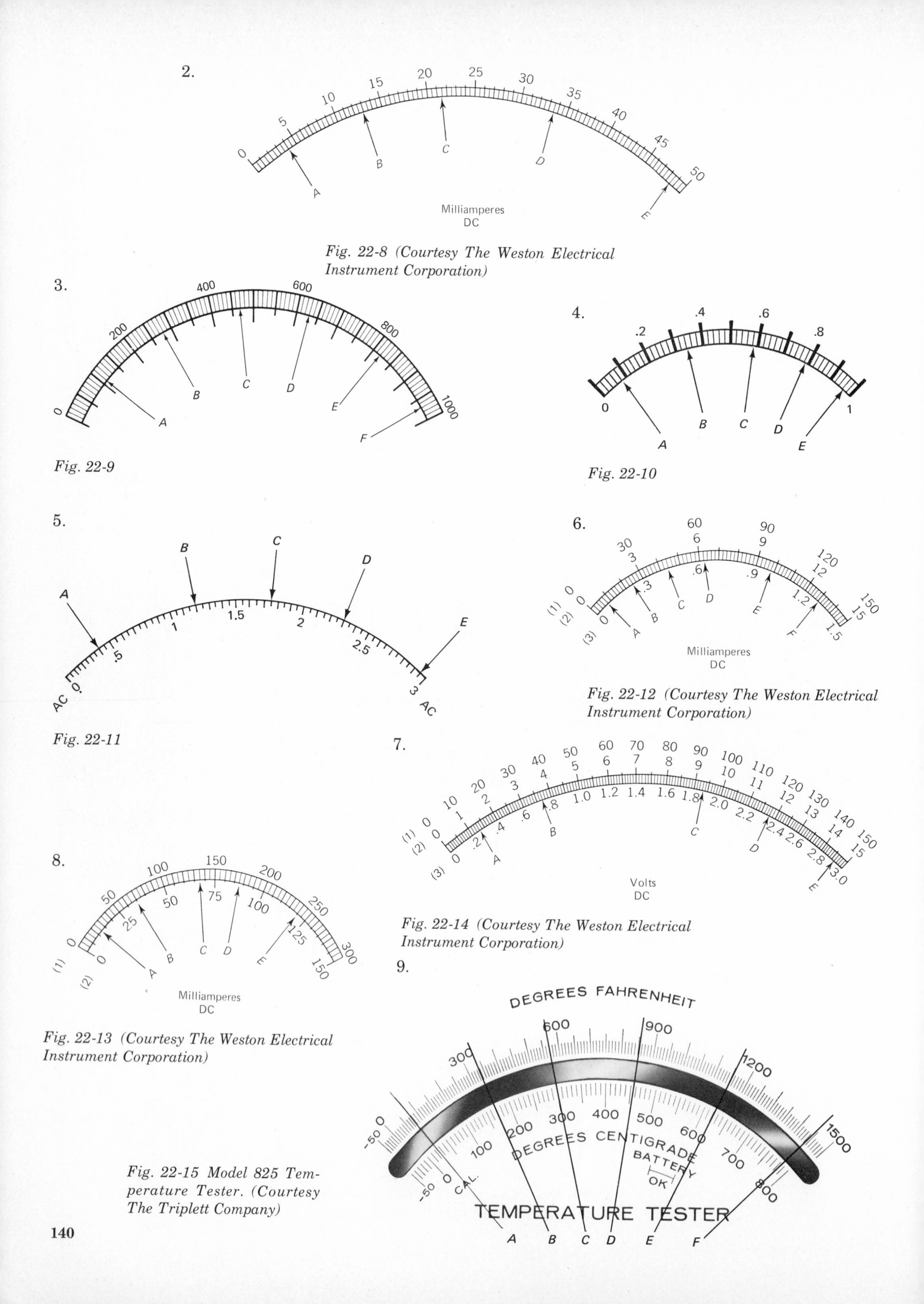

Fig. 22-8 (Courtesy The Weston Electrical Instrument Corporation)

Fig. 22-9

Fig. 22-10

Fig. 22-11

Fig. 22-12 (Courtesy The Weston Electrical Instrument Corporation)

Fig. 22-13 (Courtesy The Weston Electrical Instrument Corporation)

Fig. 22-14 (Courtesy The Weston Electrical Instrument Corporation)

Fig. 22-15 Model 825 Temperature Tester. (Courtesy The Triplett Company)

10.

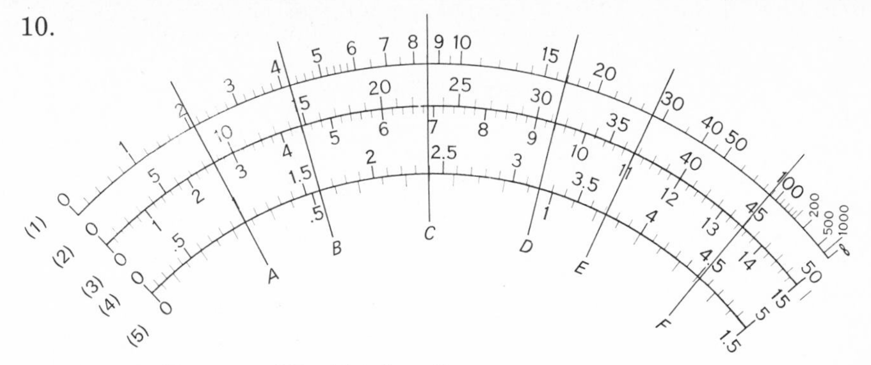

Fig. 22-16 (Courtesy The Triplett Company)

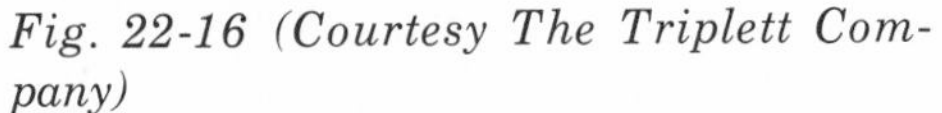

JOB 23 The Electrical Resistor Color Code

FINDING THE VALUE OF FIXED RESISTORS

Instead of the number of ohms of resistance being stamped on carbon-type resistors, the resistors are colored according to a definite system approved by the EIA (Electronics Industries Association). Each color represents a number according to the plan in Table 23-1.

Table 23-1

COLOR	NUMBER	COLOR	NUMBER
Black	0	Green	5
Brown	1	Blue	6
Red	2	Violet	7
Orange	3	Gray	8
Yellow	4	White	9

Gold—multiply by 0.1.
Silver—multiply by 0.01.

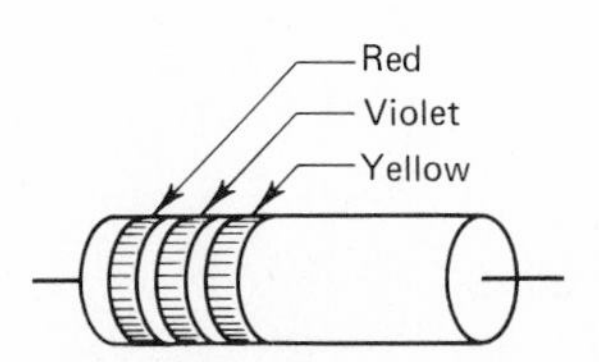

Fig. 23-1 The resistance value is indicated by three bands of color read in order from left to right.

The value of the resistor is obtained by reading the colors according to the following systems.

The three-band system. The first band represents the first number in the value. The second band represents the second number. The third band represents the number of zeros to be added after the first two numbers. If the third band is gold or silver, multiply the value indicated by the first two bands by 0.1 or 0.01, respectively, as indicated above.

Example 23-1 Find the resistance of a resistor marked red, violet, yellow, as shown in Fig. 23-1.

Solution

First band	Second band	Third band
Red	Violet	Yellow
2	7	0000

Resistance = 270,000 Ω *Ans.*

Example 23-2 A resistor is marked yellow, orange, black. What is its resistance?

Solution

First band	Second band	Third band
Yellow	Orange	Black
4	3	No zeros

Resistance = 43 Ω *Ans.*

Example 23-3 A resistor is marked green, blue, gold. What is its resistance?

Solution

First band	Second band	Third band
Green	Blue	Gold
5	6	Multiply by 0.1

The first two bands indicate a value of 56 Ω. Therefore,

$56 \times 0.1 = 5.6\ \Omega$ *Ans.*

Example 23-4 A resistor is marked gray, red, silver. What is its resistance?

Solution

First band	Second band	Third band
Gray	Red	Silver
8	2	Multiply by 0.01

The first two colors indicate a value of 82 Ω. Therefore,

$82 \times 0.01 = 0.82\ \Omega$ *Ans.*

Example 23-5 A television mechanic needs a 510,000-Ω resistor. What combination of colors will this resistor have in the three-band system?

Solution The first digit is a 5, indicating green.

The second digit is a 1, indicating brown.

The four zeros that remain indicate yellow.

The resistor will be color-coded green, brown, yellow.

Self-Test 23-6 What color combination is needed to indicate a 6.8-Ω resistor in the three-band system?

Solution The color combination needed is as follows:

The first digit is a 6, indicating ______ | blue

The second digit is an 8, indicating ______ | gray

To obtain the number 6.8 from the number 68, it is necessary to multiply 68 by ______, which indicates gold. Therefore, the resistor will be color-coded as follows: | 0.1

First band, blue second band, ______ | gray
third band, ______ *Ans.* | gold

TOLERANCE MARKINGS

A fourth band of color in the band system is used to indicate how accurately the part is made to conform to the indicated markings. Gold means that the value is not more than 5 percent away from the indicated value. Silver indicates a tolerance of 10 percent, and black a tolerance of 20 percent. If the tolerance is not indicated by a color, it is assumed to be 20 percent.

Problems

What value of resistance is indicated by each of the following color combinations?

1. Brown, black, yellow
2. Red, yellow, red
3. Gray, red, orange
4. Green, brown, red

5. Violet, green, black
6. Red, black, green
7. Yellow, violet, gold
8. Brown, green, yellow
9. Brown, red, gold
10. Brown, red, red
11. Brown, gray, green
12. Orange, white, silver
13. Blue, red, brown
14. Brown, gray, brown
15. White, brown, red
16. Orange, orange, gold
17. Yellow, violet, silver
18. Yellow, green, silver
19. Gray, red, black
20. Green, blue, gold

What color combination is needed to indicate each of the following resistances?

21. 240,000 Ω
22. 430,000 Ω
23. 51 Ω
24. 150 Ω
25. 10,000,000 Ω
26. 8.6 Ω
27. 3,900 Ω
28. 0.47 Ω
29. 12 Ω
30. 750,000 Ω
31. 0.68 Ω
32. 1.2 Ω
33. 2.2 Ω
34. 100 Ω
35. 1.8 Ω
36. 1,000,000 Ω
37. 360 Ω
38. 6,200 Ω
39. 1.5 Ω
40. 1 Ω

JOB 24 Review of Jobs 20 to 23

Complete the following list.

1 m = ______ mm	1,000
1 dm = ______ cm	10
1 cm = ______ mm	10
1 cm = ______ in	0.394
1 in = ______ cm	2.54
1 ft = ______ m	0.305
1 m = ______ ft	3.28
1 km = ______ m	1,000
1 km = ______ mi	0.62
1 mi = ______ km	1.61
1 l = ______ gal	0.264
1 gal = ______ l	3.8
1 qt = ______ l	0.95
1 l = ______ qt	1.06
1 cu in = ______ cc	16.4
1 cc = ______ cu in	0.061
1 cu ft = ______ cu m	0.0283
1 cu m = ______ cu ft	35.32
1 cu ft = ______ l	28.32

1 l	= ______	cu ft	0.0353
1 kg	= ______	lb	2.2
1 lb	= ______	kg	0.454
1 sq in	= ______	sq cm	6.45
1 sq cm	= ______	sq in	0.155
1 sq ft	= ______	sq m	0.093
1 sq m	= ______	sq ft	10.76
1 lb/sq in	= ______	kg/sq cm	0.0703
1 kg/sq cm	= ______	lb/sq in	14.22

1. To change inches to centimeters, we should (multiply)(divide) the number of inches by 2.54. — multiply

2. To change centimeters to inches, we should multiply the number of centimeters by ______. — 0.394

3. To change inches to feet, we should (multiply)(divide) the number of inches by 12. — divide

4. To change kilometers to miles, we should multiply the number of kilometers by ______. — 0.62

5. To change miles to kilometers, we should multiply the number of miles by ______. — 1.61

6. To change liters to gallons, we should multiply the number of liters by ______. — 0.264

7. To change gallons to liters, we should (multiply)(divide) the number of gallons by 3.8. — multiply

8. To change quarts to liters, we should multiply the number of quarts by ______. — 0.95

9. To change liters to quarts, we should multiply the number of liters by ______. — 1.06

10. To change cubic inches to cubic centimeters, we should multiply the number of cubic inches by ______. — 16.4

11. To change cubic centimeters to cubic inches, we should multiply the number of cubic centimeters by ______. — 0.06

12. Each line above the gage line on a metric micrometer represents a distance of ______ mm. — 1

13. Each line below the gage line on a metric micrometer represents a distance of ______ mm. — 0.5

14. Each line on the thimble of a metric micrometer represents a distance of ______ mm. — 0.01

15. (*A, M*) Read the indicated measurements in Fig. 24-1 in centimeters and millimeters.

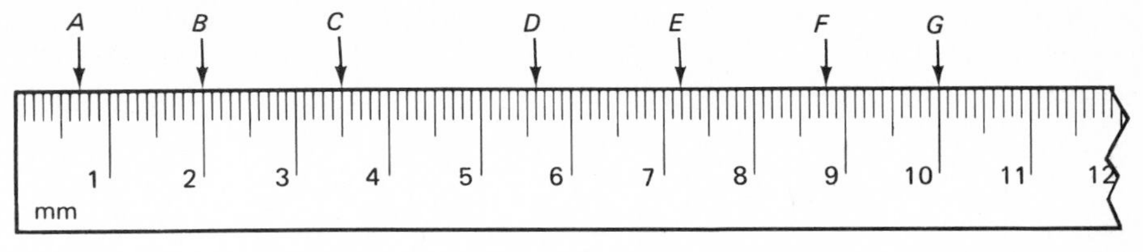

Fig. 24-1

16. (*A, M*) Using Fig. 24-2, measure the dimensions *A*, *B*, and *C* in millimeters.

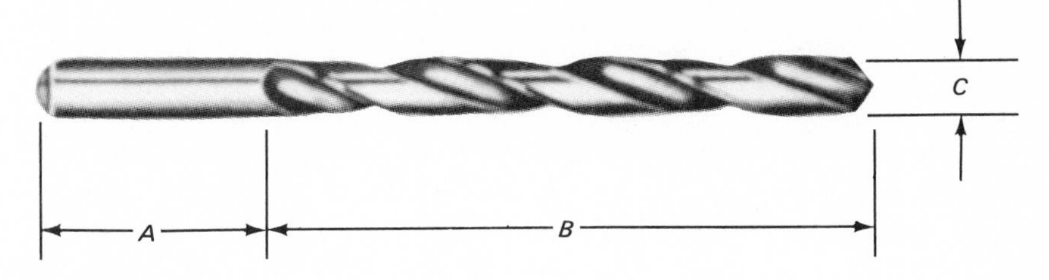

Fig. 24-2

17. (*M*) The inside diameter of the flywheel shown in Fig. 24-3 measures 75.93 cm. Express this measurement in inches.

Fig. 24-3 Measuring the diameter of a flywheel with a Starrett No. 124 Inside Micrometer.

Fig. 24-4 An abrasive belt grinder. (Courtesy The Baldor Electric Company)

18. (*M*) The diameters of the pulleys on the abrasive belt grinder shown in Fig. 24-4 are 7.5 in and 6 in. Express these measurements in centimeters.

19. (*A, M*) Change the measurements on the socket head cap screw shown in Fig. 24-5 to millimeters.

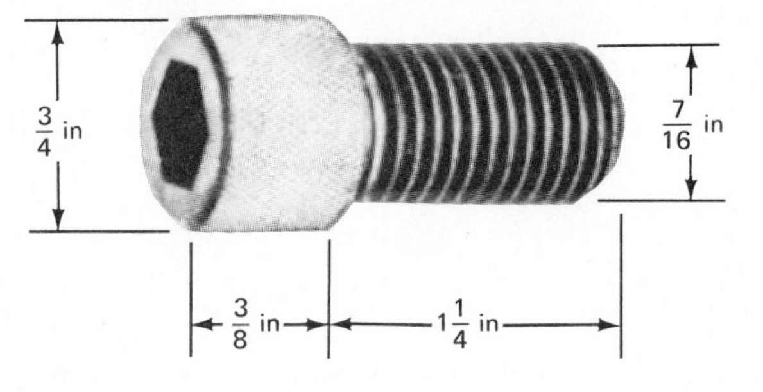

Fig. 24-5

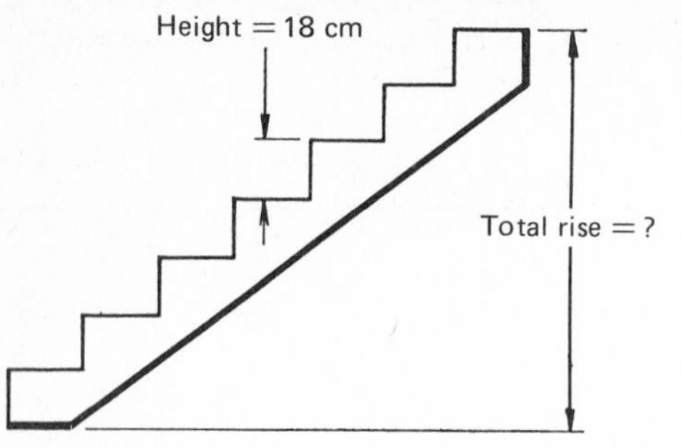

Fig. 24-6

20. (*C*) Find the total rise in the flight of steps shown in Fig. 24-6. State the answer in feet and inches.
21. (*C*) How many turns of the wood screw shown in Fig. 24-7 are required to advance it a total of 12 mm?
22. (*G*) What is the equivalent (in feet) of the 200-m dash?
23. (*G*) What is the equivalent (in yards) of the 800-m run?
24. (*A*) The cylinder bore on an engine is 3.35 in and the stroke of the piston is 2.746 in. Express these measurements as millimeters.
25. (*M, A*) Change the dimensions of the bushing shown in Fig. 24-8 to inches.

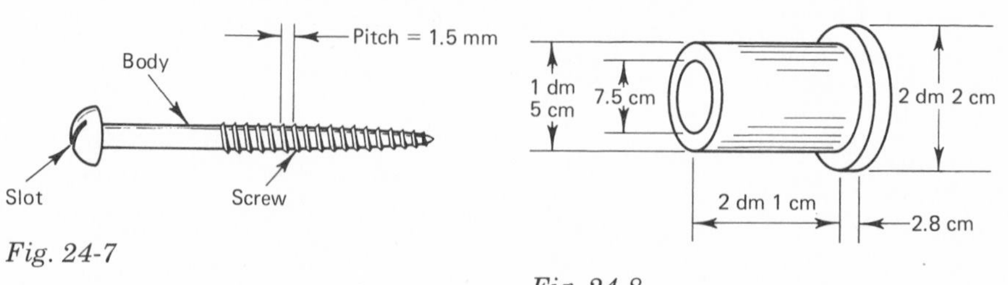

Fig. 24-7

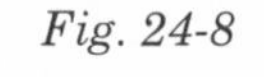

Fig. 24-8

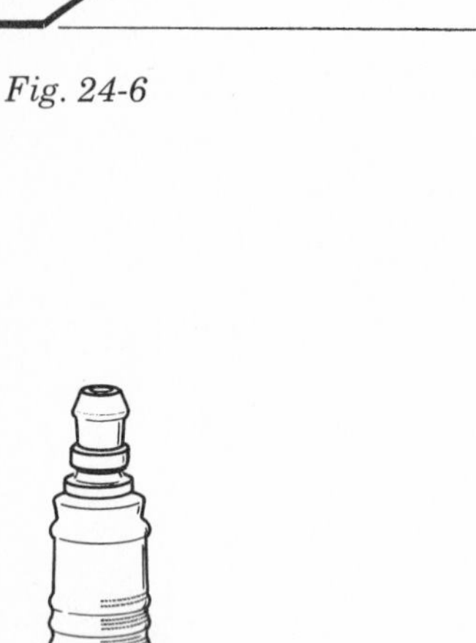

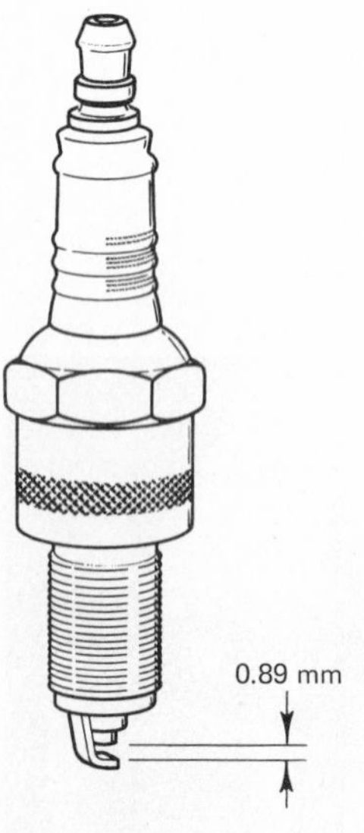

Fig. 24-9

26. (*A*) What size (in inches) of feeler gage should be used to check the spark plug gap shown in Fig. 24-9?
27. (*M*) The formula for the double depth of the thread shown in Fig. 24-10 is $DD = 1.3 \times P$. Find the double depth in millimeters. If the formula for the minor diameter is

    Minor diameter = outside diameter − DD

    find the minor diameter in inches.
28. (*F*) A strip 8.2 mi long in a field is to be plowed by the tractor shown in Fig. 24-11. Express this distance in kilometers.
29. (*A, G*) The speed limit in the United States is 55 mph. Express this speed as kilometers per hour.
30. (*G*) The airline distance between New York and Paris is 5,818 km. Express this distance in miles.
31. (*G*) The speed of sound is 1,192 km/hr. Express this speed as miles per hour.

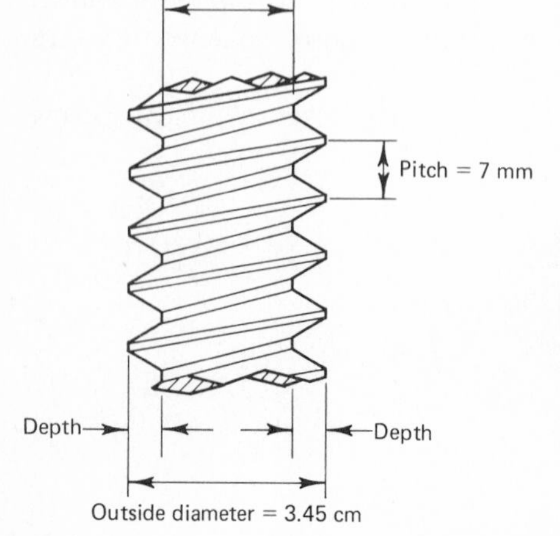

Fig. 24-10

Fig. 24-11 A tractor and plow. (Courtesy International Harvester Company)

32. (*A, M*) In the drill gage shown in Fig. 24-12, express the diameter of the following drills in millimeters. (a) $\frac{5}{32}$ in, (b) $\frac{1}{4}$ in, (c) $\frac{5}{16}$ in, (d) $\frac{3}{8}$ in, and (e) $\frac{1}{2}$ in.
33. (*C*) Express the size of the 50-in-diameter water main pipe shown in Fig. 24-13 as meters.
34. (*F, G*) A typical corn belt fertilizing program includes the broadcast of phosphate, potash and nitrogen in the form of anhydrous ammonia. If each tank on the tractor shown in Fig. 24-14 has a capacity of 200 gal, express this capacity as liters.
35. (*A, B, G*) If gasoline sells for \$0.649/gal, what is the cost per liter?
36. (*A, G*) If a 1,500-mi trip consumed 450 l of gasoline, find the gas consumption in miles per gallon.

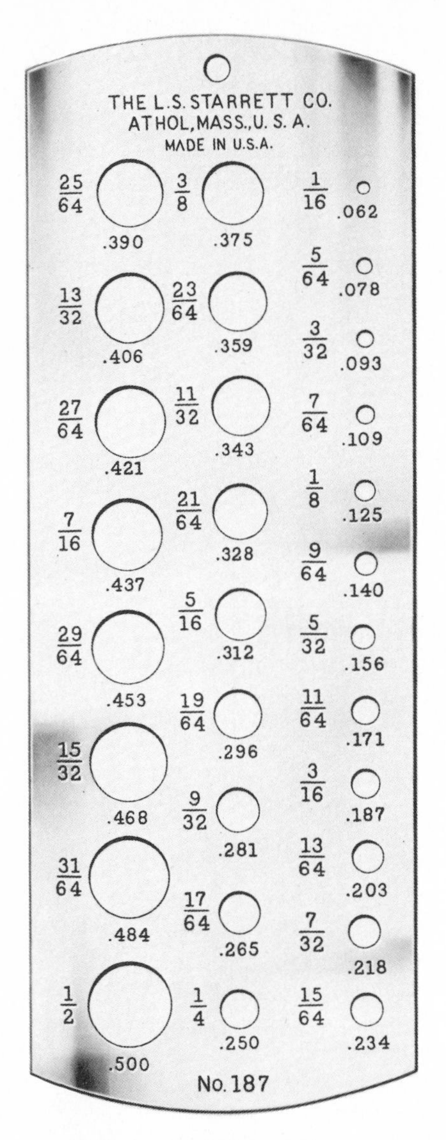

Fig. 24-12 A drill gage. (Courtesy The L.S. Starrett Company)

Fig. 24-13 Installing a water main pipe. (Courtesy The Dillon Company)

Fig. 24-14 Adding nitrogen to soil in the form of anhydrous ammonia. (Courtesy U.S. Steel)

37. (*A, G*) Express the number of liters in 5 qt of oil.
38. (*A*) If an automobile has a 315-cu in engine, what is the capacity in liters?
39. (*A*) If an automobile has a 5-l engine, express the capacity in cubic inches.
40. (*G*) If 1 cu ft = 7.5 gal, find the number of liters in 1 cu ft.
41. (*G*) A rectangular tank measures 100 cm × 50 cm × 30 cm. Find (a) the volume in liters, and (b) the volume in gallons.
42. (*A, M*) Find the values of the metric micrometer settings shown in Fig. 24-15.

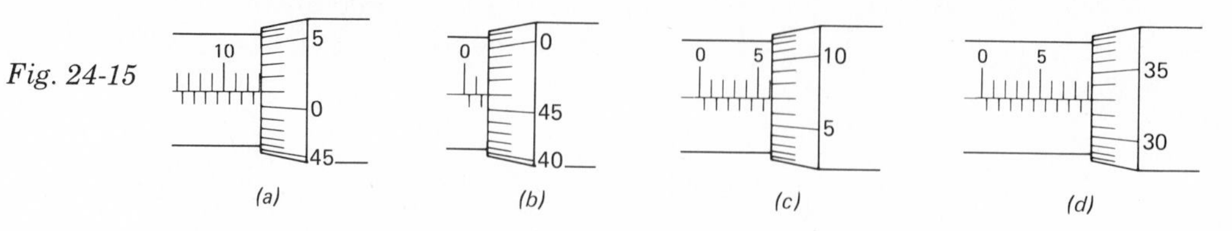

Fig. 24-15

43. (*A, M*) Read the settings on the 25-division metric vernier caliper shown in Fig. 24-16.

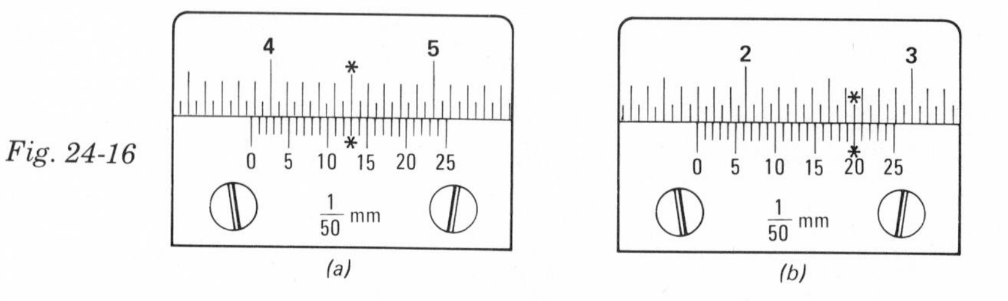

Fig. 24-16

44. (*A, M*) Read the combination verniers shown in Fig. 24-17 for the outside and inside measurements on both scales.

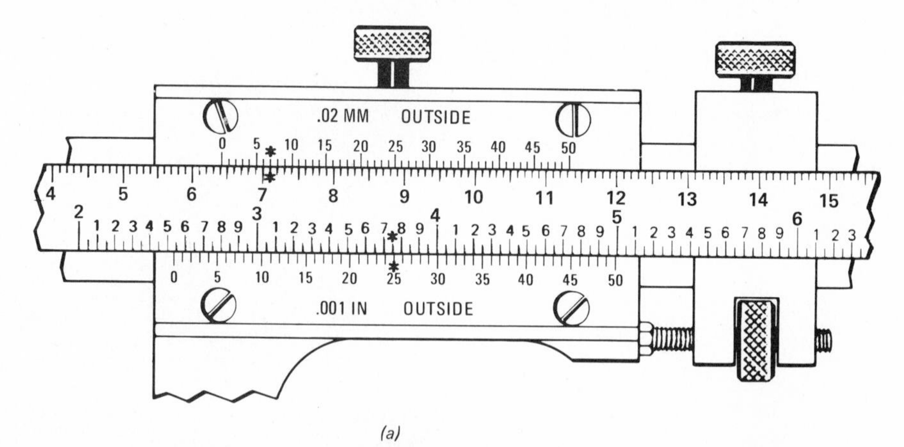

Fig. 24-17

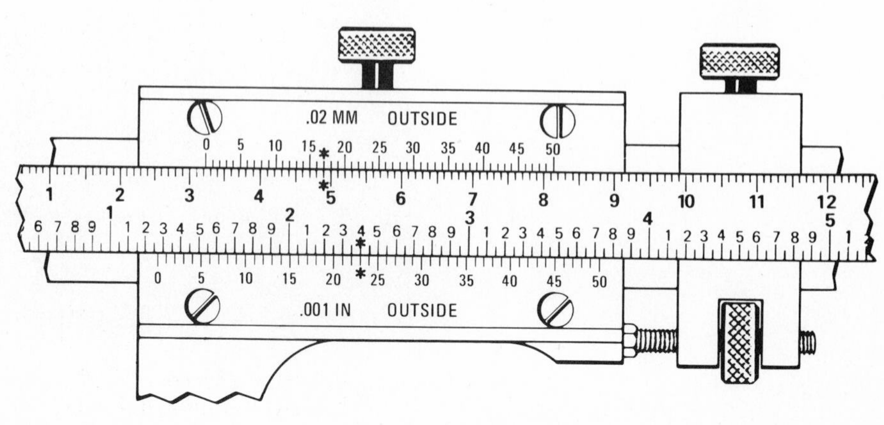

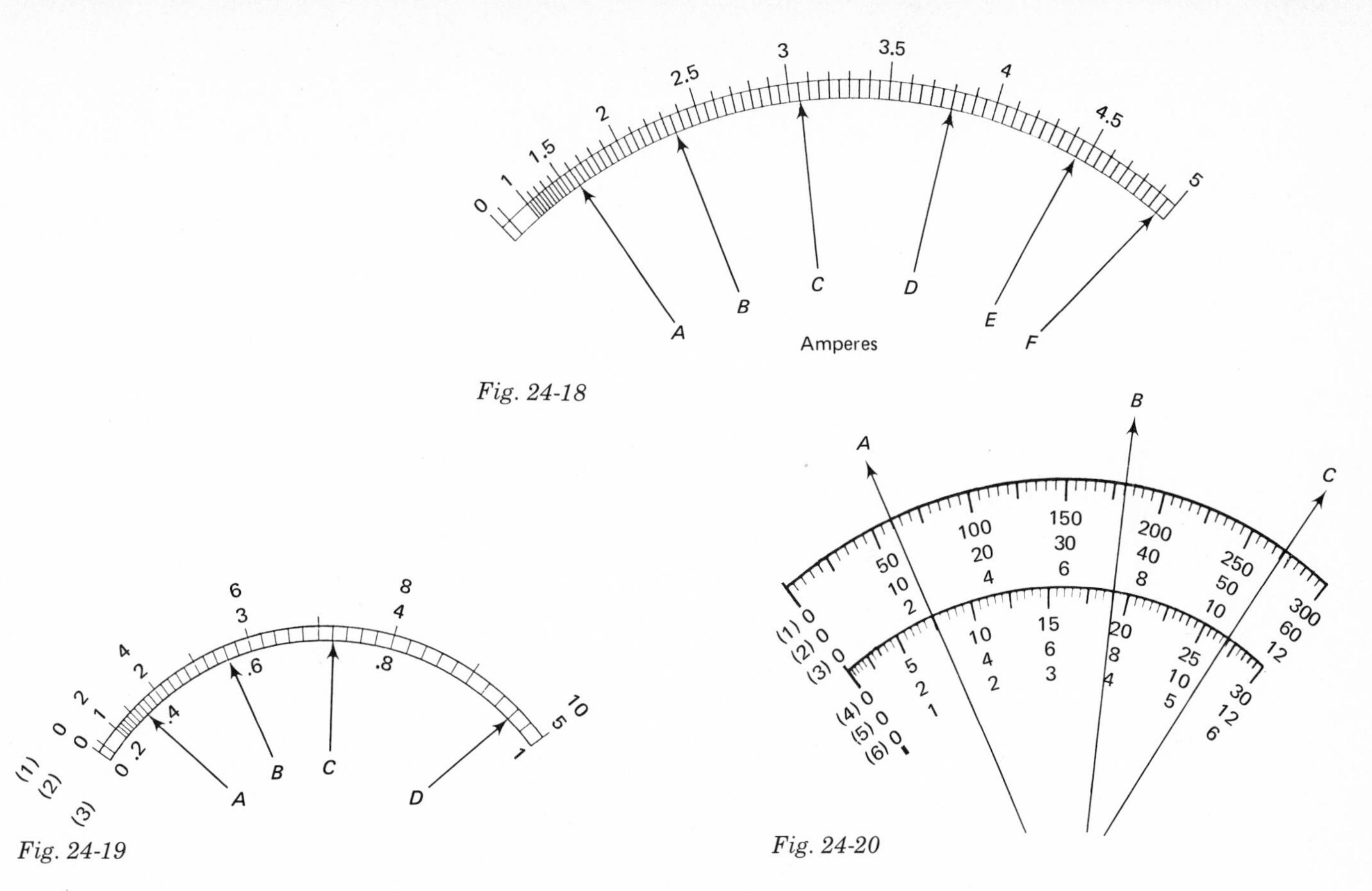

Fig. 24-18

Fig. 24-19

Fig. 24-20

45. (*E*) Read the dial at the points indicated in Fig. 24-18.
46. (*E*) Read the dial on each scale at the points indicated in Fig. 24-19.
47. (*E*) Find the readings at points *A*, *B*, and *C* for all six scales of the dial shown in Fig. 24-20.
48. (*C*) The main towers of the Mackinac Bridge, shown in Fig. 24-21, are each 552 ft high, and the center point of the bridge is 185 ft above water. Express each of these measurements as meters.

Fig. 24-21 The Mackinac Bridge nearing completion. (Courtesy Mackinac Bridge Authority and U.S. Steel)

49. (*E*) What value of electrical resistance is indicated by each of the following color combinations?
    a. Red, green, orange
    b. Brown, green, black
    c. Orange, black, yellow
    d. Yellow, black, green
    e. Violet, red, black
    f. Yellow, gray, gold

50. (*E*) What color combination is needed to indicate each of the following resistances?
    a. 350,000 Ω
    b. 2.5 Ω
    c. 4,200 Ω
    d. 9 Ω
    e. 2,000,000 Ω
    f. 180 Ω
51. (*M, G, B*) Find the cost in dollars to ship a 500-kg milling machine at a cost of 21 cents/lb.
52. (*G*) The methane content of coal is 200 cu ft/ton. Express this volume in liters per kilogram.
53. (*G, A, F*) Express a barometric pressure of 15 lb/sq in as kilograms per square centimeter.
54. (*F*) A section of land contains 640 acres. At 43,560 sq ft/acre, find the number of square meters in 1 section.
55. (*A, C*) A steel channel in the frame of a truck is stressed to 700 kg/sq cm. Express this stress in pounds per square inch.

(See Answer Key for Test 6—Review of Jobs 20 to 24)

JOB 25 Applying Decimals—Percent

Fig. 25-1 A recent advertisement in a metropolitan newspaper. Where's the error?

The advertisement shown in Fig. 25-1 appeared recently in a great metropolitan newspaper. It contains an error, and you, as a consumer, should be able to spot it. We shall return to this problem in Job 27, after the following introduction to percent.

INTRODUCTION TO PERCENT

The word *percent* is used to indicate some portion of 100. A grade of 92 percent on an examination indicates that the student got 92 points out of a possible 100 points. This may also be written as a decimal fraction (0.92) or as a common fraction ($\frac{92}{100}$), both of which are read as 92 *hundredths.* A percent, indicated by a percent sign (%), cannot be used in a calculation until it has been changed to an equivalent decimal fraction. This equivalent decimal is sometimes called the *working* decimal.

CHANGING PERCENTS TO DECIMALS

Since 92 percent (92%) means $\frac{92}{100}$ or 0.92, changing from 92 percent to 0.92 is easily done by just moving the decimal point two places to the *left.* (See Example 3-3 on page 19.)

RULE 1 **To change a percent to a working decimal, move the decimal point two places to the left and drop the percent sign or the word *percent.***

Example 25-1 Change the following percents to working decimals.

Solution Please note that whole numbers have a decimal point at the end of the number—written or not.

$32\% = 32\odot\% = .32\odot = 0.32$ *Ans.*

$8\% = .08\odot\% = .08\odot = 0.08$ *Ans.*

Example 25-2 Change $16\frac{1}{2}$ percent to a working decimal.

Solution Before we can move any decimal point, we must change the fraction $\frac{1}{2}$ itself into a decimal. Since $\frac{1}{2} = 0.5$, $16\frac{1}{2}\% = 16.5\%$.

Notice that we still have the percent sign. The number 16.5% is *still* a percent, *not* a working decimal. All that we did was to change the fraction $\frac{1}{2}$ into its equivalent 0.5. Now we can change 16.5% into a working decimal by moving the decimal point two places to the left.

$16.5\% = 0.16\odot5 = 0.165$ *Ans.*

Example 25-3 Change $33\frac{1}{3}$ percent to a working decimal.

Solution The decimal equivalent of the fraction $\frac{1}{3}$ is 0.33. Therefore, $33\frac{1}{3}\%$ may be written as 33.33%.

$33.33\% = .33\odot33 = 0.3333$ *Ans.*

Example 25-4 Change $\frac{1}{4}$ percent to a working decimal.

Solution The decimal equivalent of the fraction $\frac{1}{4}$ is 0.25. Therefore, $\frac{1}{4}\%$ may be written as 0.25%.

$0.25\% = .00\odot25 = 0.0025$ *Ans.*

Example 25-5 Many banks offer $6\frac{3}{4}\%$ interest on certain time deposits. In order to calculate the interest earned on the investment, we must first change the percent to a working decimal.

Solution The decimal equivalent of the fraction $\frac{3}{4}$ is 0.75. Therefore, $6\frac{3}{4}\%$ may be written as 6.75%.

$6.75\% = .06\odot75 = 0.0675$ *Ans.*

Problems

Change the following percents to working decimals:

1. 38%
2. 60%
3. 6%
4. 19%
5. 4%
6. 26.4%
7. 3.6%
8. 100%
9. 125%
10. 0.5%
11. $16\frac{2}{3}\%$
12. $\frac{3}{4}\%$
13. 62.5%
14. 1%
15. 12.5%
16. 0.9%
17. 2.25%
18. $5\frac{1}{2}\%$
19. $4\frac{1}{4}\%$
20. $\frac{1}{2}\%$

CHANGING DECIMALS TO PERCENTS

To change a decimal to a percent is to reverse the process of changing a percent to a decimal.

RULE 2 **To change a decimal to a percent, move the decimal point two places to the right and add a percent sign.**

Example 25-6 Change the following decimals to percents.

Solution

$0.20 = 0.20.\% = 20\%$ *Ans.*

$0.06 = 0.06.\% = 6\%$ *Ans.*

$0.125 = 0.12.5\% = 12.5\%$ *Ans.*

$1.1 = 1.10.\% = 110\%$ *Ans.*

$2 = 2.00.\% = 200\%$ *Ans.*

$0.0007 = 0.00.07\% = 0.07\%$ *Ans.*

Problems

Express the following decimals as percents:

1. 0.50	6. 0.008	11. 0.222
2. 0.04	7. 1.00	12. 0.15
3. 0.2	8. 0.87	13. 0.7
4. 0.075	9. 0.625	14. 3
5. 1.45	10. 0.092	15. 0.055

CHANGING COMMON FRACTIONS TO PERCENTS

RULE 3 **To change a common fraction to a percent, express the fraction as a decimal, then move the decimal point two places to the right and add a percent sign.**

The decimal equivalents for many common fractions may be found in Table 2-2 on page 13.

Example 25-7 Express the following fractions as percents.

Solution

$\frac{3}{4} = 0.75 = 75\%$ *Ans.*

$\frac{1}{2} = 0.50 = 50\%$ *Ans.*

$\frac{5}{8} = 0.625 = 62.5\%$ *Ans.*

If the fraction is not on the decimal equivalent chart, change the fraction into a decimal by dividing the numerator by the denominator, as shown in Job 2.

Example 25-8 Express $\frac{8}{25}$ as a percent.

Solution

$$\frac{8}{25} = 25\overline{)8.00} = 0.32 = 32\% \quad Ans.$$

```
     0.32
25)8.00
  -7 5↓
     50
    -50
      0
```

Problems

Express the following fractions as percents:

1. $\frac{1}{4}$	6. $\frac{3}{16}$	11. $\frac{13}{20}$	16. $\frac{3}{7}$
2. $\frac{3}{8}$	7. $\frac{3}{10}$	12. $\frac{3}{32}$	17. $\frac{16}{48}$
3. $\frac{2}{5}$	8. $\frac{3}{5}$	13. $\frac{2}{3}$	18. $\frac{12}{30}$
4. $\frac{4}{5}$	9. $\frac{3}{6}$	14. $\frac{5}{6}$	19. $\frac{2}{9}$
5. $\frac{5}{8}$	10. $\frac{9}{10}$	15. $\frac{5}{11}$	20. $\frac{13}{15}$

JOB 26 Finding the Part in Percent Problems

There are three parts to every problem involving percent:

1. The *base* B is the entire amount.
2. The *rate* R is the percent of the base.
3. The *part* P is the portion of the base.

These three parts are combined in the following formula:

FORMULA $B \times R = P$

Example 26-1 A mechanic earning \$180/week got a 15 percent increase in pay. Find the amount of the increase.

Solution Given: $B = \$180$ Find: $P = ?$
$R = 15\% = 0.15$

The rate of 15 percent must be changed to a working decimal, as shown above. In the formula, R will equal 0.15.

$$P = B \times R$$
$$= 180 \times 0.15$$
$$= \$27.00 \quad \textit{Ans.}$$

Example 26-2 A bronze bearing for a propeller shaft is made of 75 percent copper, 17 percent lead, and 8 percent tin. If the bearing weighs 400 lb, find the weight of each material in the bearing.

Solution Given: $B = 400$ lb
For copper, $R = 75\% = 0.75$
For lead, $R = 17\% = 0.17$
For tin, $R = 8\% = 0.08$

Find: lb of copper, $P = ?$
lb of lead, $P = ?$
lb of tin, $P = ?$

Find the weight of copper.

$$P = B \times R$$
$$= 400 \times 0.75$$
$$= 300 \text{ lb} \quad \textit{Ans.}$$

Find the weight of lead.

$P = B \times R$

$= 400 \times 0.17$

$= 68$ lb *Ans.*

Find the weight of tin.

$P = B \times R$

$= 400 \times 0.08$

$= 32$ lb *Ans.*

Check. The sum of the weights of the materials should equal the total weight of 400 lb.

$300 + 68 + 32 \stackrel{?}{=} 400$

$400 = 400$ *Check*

Example 26-3 The resistance of a resistor depends on the accuracy with which it is manufactured. This is shown on the resistor as a fourth band of color called the *tolerance value*. What are the upper and lower values of resistance that might be expected on a resistor marked 3,000 Ω and 5 percent tolerance?

Note: Gold = 5 percent, silver = 10 percent, and black (or no color) = 20 percent.

Solution Given: $B = 3{,}000\ \Omega$ Find: Tolerance $= P = ?$

$R =$ tolerance $= 5\% = 0.05$ Upper value = ?

Lower value = ?

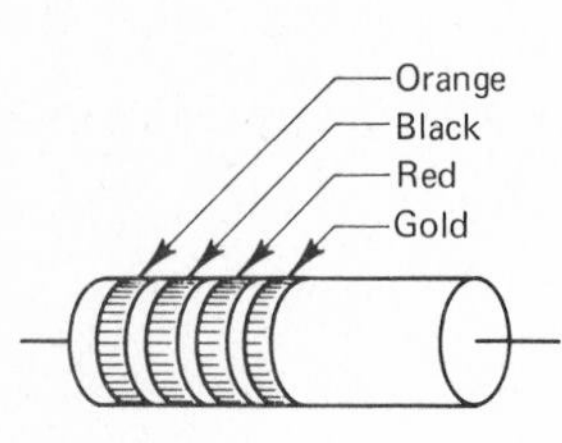

Fig. 26-1

1. Find the permitted tolerance.

$P = B \times R$

$= 3{,}000 \times 0.05$

$= 150\ \Omega$ *Ans.*

2. Find the upper and lower values.

Upper value $= 3{,}000 + 150 = 3{,}150\ \Omega$ *Ans.*

Lower value $= 3{,}000 - 150 = 2{,}850\ \Omega$ *Ans.*

Example 26-4 Molten iron shrinks $1\frac{3}{4}$ percent while cooling. A pattern 6.5 cm long is used to cast a piece. What is the length of the cooled cast piece?

Solution Given: $B = 6.5$ cm Find: Shrinkage $= P = ?$

$R = 1\frac{3}{4}\%$ Final length = ?

1. Change the rate into a working decimal.

$1\frac{3}{4}\% = 1.75\% = 0.0175$

2. Find the amount of shrinkage P.

$P = B \times R$

$= 6.5 \times 0.0175$

$= 0.11$ cm *Ans.*

Fig. 26-2 Charging molten metal into a basic oxygen furnace. (Courtesy Allegheny Ludlum Industries, Inc.)

3. Find the cooled length.

Cooled length = pattern length − shrinkage

$$= 6.5 - 0.11 = 6.39 \text{ cm} = 63.9 \text{ mm} \qquad \textit{Ans.}$$

A contractor must be able to estimate the cost of flooring, sheathing, siding, etc. The estimate is usually based on the area to be covered, but an additional allowance must be made for waste, short ends, matching, and the fact that the boards are not always as wide as their nominal measurements indicate. That is, an 8-in board will often cover *less* than 8 in of width. This is called a *scant* width.

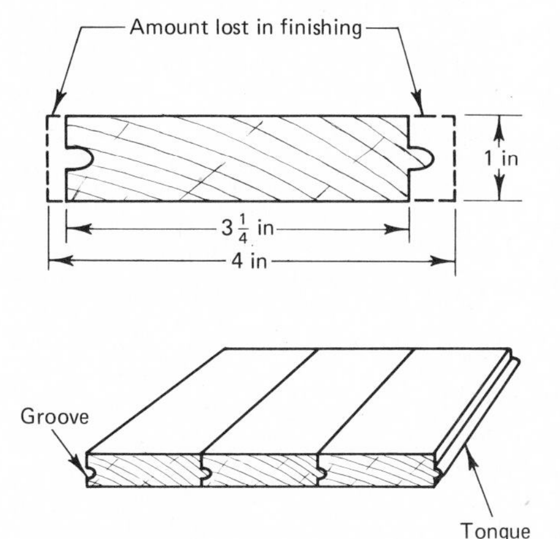

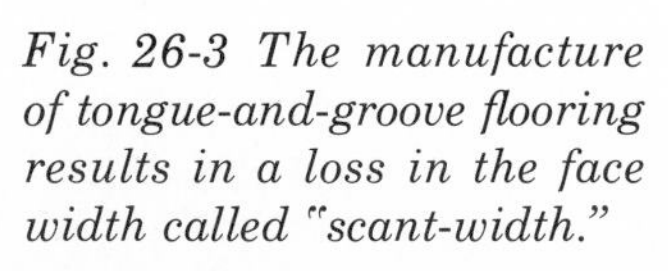

Fig. 26-3 The manufacture of tongue-and-groove flooring results in a loss in the face width called "scant-width."

Hardwood flooring is laid directly over the rough flooring at right angles to the direction of the floor joists. Because the flooring is tongue-and-groove, there is a large scant-width loss that varies according to the face width of the flooring. For example, ordinary 1 in × 4 in flooring is reduced to about $3\frac{1}{4}$ in net width, as shown in Fig. 26-3.

The following allowances are recommended by the National Oak Flooring Manufacturer's Association:

FACE WIDTH, in	ALLOWANCE, PERCENT
$1\frac{1}{2}$	50
2	$37\frac{1}{2}$
$2\frac{1}{4}$	33.3
$3\frac{1}{4}$	24

Self-Test 26-5 A room requiring 108 board feet of ordinary lumber is to be covered with matched flooring $\frac{1}{2}$ in thick and 2 in wide. Find the total number of board feet required if extra allowance is made for scant width and 5 percent extra is allowed for waste.

Solution **1.** Find the allowance to be added.

Waste allowance = ______ %	5
Scant width allowance = ______ %	$37\frac{1}{2}$
Total allowance = $42\frac{1}{2}$% = (decimal value)	0.425

2. Convert these allowances into board feet.

$P = B \times$ ______	*R*
= ______ × 0.425	108
= ______ board feet	45.9

3. Find the total number of board feet needed.

Total = 108 + ______ = ______ board feet *Ans.* 45.9, 153.9

Problems

1. (*G*) How much is 22 percent of 150?
2. (*G, B*) How much is 6 percent of $500?
3. (*G, B*) How much is 2.5 percent of $300?
4. (*E, B*) A television benchman earned $225 per week. He received an increase of 12 percent. Find the amount of the increase.
5. (*M, C*) Brass is composed of 65 percent copper and 35 percent zinc. Find the amount of each metal in a casting weighing 7.4 lb.
6. (*E*) What is the possible error in a 500-Ω resistor if it is marked with a silver band, indicating only 10 percent accuracy?
7. (*M*) A certain alloy shrinks 2 percent when cooled. Find the cooled length of a piece if the cast length is 9.75 cm long.
8. (*F, B*) Ms. Ennis shipped 350 crates of apples. If 8 percent were spoiled due to freezing, how many crates were spoiled?
9. (*A, G*) If 9 percent of the weight of a car is aluminum, how many pounds of aluminum are there in a 3,500-lb car?
10. (*F, B*) How much butterfat is there in 95 lb of cream testing 38 percent butterfat?
11. (*M, C, B*) Steel sheets with an area of 600 sq cm are used in a punch press. If 88 percent of the material is used, find the number of square centimeters of scrap.
12. (*M, A, G*) A belt (Fig. 26-4) is designed to transmit 4 hp. If $1\frac{1}{2}$ percent of the power is lost due to slippage, how many horsepower does the belt actually transmit?
13. (*E*) A 15-A fuse carried a 15 percent overload for 2 sec. What current flowed through the fuse during this time?
14. (*E*) In a brightness control circuit, 6 percent of the 525 horizontal lines were blanked out. How many lines were blanked out?
15. (*M, G*) If the weight of a wood pattern is 8 percent of that of a 425-kg casting, what is the weight of the pattern?
16. (*A*) A radiator of 20 qt capacity requires 25 percent antifreeze for protection down to 10°F. How many quarts of antifreeze should be used to get this protection?

Fig. 26-4 Triple, matched, V-belt drive. (Courtesy Rockwell International)

17. (*C*) A contractor ordered 2,150 board feet of lumber. Because of knots and cracks, 16 percent of the lumber could not be used. How many board feet were used?
18. (*C*) In laying asphalt shingles on a hip roof, 8 percent must be added for waste. How many square feet should a roofer figure on covering if the roof has an area of 2,500 sq ft?
19. (*F*) Which of the following provides the most protein: 100 lb of barley (6.9 percent protein), 80 lb of wheat (9.5 percent protein), or 180 lb of alfalfa (4.0 percent protein)?
20. (*A*) Due to the efficiency of the clutch linkage, a driver has to exert only 3 percent of the 900 lb holding the pressure plate in order to release the plate. Find the force exerted.
21. (*M, G*) Standard type metal contains $77\frac{1}{2}$ percent lead, $6\frac{1}{2}$ percent tin, and 16 percent antimony. How many pounds of each metal are in 200 lb of type metal?
22. (*M, C*) A 15.5-lb ingot of specialty steel (Fig. 26-5) contains $2\frac{3}{10}$ percent vanadium. How many pounds of vanadium are in the ingot?
23. (*A*) An engine used 2,500 lb of a mixture of fuel and air. If 9 percent of this mixture was fuel, how many pounds of fuel were used?
24. (*F*) A farm of 260 acres has 87 percent of the land under cultivation. How many acres are *not* being cultivated?
25. (*A*) A 350-hp engine lost 14 percent of its power in the automatic transmission. How many horsepower were lost?
26. (*M*) A sample of iron ore assays $4\frac{1}{4}$ percent iron. Find the number of pounds of iron per ton of ore.
27. (*B, F*) A fruit dealer bought 144 baskets of strawberries at $0.39 per basket. If 6 baskets spoiled, at what price must she sell the rest to make a profit of 40 percent on the total cost?
28. (*E*) How many watts of power are lost in a transformer rated at 250 W if 3 percent of the energy is lost as heat?

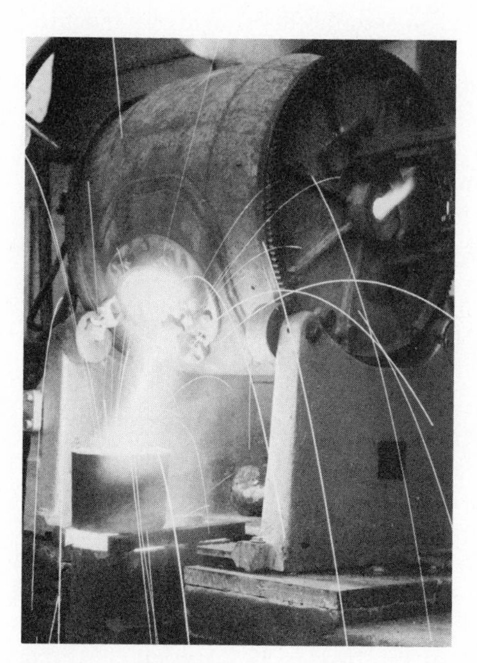

Fig. 26-5 Pouring a small test ingot of specialty steel. (Courtesy Allegheny Ludlum Industries, Inc.)

29. (*A, M*) Out of a lot of 1,000 aluminum engine block castings, 5 percent were defective. How many were defective? Of the remainder, 6 percent were spoiled in machining. How many finished castings were made?
30. (*F*) If 45 percent of 1,200 acres of timber were clear-cut, how many acres were cut?
31. (*E*) The voltage lost in the line wires supplying a motor is 5 percent of the generator voltage of 220 V. Find (a) the voltage lost, and (b) the voltage supplied to the motor.
32. (*C*) A bundle of white cedar shingles contained 250 shingles. If 30 percent were used on a job, how many shingles were used?
33. (*C, B*) An estimator added 5 percent for waste to the 625 sq ft of area to be covered with wire lath. On how many square feet did he base his cost?
34. (*A*) The stem-to-guide clearance on an automobile engine is 0.8 percent of the stem diameter of 0.3104 in. Find the clearance.
35. (*A, B*) The owner of an auto body repair shop borrowed $2,000.00 at $7\frac{1}{2}$ percent for 3 years. Find the amount of interest that she paid.
36. (*A*) An engine developed its maximum torque of 220 lb-ft at 1,500 rpm. Find the torque developed at 3,500 rpm if it is only 70 percent of the maximum torque.
37. (*C, B*) Roof sheathing for hip roofs is calculated on the basis of base area plus an allowance for the pitch of the roof. Using the following table, find the area to be sheathed if the base area is 1,000 sq ft and the roof is (a) $\frac{3}{8}$ pitch, and (b) $\frac{1}{2}$ pitch.

PITCH	ALLOWANCE	PITCH	ALLOWANCE
$\frac{1}{4}$	12%	$\frac{1}{2}$	42%
$\frac{1}{3}$	20%	$\frac{5}{8}$	60%
$\frac{3}{8}$	25%	$\frac{3}{4}$	80%

38. (*F, B*) A cotton picker (Fig. 26-6) costing $28,000 was estimated to depreciate (lose value) an amount equal to 9 percent of its original cost each year. What would be the estimated value of the machine after 5 years of use?
39. (*A*) An automobile valve spring should produce a pressure of 160 lb with a tolerance of 10 percent. Find the maximum and minimum acceptable pressures.
40. (*M*) In punching sheet aluminum, the clearance between the punch and the die should be 6 percent of the thickness of the metal. Find the clearance needed to punch sheets whose thicknesses are (a) 2.65 cm, and (b) 3.57 cm.
41. (*E*) A TV repairer tested a 120-Ω, 2 percent resistor and found the resistance to be 117 Ω. Is this resistor within the acceptable tolerance?
42. (*M*) A machinist is using a high-speed cutter with a recommended cutting speed of 80 ft/min. A carbide cutter will permit an increase in the cutting speed of 35 percent. What is the permissible cutting speed using a carbide cutter?
43. (*M*) The minor diameter of a $\frac{7}{8}''$-9 American Standard thread is 0.7307 in. Find the tap drill size if it is equal to 105 percent of the minor diameter.
44. (*C*) An area requiring 200 board feet of lumber is to be rough-floored. Find the total number of board feet needed if 4 percent extra is allowed for waste.
45. (*C*) An area requiring 240 board feet of lumber is to be sheathed with $\frac{3}{4}$-in stock. Find the total number of board feet needed if 8 percent is added for waste and 5.3 percent is added for scant widths.

Fig. 26-6 A cotton stripper. (Courtesy International Harvester Company)

46. (*C*) The floor of a room is estimated to require 280 board feet of lumber. How many board feet are actually needed if it is to be covered with matched flooring $3\frac{1}{4}$ in wide and 5 percent extra is allowed for waste?
47. (*C*) A room is estimated to require 180 board feet of lumber. How many board feet of $2\frac{1}{4}$-in-wide matched flooring are required if 5 percent extra is allowed for waste?
48. (*F, B*) Which is the better buy: (a) Barley containing 79.2 percent nutrients @ $5.02/cwt or corn containing 83.9 percent nutrients @ $5.12/cwt?
49. (*C, B*) A contractor estimated that he would need 425 concrete blocks for a foundation. If he allows an extra 3 percent for breakage, how many blocks should he order?
50. (*F, C*) A log was estimated to contain 40 cu ft of wood. If 36 percent was wasted in trimming off the four slabs, how much was wasted?
51. (*C, A*) A screw jack is only 55 percent efficient because of the tremendous friction losses. Find the actual lift of the jack if the theoretical lift is 6 tons.
52. (*F*) Timbers over 8 ft long are called logs. Shorter pieces are called sticks. If the bark averages 25 percent of the total stick volume, find the amount of wood in 300 cu ft of sticks.
53. (*F, B*) A farmer averaged 8,000 lb of strawberries per acre on each of 5 acres. If the dockage rate (the deduction for spoilage) was $4\frac{1}{2}$ percent, for how many pounds was she paid?
54. (*F*) Spoilage in hand-picked raspberries averages 0.3 percent. Find the amount of spoilage in 1,500 lb of hand-picked raspberries.
55. (*M, G*) The brass pendulum of a clock is 35.5 cm long. If brass expands at the rate of 0.02 percent in the heat of the summer, find the increase in length of the pendulum.
56. (*F, B*) In preparing packs of frozen asparagus, the packer estimates a loss of 55 percent of the original weight. (a) What percent is useful? (b) How many 12-oz packages can be produced from 1,000 lb of raw asparagus?

JOB 27 Taxes, Discounts, and Commissions

A *discount* is a *reduction* in the price of an article as a result of a sale, large purchase, etc. The amount of the discount is *subtracted* from the total price to get the net price to be paid. The sale offered in Fig. 25-1 promised a discount of 20 percent off the regular price of $20.00.

Example 27-1 Given: $B = \$20.00$ Find: Discount = ?
$R = 20\% = 0.20$ Net price = ?

Solution **1.** Find the discount.

$$P = B \times R$$
$$= 20 \times 0.20$$
$$= \$4.00 \quad \textit{Ans.}$$

2. Find the net price.

$$\text{Net price} = \text{total price} - \text{discount}$$
$$= \$20.00 - \$4.00$$
$$= \$16.00 \qquad \textit{Ans.}$$

Here, then, is the error. The advertisement listed the price after the discount as $17.00 when it should have been $16.00.

Example 27-2 A plumber earns $210/week. Find his take-home pay if the combined withholding tax and Social Security tax is 23.4 percent of his gross earnings.

Solution Given: $B = \$210$ Find: Take-home pay = ?
$R = 23.4\% = 0.234$

1. Find the total tax to be paid.

$$P = B \times R$$
$$= 210 \times 0.234$$
$$= \$49.14$$

2. The take-home pay equals the gross earnings minus the taxes.

$$\text{Take-home pay} = \$210.00 - \$49.14$$
$$= \$160.86 \qquad \textit{Ans.}$$

Example 27-3 Find the net price paid for 10 dozen flat head socket cap screws (Fig. 27-1) at $1.05/dozen if the discount is 15 percent.

Fig. 27-1

Solution 1. Find the total cost.

10 dozen at $1.05 per dozen = 10 × $1.05 = $10.50

2. Find the discount.

$$P = B \times R$$
$$= \$10.50 \times 0.15$$
$$= \$1.58$$

3. Since a discount means an amount which is *subtracted* from the total,

$$\text{Net price} = \text{total} - \text{discount}$$
$$= \$10.50 - \$1.58 = \$8.92 \qquad \textit{Ans.}$$

Self-Test 27-4 A machine tool salesperson earns a salary of $600 a month plus an 8 percent commission on all sales above $10,000. One month, her sales were $16,000. How much did she earn that month?

Solution 1. Find the amount of sales above $10,000.

Difference = total sales − required sales
= $16,000 − ______
= ______

$10,000
$6,000

2. Find the commission on this amount.

$P =$ ______ $\times R$

$= 6{,}000 \times$ ______

$=$ ______

B
0.08
$480.00

3. The commission is to be (added to)(subtracted from) her salary to find the total earned.

Total = $600 + ______ = ______ earned *Ans.*

added to
$480, $1,080

Problems

1. (*C, B*) A plumber bought 100 ft of 1-in galvanized pipe at $0.21/ft. If the sales tax is 6 percent, find (a) the tax, and (b) the total bill.
2. (*M, B*) A machine shop ordered 1 dozen micrometers at $31.80 each. If the discount was 15 percent, find (a) the amount of the discount, and (b) the net cost of the order.
3. (*C, B*) A contractor bought 3,000 board feet of lumber at $125 per thousand board feet. If the discount was 6 percent, find the net cost.
4. (*E, B*) An electrician earned $180 per week. Social Security taxes are 5.8 percent, withholding taxes are 20 percent, state taxes are 4.5 percent, and city taxes are 1.4 percent. Find (a) the total taxes paid, and (b) the electrician's net take-home pay.
5. (*F, B*) A pear grower shipped 540 boxes of pears to a broker who received a 15 percent commission. If the pears were sold at $6.20 per box, find the grower's net proceeds.
6. (*C, B*) A paint wholesaler offered a discount of 20 percent plus 2 percent for cash. Find the net cash cost of 120 gal of white paint at $6.95/gal and 250 gal of latex paint at $7.15/gal.
7. (*A, M, B*) A welding shop bought 1,000 ft of bar stock at an 8 percent discount from the list price of $0.32/lb. If the bar weighs 30 lb/20-ft length, find (a) the total weight of 1,000 ft of bar stock, (b) the list price for this weight, (c) the amount of the discount, and (d) the net price paid.
8. (*C, B*) A building-supply salesperson was paid a 4 percent commission on his first $800 of sales, 6 percent on his next $900 of sales, and 8 percent on all sales over $1,700. Last week, his sales were $2,960. Find his total commission earnings.
9. (*A, B*) An automobile mechanic bought a set of brake linings for $13.65. If her mark-up is 25 percent, what should she charge for the linings?
10. (*C, B*) A carpenter's supply house ordered 1 dozen planes at $6.25 each (6$\frac{1}{4}$ percent discount), 2 dozen crosscut saws at $7.50 each (15 percent discount), and 3 dozen hammers at $2.40 each (30 percent discount). Find the net cost of the order.
11. (*G, B*) A printing salesperson earns a salary of $130/week and 9 percent commission on all sales over $3,500/week. One week, her sales were $5,200. How much did she earn that week?
12. (*C, B*) A salesperson for a builder sold 2 houses at $18,500 each during June. His commission was 5 percent. If his total withholding taxes were 23 percent, find his net earnings for the month.
13. (*M, B*) A machinist (Fig. 27-2) worked a 40-hr week at $4.65/hr. His federal withholding tax was 20 percent, and his Social Security tax was 5.8 percent. Find his net pay for the week.
14. (*E, B*) An electrician purchased 500 ft of 2-in conduit at $55.00/100 ft. She got successive discounts of 25 percent and 5 percent. Find her net cost.

Fig. 27-2 A machinist operating a Clausing engine lathe. (Courtesy Clausing, Inc.)

15. (*E, C, B*) An electrical contractor bought 750 ft of BX cable at $43.00/100 ft. Find his net cost if he obtained successive discounts of 15 percent and 4 percent.
16. (*F, B*) A farmer bought a new truck for $5,850.00 plus 7 percent sales tax. If she made a down payment of $750.00, find her finance charge at $7\frac{1}{2}$ percent interest on the balance for 1 year.
17. (*A, B*) An auto parts dealer filled the following order:

    8 spark plugs @ $1.75 each
    1 set gaskets @ $23.50
    1 set rings @ $35.50
    8 valves @ $4.80 each
    16 valve springs @ $1.40 each

    Find (a) the cost for each item, (b) the total cost, (c) the total discount at 40 percent, and (d) the net price to be charged.
18. (*F, B*) A farmer sold 3,000 bushels of potatoes at $3 per bushel. The commission charge was 4 percent of the total selling price. How much did the farmer get if the trucking charges were $48.50?
19. (*B, F*) A commercial baker purchased 5,000 bushels of wheat through a commission merchant. The merchant got 3,500 bushels at $3.50 and the remainder at $3.25 per bushel. If the commission was 5 percent and the freight charges were $9 per 100 bushels, find the total amount paid by the baker.
20. (*F, B*) A commission agent bought 350 bales of cotton (500 lb/bale) at $0.48/lb. Find the total cost to the buyer if the commission was 4 percent, insurance $1\frac{1}{2}$ percent, and freight charges a total of $645.75.

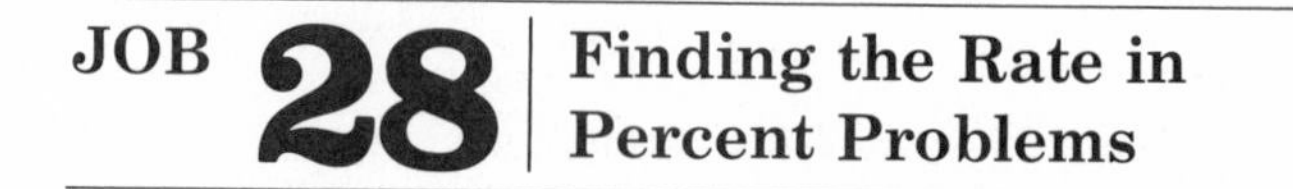

JOB 28 Finding the Rate in Percent Problems

Example 28-1 20 is what percent of 80?

Solution (Review Job 14 on page 82.)

Given: The base B is the total amount. $B = 80$
The part P is some portion of the base. $P = 20$
Find: The rate $R = ?$

$$B \times R = P$$

$$80 \times R = 20$$

$$R = \frac{20}{80} = \frac{1}{4}$$

The rate of $\frac{1}{4}$ must now be expressed first as a decimal and then as a percent.

$$\frac{1}{4} = 0.25 = 25\% \qquad \textit{Ans.}$$

Example 28-2 In transmitting 50 hp by a belt system, 1.2 hp was lost due to slippage. What percent of the power was lost?

Solution Given: $B = 50$ hp Find: $R = ?$
$P = 1.2$ hp

$B \times R = P$

$50 \times R = 1.2$

$$R = \frac{1.2}{50} = 0.024 = 2.4\% \quad \textit{Ans.}$$

Example 28-3 A pattern maker used a shrink rule allowing for a shrinkage in the cooling metal of $\frac{3}{16}$ in/ft of length. Express this shrinkage as a percent.

Solution Given: $B = 1$ ft $= 12$ in Find: $R = ?$
$P = \frac{3}{16}$ in

$B \times R = P$

$$12 \times R = \frac{3}{16}$$

$$R = \frac{3}{16} \div 12$$

$$= \frac{3}{16} \times \frac{1}{12} = \frac{1}{64}$$

Since $\frac{1}{64} = 0.0156$, the rate of shrinkage $= 1.56\%$. *Ans.*

Self-Test 28-4 A feed mixture was prepared consisting of 42 lb of alfalfa and 28 lb of corn silage. What percent of the mixture is alfalfa?

Solution

1. First find the total weight of the mixture.

   Total weight $= 42 +$ _______ $=$ _______ lb   28, 70

2. Find the percent of alfalfa.

   Given: $B =$ total weight $=$ _______ lb   70
   $P =$ weight of alfalfa $=$ _______ lb   42

   Find: $R =$ percent of alfalfa $= ?$

   $B \times R = P$

   _______ $\times R =$ _______   70, 42

   $$R = \frac{42}{?}$$ 70

   $= 0.60$

   $=$ _______ % alfalfa *Ans.*   60

Problems

1. (*G*) 30 is what percent of 120?
2. (*G*) 30 is what percent of 80?
3. (*G*) What percent of 25 is 5?
4. (*G*) What percent of 30 is 50?
5. (*G*) 24.5 is what percent of 70?
6. (*G*) What percent of 18.5 is 3.5?
7. (*B*) A discount of \$2 was given on a bill of \$16. Find the rate of discount.

8. (*E*) If 3 transistors out of a lot of 120 transistors are defective, what percent are defective?
9. (*F, E, C*) If 100 m of a 250-m roll of wire has been used, what percent has been used?
10. (*A, B*) In early 1975, Mr. Weber bought a car for $4,000 and received a rebate of $200. What percent of the cost was the rebate?
11. (*E*) The National Electrical Code specifies that #14 rubber-covered wire can safely carry only 15 A. What percent of the wire's capacity is used by a broiler drawing 10 A?
12. (*F*) If a section of land equals 640 acres, what percent of a section is represented by a farm of 80 acres?
13. (*A, M*) The auto wrench shown in Fig. 28-1 weighs 4.5 lb. What percent is this of a carton weighing 45 lb?

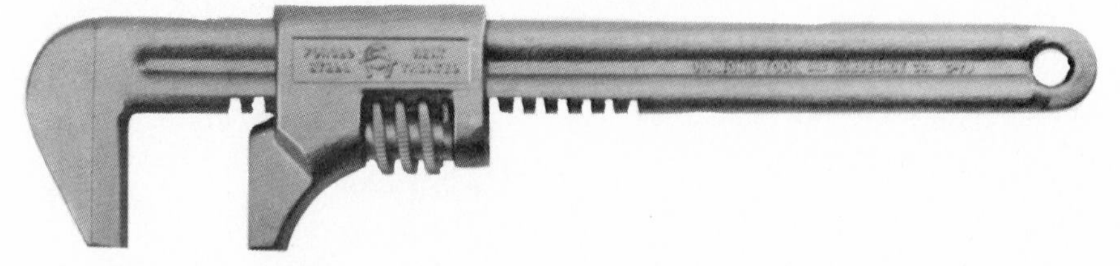

Fig. 28-1 An auto wrench. (Courtesy Diamond Tool and Horseshoe Company)

14. (*G, C*) The capacity of the fork lift shown in Fig. 28-2 is 3,000 lb. What percent of the capacity is used in lifting 2,500 lb?

Fig. 28-2 A fork lift at work. (Courtesy Clark Equipment Company)

15. (*E*) The voltage loss in a supply line is 5.5 V. If the generator voltage is 110 V, find the rate of loss.
16. (*C*) A riveting team on a construction job spoiled 60 rivets out of a lot of 1,000. What percent were spoiled?
17. (*M, G*) A 90-lb piece of naval brass contained 55.8 lb of copper, 33.3 lb of zinc, and 0.9 lb of tin. Find the percent of each metal in the piece.
18. (*C, B*) Only 750 board feet out of 800 board feet ordered for a job could be used because of imperfections. What percent was wasted?

19. (*B, F*) In 1975, the International Harvester Corporation earned \$1.70 per share of stock. If the stock cost \$22.50 per share, what was the percent return on the investment?
20. (*C*) Rough 2-in-thick lumber is finished to a thickness of $1\frac{5}{8}$ in. What percent of the original thickness was removed in the finishing process?
21. (*C, B*) The cost of 3-ply Douglas fir panels increased from \$0.18/sq ft to \$0.24/sq ft. Find the percent of increase.
22. (*E*) A 15-A fuse carried 18 A for 2 sec. What was the percent of overload?
23. (*A*) A safety inspector measured the thickness of the brake lining on a car as $\frac{7}{64}$ in. What percent of the original $\frac{1}{4}$-in thickness was worn away?
24. (*M*) If a casting shrinks $\frac{1}{8}$ in/ft of length, express the shrinkage as a percent.
25. (*M*) In Fig. 28-3, what percent of metal is allowed for cutoff on each 6-cm length of stock?

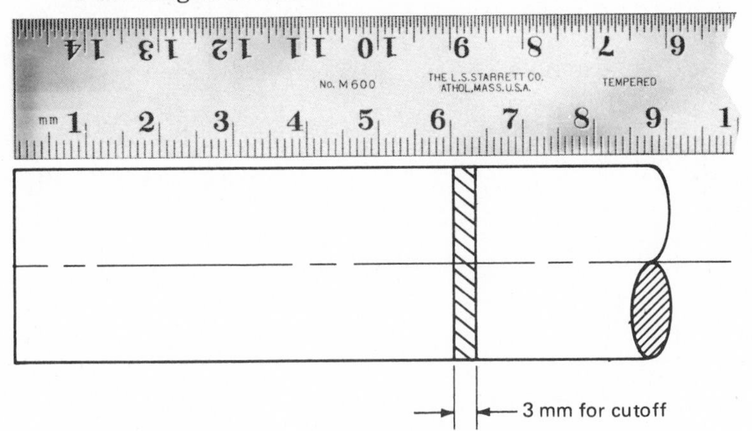

Fig. 28-3

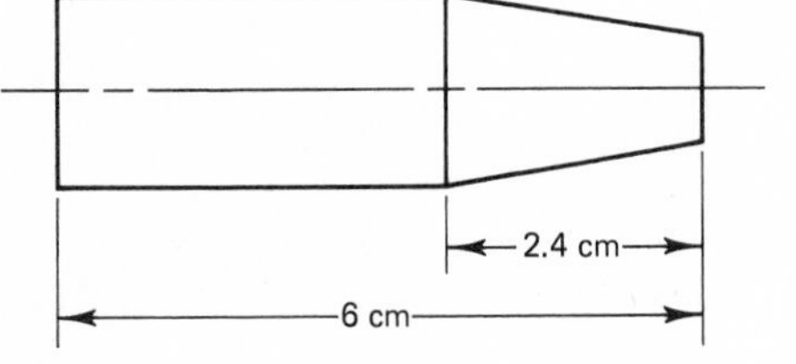

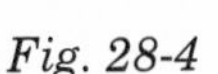
Fig. 28-4

26. (*B, A*) A total of \$51.45 in taxes was withheld from an automobile mechanic's salary of \$175.00. Find the total tax rate.
27. (*M*) In Fig. 28-4, what percent of the total length is tapered?
28. (*C, A*) Each tire on the International PAY loader axle shown in Fig. 28-5 weighs 5,830 lb. What percent of the 45,000-lb axle assembly is the weight of 1 tire?

Fig. 28-5 The axle of the International 580 PAY loader. (Courtesy The International Harvester Company)

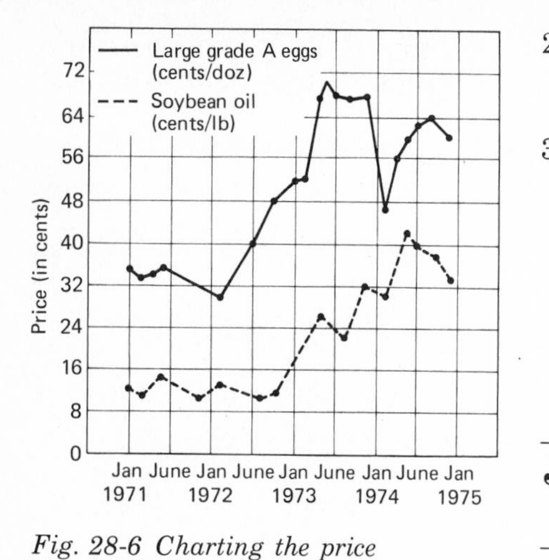

Fig. 28-6 Charting the price of eggs and soybean oil from 1971 to 1975.

29. (A, B) An automobile repair shop paid $28.15 for parts, but charged the list price of $39.41. (a) Find the profit. (b) What percent of the cost was the profit?
30. (F, B) Using the graph shown in Fig. 28-6, find the price of eggs in (a) June 1972 and (b) June 1973. (c) Find the amount of increase and the percent of increase. Find the price of soybean oil in (d) January 1971 and (e) June 1974. (f) Find the amount of increase and the percent of increase.

JOB 29 Finding the Base in Percent Problems

Example 29-1 5 is 25 percent of what number?

Solution Given: The part of the base $= P = 5$ Find: Total $= B = ?$

$R = 25\% = 0.25$

$$B \times R = P$$

$$B \times 0.25 = 5$$

$$B = \frac{5}{0.25} = 20 \qquad \textit{Ans.}$$

Example 29-2 A mechanic is able to save $18.50 each week. This amount is equal to $12\frac{1}{2}$ percent of his weekly wages. How much does he earn each week?

Solution Given: P = amount saved = $18.50

R = percent saved = $12\frac{1}{2}\% = 0.125$

Find: B = total wages = ?

$$B \times R = P$$

$$B \times 0.125 = 18.50$$

$$B = \frac{18.50}{0.125}$$

$$= \$148.00 \qquad \textit{Ans.}$$

Self-Test 29-3 A carpenter used 450 board feet of lumber on a job. This was 90 percent of the number of board feet delivered to the job. How many board feet were delivered?

Solution The problem is finding the number of board feet delivered.

Given: Part of the total =

P = ______ Find: Total $= B = ?$

$R = 90\% =$ (decimal)

$$B \times R = P$$

$$B \times 0.9 = \_\_\_\_\_\_$$

$$B = \frac{450}{?}$$

$= \_\_\_\_\_\_$ board feet *Ans.*

450
0.9
450
0.9
500

Problems

1. (*G*) 10 is 50 percent of what number?
2. (*G*) 16 is 40 percent of what number?
3. (*G*) 25 is 2.5 percent of what number?
4. (*G*) 70 is $3\frac{1}{2}$ percent of what number?
5. (*E*) The voltage loss in a line is 3 V. If this is 2 percent of the generator voltage, what is the generator voltage?
6. (*M, G*) To make a bronze casting, 232 lb of copper were used. If this is 80 percent of the total weight of the casting, find the weight of the casting.
7. (*F*) A feed lot operator figures that each animal requires 2.5 lb of protein nutrients per day. How many pounds of 10 percent protein food should be provided per day?
8. (*B, F, A*) A tractor depreciates (loses value) at the rate of $450/yr, which is equal to 9 percent of the cost. What was the original cost of the tractor?
9. (*C*) A woman wasted $3\frac{1}{2}$ percent of the tiles in laying a vinyl tile floor. If 7 tiles were wasted, with how many did she start?
10. (*M*) The wooden pattern for a casting weighs $10\frac{1}{2}$ lb. If this is $7\frac{1}{2}$ percent of the casting weight, find the weight of the casting.
11. (*A*) If $5\frac{1}{2}$ l of antifreeze represent $27\frac{1}{2}$ percent of the capacity of an automobile radiator, what is the capacity of the radiator?
12. (*C*) The safe compressive stress of wood is 750 lb/sq in. This is only 5 percent of the safe compressive stress of steel. Find the safe compressive stress of steel.
13. (*F*) How many pounds of nitrate fertilizer averaging 25 percent available nitrogen are required to provide 50 lb/acre of nitrogen on a 100-acre field?
14. (*B*) A salesperson earned $220 in commissions. If his commission was 8 percent of his sales, what was the amount of his sales?
15. (*M, B*) A lathe was sold at a discount of 12 percent. This represented a saving of $48. What was the original price of the lathe?
16. (*C, B*) A dealer sold a saw at a profit of $0.87, which represents a 15 percent profit on the cost. What did the saw cost her?
17. (*C, B*) A stonemason received a 15 percent increase in wages, amounting to $31.50. What were his wages before the increase?
18. (*E*) A fuse carried 16.5 A for a few seconds. This was 110 percent of the fuse rating. What is the fuse rating?
19. (*E*) What must be the generator voltage if only 98 percent of the generator voltage is delivered to a 117-V line?
20. (*A*) In the cylinder shown in Fig. 29-1, the volume at top dead center (TDC) of 37.8 cu in is only 12 percent of the volume at bottom dead center (BDC). Find the volume at BDC.

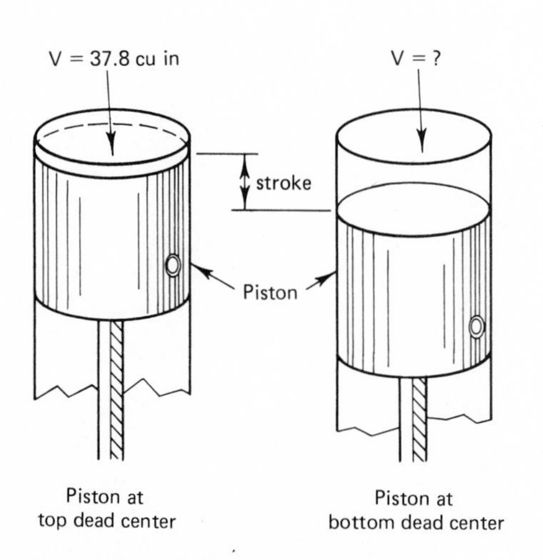

Fig. 29-1 Full compression of the air-fuel mixture is obtained at top dead center.

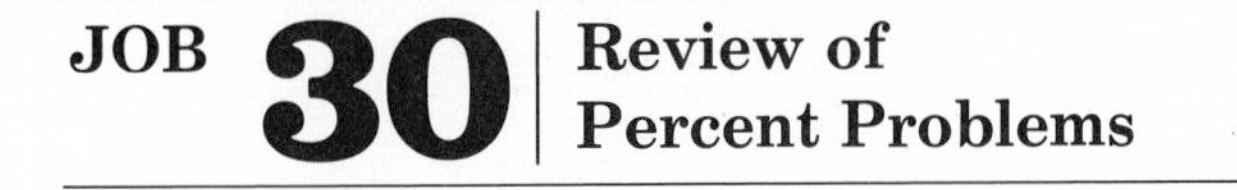

JOB 30 Review of Percent Problems

1. To change a percent into a decimal, move the decimal point ______ places to the ______ and drop the percent sign.

   two, left

2. To change a decimal into a percent, move the decimal point ______ places to the ______ and add a percent sign.

   two, right

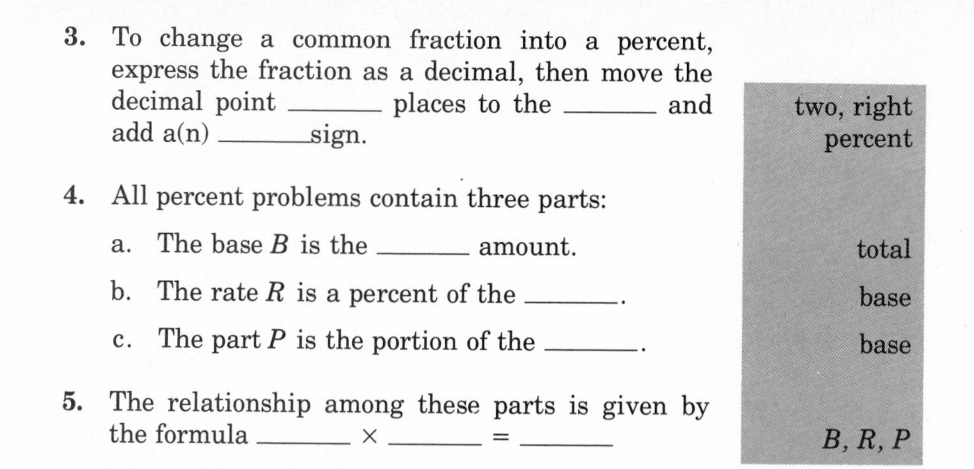

3. To change a common fraction into a percent, express the fraction as a decimal, then move the decimal point ______ places to the ______ and add a(n) ______ sign.

two, right
percent

4. All percent problems contain three parts:

a. The base B is the ______ amount.

b. The rate R is a percent of the ______.

c. The part P is the portion of the ______.

total
base
base

5. The relationship among these parts is given by the formula ______ × ______ = ______

B, R, P

Problems

1. (*G*) Change the following percents to decimals: (a) 62 percent, (b) 3 percent, (c) 5.6 percent, (d) 0.8 percent, (e) 116 percent, (f) $4\frac{1}{2}$ percent, and (g) $6\frac{1}{4}$ percent.
2. (*G*) Change the following decimals to percents: (a) 0.4, (b) 0.08, (c) 0.6, (d) 2.00, (e) 0.625, (f) 0.045, and (g) 5.
3. (*G*) Change the following fractions to percents: (a) $\frac{3}{4}$, (b) $\frac{3}{5}$, (c) $\frac{3}{7}$, (d) $\frac{3}{10}$, (e) $\frac{3}{13}$, (f) $\frac{1}{6}$, and (g) $\frac{24}{52}$.
4. (*G*) How much is 18 percent of 96?
5. (*G*) How much is 4.5 percent of 120?
6. (*G*) 20 is what percent of 120?
7. (*G*) 3.3 is 5.5 percent of what number?
8. (*C*) A plumber estimated that he would need 48 m of galvanized iron pipe for a job. If he allowed 5 percent for waste in cutting, how many additional meters should he order?
9. (*B, M*) A lathe hand earning $165 per week received an increase of 8 percent. Find the amount of the increase and her new salary.
10. (*C*) "Clear" lumber is free from knots and other defects. How much clear lumber would you expect to find in a load of 2,500 board feet that was 80 percent clear?
11. (*A, M*) In the transmission of 6 hp by a belt system, 0.09 hp was lost owing to slippage. What percent was lost?
12. (*E*) The plate voltage on an amplifier tube may vary up to 10 percent of its rated value and still be satisfactory. A tube rated at 200 V has a plate voltage of 185 V. Is this within the accepted variation?
13. (*C, B*) A jobber ordered twenty-four 12-in combination squares of the type shown in Fig. 30-1. If each square cost $37.55 and he received a discount of 15 percent, find the net cost of the order.
14. (*G*) Of the 60 workers in a shop, 80 percent got production bonuses. How many workers got bonuses?
15. (*M*) White metal is composed of 4 percent copper, 88 percent tin, and 8 percent antimony. How many pounds of each metal are in an ingot weighing 150 lb?
16. (*C*) An architect designed the window area of a school room to be 15 percent of the floor area. How many square feet of glass area should be provided for a school room 25 ft by 30 ft?
17. (*E*) If only a 2 percent voltage drop is permitted in a line, what is the maximum loss in voltage permitted at the end of a line supplied by a 120-V source? What is the voltage available at the end of the line?
18. (*B, M*) Find the net cost for 2 dozen drill gages (Fig. 30-2) listed at $2.65 each if the discount is 20 percent.

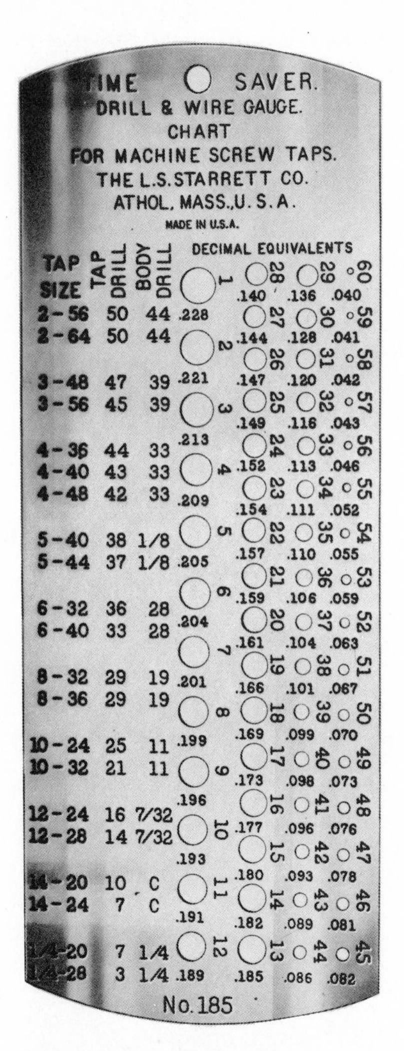

Fig. 30-2 A drill and wire gage. (Courtesy The L.S. Starrett Company)

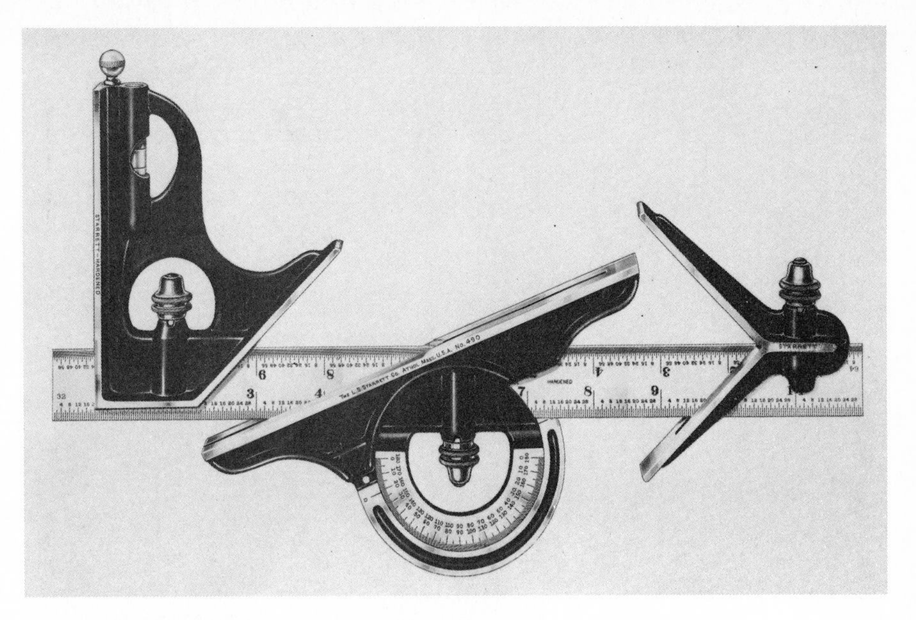

Fig. 30-1 A combination square. (Courtesy The L.S. Starrett Company)

Fig. 30-3 Slotting a screwdriver handle on a milling machine. (Courtesy Hamilton Associates)

19. (*B*) A salesperson earns a salary of $120 per week plus 7 percent commission on all sales over $2,000. One week, her sales amounted to $3,100. How much did she earn that week?
20. (*M*) A certain metal can be expected to shrink $\frac{1}{8}$ in/ft of length on cooling. Find the percent shrinkage. If a pattern length is 10 in, what will be the length of a cooled piece of this metal?
21. (*C*) A wall area to be sheathed calls for 625 board feet of lumber. How many board feet should be ordered if 10 percent extra is allowed for waste and 6.6 percent extra is allowed for scant widths?
22. (*M*) The width of the slot in the tang of the screwdriver handle shown in Fig. 30-3 should not be more than 15 percent of the width of the tang. What thickness of cutter should be used if the tang is 1.4 cm in diameter?
23. (*C*) A room requires 530 board feet of rough flooring. It is to be finished with matched hardwood flooring $\frac{5}{8}$ in thick and $3\frac{1}{4}$ in wide. How many board feet will be needed if 5 percent extra is allowed for waste and scant widths are taken into account?
24. (*E*) Emitter bias resistors usually have a wattage rating about 80 percent higher than the calculated wattage. What should be the wattage rating of an emitter bias resistor developing 0.5 W?
25. (*A*) An automobile has a power train that transmits 162 hp to the rear wheels at an efficiency of 72 percent. How many horsepower does the engine develop?
26. (*A*) The mechanical efficiency of an engine (in percent) is obtained by dividing the brake horsepower by the rated horsepower. Find the mechanical efficiency of an engine if the brake horsepower = 120 and the rated horsepower = 160.
27. (*F*) A farmer planted 1,500 strawberry plants, but only 1,275 bore fruit. What percent of the planting bore fruit?
28. (*C*) The grade of a highway is the ratio of the vertical distance to the horizontal distance. If the grade of a highway is 0.043, express this grade as a percent.
29. (*C*) The pressure in a steam boiler was increased from 56 lb/sq in to 63 lb/sq in. Find the percent increase in pressure.

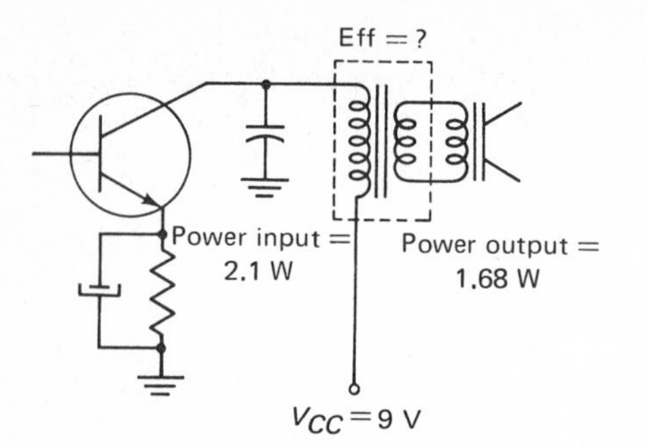

Fig. 30-4 Finding the efficiency of an impedance-matching transformer.

30. (*C*) A surveyer made an error of 0.18 ft in a distance of 577.8 ft. Find the percentage of error.
31. (*E*) In Fig. 30-4, an impedance-matching transformer couples an output transistor delivering 2.1 W to a voice coil which receives 1.68 W. Find the efficiency of the transformer by dividing the power output by the power input.
32. (*C*) A carpenter using a Portanailer (Fig 30-5) increased the number of square feet of subflooring laid in a given time from 200 to 330 sq ft. Find the percent of increase.

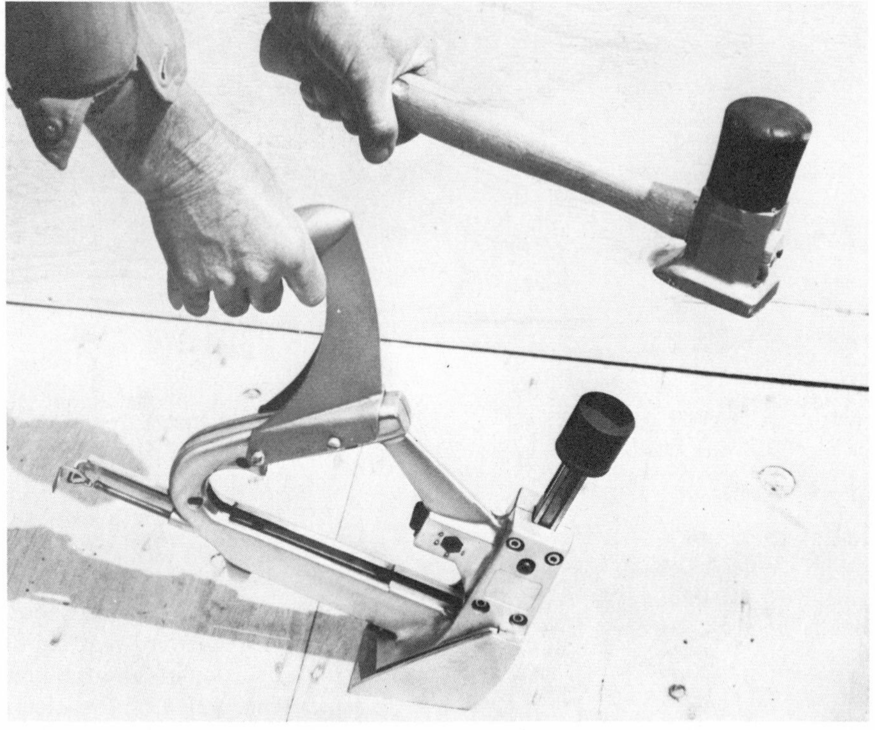

Fig. 30-5 Installing a floor with a Rockwell Delta Portanailer. (Courtesy Rockwell International)

33. (*C*) An enamel paint is thinned at the rate of 1 pint of thinner per quart of enamel. Find the percent of thinner in the final mixture.
34. (*M, G*) "Fool's gold," or iron pyrite, is a compound of iron and sulfur. Find the percent of iron in a 7.51-g sample if it contains 4.73 g of iron by weight.
35. (*C*) The number of concrete reinforcing bars is based on the number of eighths of an inch in the diameter. Thus,

    #5 bar = *five*-eighths inch in diameter

    #4 bar = *four*-eighths inch in diameter

    Using the table below, what percent of the weight of a $\frac{3}{4}$-in-diameter bar is the weight of a $\frac{3}{8}$-in-diameter bar?

BAR NO.	lb/ft
2	0.167
3	0.376
4	0.668
5	1.043
6	1.502

Fig. 30-7 An LP-fueled lift truck. (Courtesy Allis-Chalmers Corporation)

Fig. 30-6 The foundation of the Louisiana Super Dome used 11,000 tons of reinforcing bar. (Courtesy U.S. Steel)

36. (*C*) In Fig. 30-7, if the crate weighs 800 lb and the lift has a capacity of 2,500 lb, what percent of the lift's capacity is being used?

(See Answer Key for Test 7—Percent Problems)

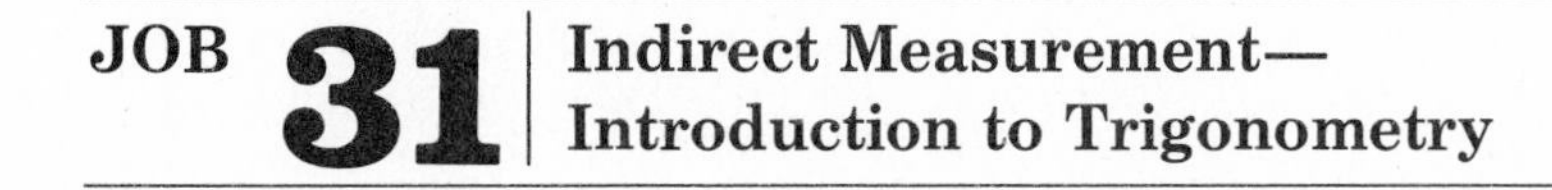

JOB 31 Indirect Measurement—Introduction to Trigonometry

TRIGONOMETRIC FUNCTIONS OF A RIGHT TRIANGLE

Trigonometry is the study of the relationships that exist among the sides and angles of a triangle. These relationships will form a basic mathematical tool used in the solution of many problems.

ANGLES

An angle is a figure formed when two lines meet at a point. In Fig. 31-1, the lines *OA* and *OC* are the sides of the angle. They meet at point *O*, which is the vertex of the angle.

Fig. 31-1 Naming an angle: angle AOC or angle 0 or angle 1.

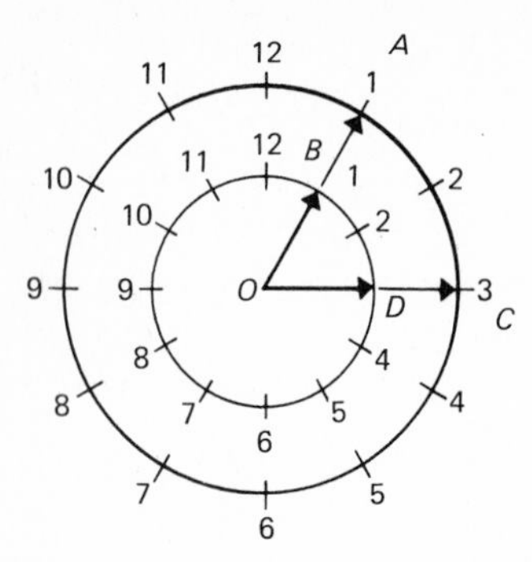

Fig. 31-2 The angle formed by the rotation from 1 to 3 o'clock is the same on any clock face. $\angle BOD = \angle AOC$.

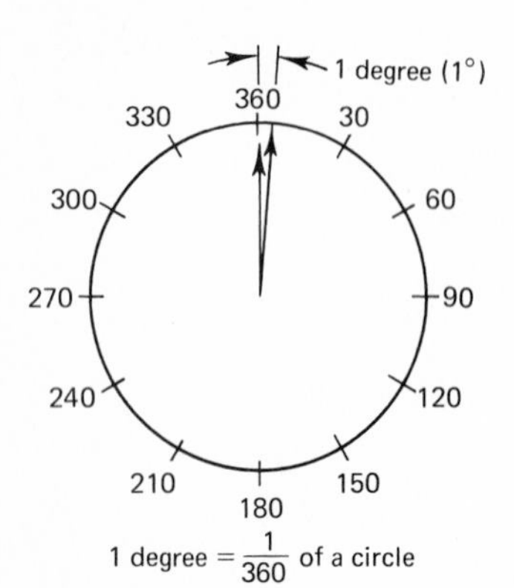

Fig. 31-3 An angle is measured by the amount of rotation about a fixed point.

NAMING ANGLES

The angle shown in Fig. 31-1 may be named in three ways. (1) Use three capital letters, setting them down in order from one end of the angle to the other. Thus the angle may be named $\angle AOC$ or $\angle COA$. (2) Use only the capital letter at the vertex of the angle, as $\angle O$. (3) Use a small letter or number inside the angle, as $\angle 1$.

MEASURING ANGLES

When a straight line *turns* about a point from one position to another, an angle is formed. When the minute hand of a clock turns from the 1 to the 3, the hand has turned through an angle approximately equal to $\angle AOC$. The size of an angle is measured only by the *amount of rotation of a line about a fixed point.* Suppose we place a small watch on the face of a large clock, as shown in Fig. 31-2. When the hand of the large clock turns from 1 to 3, it forms $\angle AOC$. But in this *same time,* the hands of the small watch have formed $\angle BOD$, which is exactly equal to $\angle AOC$. Thus we see that the lengths of the sides have no effect on the size of the angle. The size of an angle depends *only* on the amount of rotation between the sides of the angle. The end of a line that makes one complete turn will describe a circle. In Fig. 31-3, a circle is divided into 360 parts. The angle formed by a rotation through $\frac{1}{360}$ part of the circle is called one degree (°).

A *right angle* is an angle formed by a rotation through one-fourth of a circle, as in $\angle ABC$ (Fig. 31-4). Since a complete circle contains 360°, a right angle contains $\frac{1}{4} \times 360 = 90°$.

An *acute angle* is an angle containing *less* than 90°, as in $\angle AOC$ (Fig. 31-1).

TRIANGLES

A *triangle* is a closed plane figure with three sides. Triangles are named by naming the three vertices of the triangle, using capital letters. The letters are then given in order around the triangle. Thus, in Fig. 31-4, the triangle can be named $\triangle ACB$ or $\triangle CBA$ or $\triangle BAC$ or $\triangle ABC$ or $\triangle BCA$ or $\triangle CAB$.

A *right triangle* is a triangle which contains a right angle, like $\triangle ABC$ in Fig. 31-4. The little square at $\angle B$ is used to indicate a right angle.

NAMING THE SIDES OF A RIGHT TRIANGLE

In trigonometry, the names for the sides of a right triangle depend on which of the *acute* angles are used as the reference point. In Fig. 31-4,

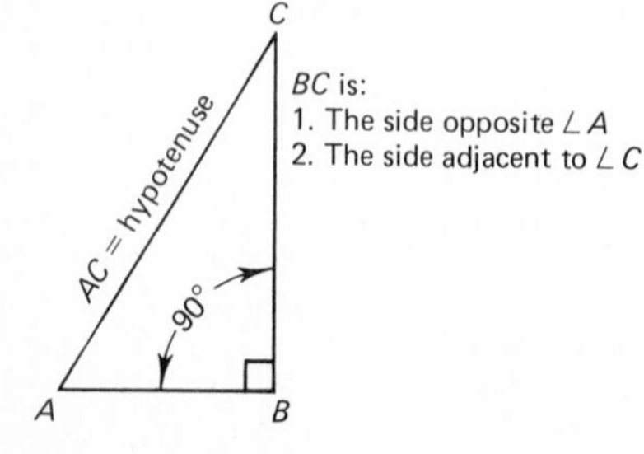

Fig. 31-4 Naming the sides of a right triangle.

For $\angle A$:

1. The side opposite the right angle is called the *hypotenuse* (AC).
2. The side opposite $\angle A$ is called the *opposite* side (BC).
3. The side of $\angle A$ which is *not* the hypotenuse is called the *adjacent* side (AB).

For $\angle C$:

1. The side opposite the right angle is called the *hypotenuse* (AC).
2. The side opposite $\angle C$ is called the *opposite* side (AB).
3. The side of $\angle C$ which is *not* the hypotenuse is called the *adjacent* side (BC).

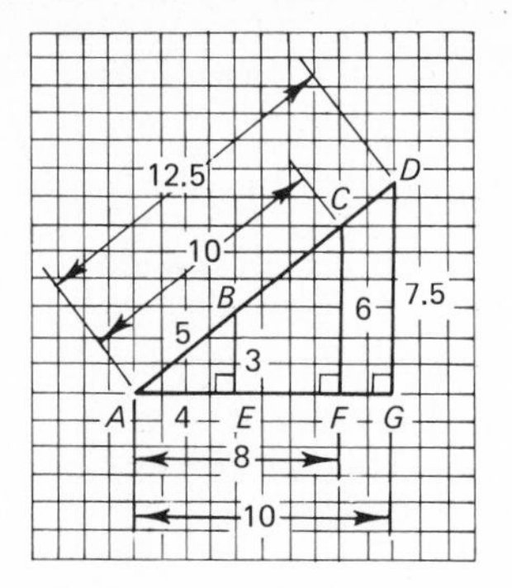

Fig. 31-5 Diagram used to obtain the trigonometric formulas.

THE TANGENT OF AN ANGLE

In addition to degrees, the size of an angle may also be described in terms of the lengths of the sides of a right triangle formed from the angle. Refer to Fig. 31-5, in which each space represents one unit of length. From various points on one side of angle A, lines have been drawn making right angles with the other side. These lines have formed the right triangles ABE, ACF, and ADG. Notice that all these triangles contain the same $\angle A$.

	IN $\triangle ABE$	IN $\triangle ACF$	IN $\triangle ADG$
Hypotenuse	$= AB = 5$	$AC = 10$	$AD = 12.5$
Side opposite $\angle A$	$= BE = 3$	$CF = 6$	$DG = 7.5$
Side adjacent $\angle A$	$= AE = 4$	$AF = 8$	$AG = 10$

In each triangle, let us divide the side *opposite* $\angle A$ by the side *adjacent* to $\angle A$.

In $\triangle ABE$,

$$\frac{\text{Side opposite } \angle A}{\text{Side adjacent } \angle A} = \frac{BE}{AE} = \frac{3}{4} = 0.75$$

In $\triangle ACF$,

$$\frac{\text{Side opposite } \angle A}{\text{Side adjacent } \angle A} = \frac{CF}{AF} = \frac{6}{8} = 0.75$$

In $\triangle ADG$,

$$\frac{\text{Side opposite } \angle A}{\text{Side adjacent } \angle A} = \frac{DG}{AG} = \frac{7.5}{10} = 0.75$$

Notice that this particular division of the opposite side by the adjacent side always resulted in the same answer, *regardless* of the size of the triangle. This is true because all the triangles contain the same $\angle A$. The ratio of these sides remains constant because they all describe the same $\angle A$. This constant number describes the size of $\angle A$ and is called the *tangent* of angle A (tan $\angle A$). Therefore, if the tangent of an unknown angle were calculated to be 0.75, then the angle would be equal to $\angle A$, or about 37°. If the angle changes, then the number for the tangent of the angle will also change. However, the tangent of every angle is a specific number that never changes. These numbers are found in Table 32-1.

RULE 1 **The tangent of an angle** $= \dfrac{\textbf{side opposite the angle}}{\textbf{side adjacent to the angle}}$

FORMULA $\tan \angle = \dfrac{o}{a}$

THE SINE OF AN ANGLE

In each triangle shown in Fig. 31-5, let us divide the side *opposite* $\angle A$ by the side called the *hypotenuse.*

In $\triangle ABE$,

$$\frac{\text{Side opposite } \angle A}{\text{Hypotenuse}} = \frac{BE}{AB} = \frac{3}{5} = 0.6$$

In $\triangle ACF$,

$$\frac{\text{Side opposite } \angle A}{\text{Hypotenuse}} = \frac{CF}{AC} = \frac{6}{10} = 0.6$$

In $\triangle ADG$,

$$\frac{\text{Side opposite } \angle A}{\text{Hypotenuse}} = \frac{DG}{AD} = \frac{7.5}{12.5} = 0.6$$

Once again, the resulting quotients are the same, regardless of the size of the triangles. This number is called the *sine* of angle A (sin $\angle A$) and is *another* way to describe the size of the angle. If the angle changes, the number for the sine of the angle will also change. However, the sine of every angle is a specific number that never changes. These numbers are also found in Table 32-1.

RULE 2 **The sine of an angle** $= \dfrac{\textbf{side opposite the angle}}{\textbf{hypotenuse}}$

FORMULA $\sin \angle = \dfrac{o}{h}$

THE COSINE OF AN ANGLE

In each triangle shown in Fig. 31-5, let us divide the side *adjacent* to $\angle A$ by the *hypotenuse*.

In $\triangle ABE$,

$$\frac{\text{Side adjacent } \angle A}{\text{Hypotenuse}} = \frac{AE}{AB} = \frac{4}{5} = 0.8$$

In $\triangle ACF$,

$$\frac{\text{Side adjacent } \angle A}{\text{Hypotenuse}} = \frac{AF}{AC} = \frac{8}{10} = 0.8$$

In $\triangle ADG$,

$$\frac{\text{Side adjacent } \angle A}{\text{Hypotenuse}} = \frac{AG}{AD} = \frac{10}{12.5} = 0.8$$

Here, too, the resulting quotients are the same, regardless of the size of the triangles. This number is called the *cosine* of angle A (cos $\angle A$) and is a third way to describe the size of the angle. If the angle changes, then the number for the cosine of the angle will also change. However, the cosine of every angle is a specific number that never changes. These numbers are found in Table 32-1.

RULE 3 **The cosine of an angle** $= \dfrac{\textbf{side adjacent to the angle}}{\textbf{hypotenuse}}$

FORMULA $\cos \angle = \dfrac{a}{h}$

Example 31-1 Find the sine, cosine, and tangent of $\angle A$ in the triangle shown in Fig. 31-6.

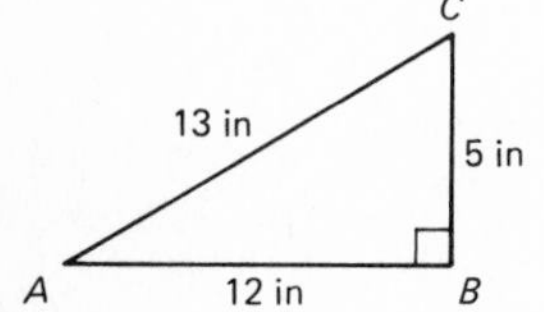

Fig. 31-6

Solution **1.** Name the sides, using $\angle A$ as the reference angle.

Hypotenuse $= AC = 13$ in

Opposite side $= BC = 5$ in

Adjacent side $= AB = 12$ in

2. Find the values of the three functions.

$$\sin \angle A = \frac{o}{h} \qquad \cos \angle A = \frac{a}{h} \qquad \tan \angle A = \frac{o}{a}$$

$$\sin \angle A = \frac{5}{13} \qquad \cos \angle A = \frac{12}{13} \qquad \tan \angle A = \frac{5}{12}$$

$$\sin \angle A = 0.384 \qquad \cos \angle A = 0.923 \qquad \tan \angle A = 0.417 \qquad \textit{Ans.}$$

Self-Test 31-2 Find the sine, cosine, and tangent of $\angle B$ in the triangle shown in Fig. 31-7.

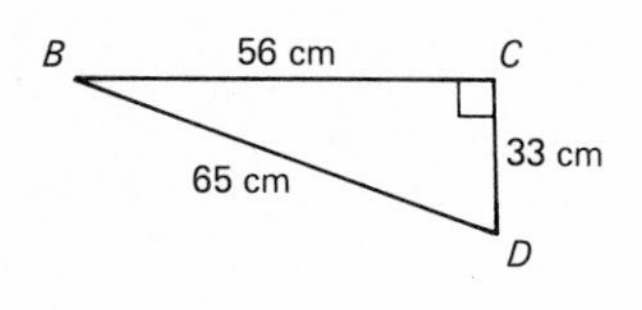

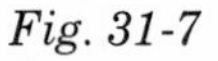

Fig. 31-7

Solution

1. The first step in finding the sine, cosine, and tangent is to name the sides, using $\angle$ ______ as *the reference angle.*

   Hypotenuse = side BD = ______ cm

   Opposite side = side ______ = ______ cm

   Adjacent side = side ______ = ______ cm

2. Find the values of the three functions.

   $$\sin \angle B = \frac{?}{h}$$

   $$\sin \angle B = \frac{33}{?}$$

   $$= \underline{\qquad}$$

   $$\cos \angle B = \frac{a}{?}$$

   $$\cos \angle B = \frac{?}{65}$$

   $$= \underline{\qquad}$$

   $$\tan \angle B = \frac{?}{?}$$

   $$\tan \angle B = \frac{?}{56}$$

   $$= \underline{\qquad}$$

Answers (shaded column):

- B
- 65
- CD, 33
- BC, 56
- o
- 65
- 0.5077
- h
- 56
- 0.8615
- $\frac{o}{a}$
- 33
- 0.5893

Problems

1. State the value of (a) the hypotenuse, (b) the opposite side, and (c) the adjacent side for each of the angles named in Fig. 31-8.

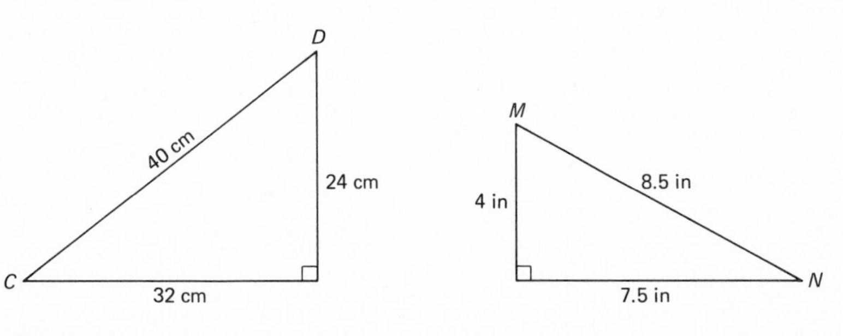

Fig. 31-8

Calculate the sine, cosine, and tangent of the angles named in each problem below:

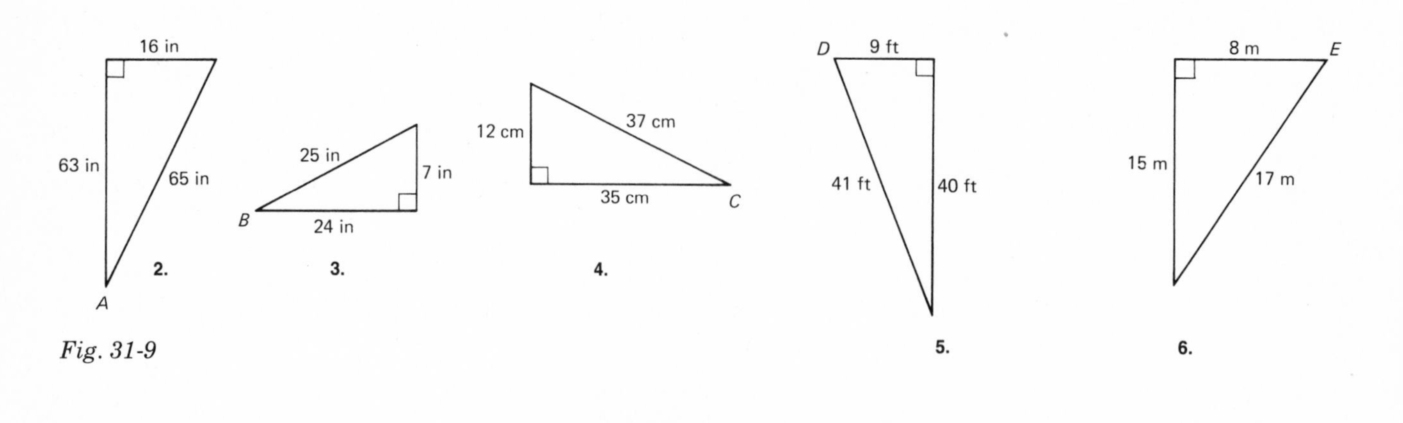

Fig. 31-9

JOB 32 Using a Table of Trigonometric Functions

Since the values of the trigonometric functions of an angle never change, it is possible to set these values down in a table like Table 32-1.

Example 32-1 Find the value of sin 50°.

Solution Follow down the column marked "angle" until you reach 50°. Read across to the right to find the number 0.7660 in the sine column. Therefore,

$\sin 50° = 0.7660$ *Ans.*

Example 32-2 Find the value of tan 30°.

Solution Follow down the column marked "angle" until you get to 30°. Read across to the right to find the number 0.5774 in the tangent column. Therefore,

$\tan 30° = 0.5774$ *Ans.*

FINDING THE ANGLE WHEN THE FUNCTION IS GIVEN

Example 32-3 Find $\angle A$ if $\cos \angle A = 0.7314$.

Solution Follow down the column marked "cos" until you find the number 0.7314. Read across to the left to find 43° in the column marked "angle." Therefore,

$\angle A = 43°$ *Ans.*

Example 32-4 Find $\angle B$ if $\tan \angle B = 1.8807$.

Solution Follow down the column marked "tan" until you find the number 1.8807. Read across to the left to find 62° in the column marked "angle." Therefore,

$\angle B = 62°$ *Ans.*

Table 32-1. Values of the Trigonometric Functions

ANGLE	SIN	COS	TAN	ANGLE	SIN	COS	TAN
0°	0.0000	1.0000	0.0000	46°	0.7193	0.6947	1.0355
1°	0.0175	0.9998	0.0175	47°	0.7314	0.6820	1.0724
2°	0.0349	0.9994	0.0349	48°	0.7431	0.6691	1.1106
3°	0.0523	0.9986	0.0524	49°	0.7547	0.6561	1.1504
4°	0.0698	0.9976	0.0699	50°	0.7660	0.6428	1.1918
5°	0.0872	0.9962	0.0875	51°	0.7771	0.6293	1.2349
6°	0.1045	0.9945	0.1051	52°	0.7880	0.6157	1.2799
7°	0.1219	0.9925	0.1228	53°	0.7986	0.6018	1.3270
8°	0.1392	0.9903	0.1405	54°	0.8090	0.5878	1.3764
9°	0.1564	0.9877	0.1584	55°	0.8192	0.5736	1.4281
10°	0.1736	0.9848	0.1763	56°	0.8290	0.5592	1.4826
11°	0.1908	0.9816	0.1944	57°	0.8387	0.5446	1.5399
12°	0.2079	0.9781	0.2126	58°	0.8480	0.5299	1.6003
13°	0.2250	0.9744	0.2309	59°	0.8572	0.5150	1.6643
14°	0.2419	0.9703	0.2493	60°	0.8660	0.5000	1.7321
15°	0.2588	0.9659	0.2679	61°	0.8746	0.4848	1.8040
16°	0.2756	0.9613	0.2867	62°	0.8829	0.4695	1.8807
17°	0.2924	0.9563	0.3057	63°	0.8910	0.4540	1.9626
18°	0.3090	0.9511	0.3249	64°	0.8988	0.4384	2.0503
19°	0.3256	0.9455	0.3443	65°	0.9063	0.4226	2.1445
20°	0.3420	0.9397	0.3640	66°	0.9135	0.4067	2.2460
21°	0.3584	0.9336	0.3839	67°	0.9205	0.3907	2.3559
22°	0.3746	0.9272	0.4040	68°	0.9272	0.3746	2.4751
23°	0.3907	0.9205	0.4245	69°	0.9336	0.3584	2,6051
24°	0.4067	0.9135	0.4452	70°	0.9397	0.3420	2.7475
25°	0.4426	0.9063	0.4663	71°	0.9455	0.3256	2.9042
26°	0.4384	0.8988	0.4877	72°	0.9511	0.3090	3.0777
27°	0.4540	0.8910	0.5095	73°	0.9563	0.2924	3.2709
28°	0.4695	0.8829	0.5317	74°	0.9613	0.2756	3.4874
29°	0.4848	0.8746	0.5543	75°	0.9659	0.2588	3.7321
30°	0.5000	0.8660	0.5774	76°	0.9703	0.2419	4.0108
31°	0.5150	0.8572	0.6009	77°	0.9744	0.2250	4.3315
32°	0.5299	0.8480	0.6249	78°	0.9781	0.2079	4.7046
33°	0.5446	0.8387	0.6494	79°	0.9816	0.1908	5.1446
34°	0.5592	0.8290	0.6745	80°	0.9848	0.1736	5.6713
35°	0.5736	0.8192	0.7002	81°	0.9877	0.1564	6.3138
36°	0.5878	0.8090	0.7265	82°	0.9903	0.1392	7.1154
37°	0.6018	0.7986	0.7536	83°	0.9925	0.1219	8.1443
38°	0.6157	0.7880	0.7813	84°	0.9945	0.1045	9.5144
39°	0.6293	0.7771	0.8098	85°	0.9962	0.0872	11.4300
40°	0.6428	0.7660	0.8391	86°	0.9976	0.0698	14.3010
41°	0.6561	0.7547	0.8693	87°	0.9986	0.0523	19.0810
42°	0.6691	0.7431	0.9004	88°	0.9994	0.0349	28.6360
43°	0.6820	0.7314	0.9325	89°	0.9998	0.0175	57.2900
44°	0.6947	0.7193	0.9657	90°	1.0000	0.0000	
45°	0.7071	0.7071	1.0000				

Problems

Find the number of degrees in each of the following angles:

1. $\tan \angle A = 0.3249$
2. $\sin \angle B = 0.6428$
3. $\cos \angle C = 0.1736$
4. $\cos \angle D = 0.9877$
5. $\sin \angle E = 0.5000$
6. $\tan \angle F = 1.0724$
7. $\tan \angle G = 0.5774$
8. $\cos \angle H = 0.7071$
9. $\sin \angle J = 0.8660$

FINDING THE ANGLE WHEN THE FUNCTION IS NOT IN THE TABLE

Example 32-5 Find $\angle A$ if $\tan \angle A = 0.5120$.

Solution The number 0.5120 is not in the table under the column "tan" but lies between 0.5095 (27°) and 0.5317 (28°). Choose the number closest to 0.5120.

$$\left.\begin{array}{l}\tan 27^\circ = 0.5095 \\ \tan \angle A = 0.5120 \\ \tan 28^\circ = 0.5317\end{array}\right\} \begin{array}{l}\text{difference} = 0.0025 \\ \text{difference} = 0.0197\end{array}$$

The smaller difference indicates the closer number. Therefore,

$\angle A = 27^\circ$ (nearest angle) *Ans.*

Self-Test 32-6 Find $\angle B$ if $\sin \angle B = 0.7699$.

Solution The number 0.7699 is not in the table under the column "sin" but lies between ______ (50°) and ______ (51°). Choose the number closest to 0.7699.

$$\left.\begin{array}{l}\sin 50^\circ = 0.7660 \\ \sin \angle B = 0.7699 \\ \sin 51^\circ = 0.7771\end{array}\right\} \begin{array}{l}\text{difference} = \_\_\_\_\_\_ \\ \text{difference} = \_\_\_\_\_\_\end{array}$$

The (smaller)(larger) difference indicates the closer number. Therefore,

$\angle B =$ ______° (nearest angle) *Ans.*

0.7660
0.7771

0.0039
0.0072

smaller

50

Problems

Find the number of degrees in each of the following angles correct to the nearest degree:

1. $\tan \angle A = 0.2700$
2. $\cos \angle B = 0.7500$
3. $\sin \angle C = 0.8500$
4. $\sin \angle D = 0.2350$
5. $\cos \angle E = 0.4172$
6. $\tan \angle F = 1.9120$
7. $\tan \angle G = 0.2783$
8. $\cos \angle H = 0.1645$
9. $\sin \angle J = 0.7250$

JOB 33 Finding the Acute Angles of a Right Triangle

In order to find the value of an angle in any problem, it is only necessary to find the number for the sine *or* the cosine *or* the tangent of the angle. If we know *any one* of these values, we can determine the angle by finding the number in the appropriate column of the table.

Example 33-1 Find $\angle A$ and $\angle B$ in the triangle of Fig. 33-1.

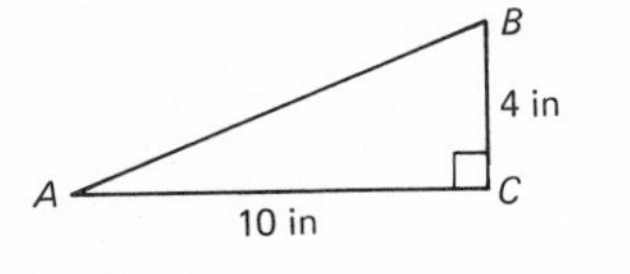

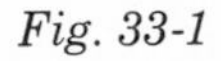
Fig. 33-1

Solution

1. Name the sides which have values, using $\angle A$ as the reference angle.

4 in = the side *opposite* $\angle A$

10 in = the side *adjacent* $\angle A$

2. Choose the trigonometric formula which uses these *opposite* and *adjacent* sides. The only formula that uses these two sides is the *tangent* formula.

$$\tan \angle A = \frac{o}{a} = \frac{4}{10} = 0.4000$$

3. Find the number in the tangent table closest to 0.4000.

$\angle A = 22°$ (nearest angle) *Ans.*

4. The two acute angles of any right triangle always total 90°, so we can use the following formula:

FORMULA $\angle B = 90° - \angle A$

$\angle B = 90° - 22° = 68°$ *Ans.*

Example 33-2 The phase angle in an ac electrical circuit may be represented by $\angle \theta$ (angle theta), as shown in Fig. 33-2. Find $\angle \theta$ and $\angle B$.

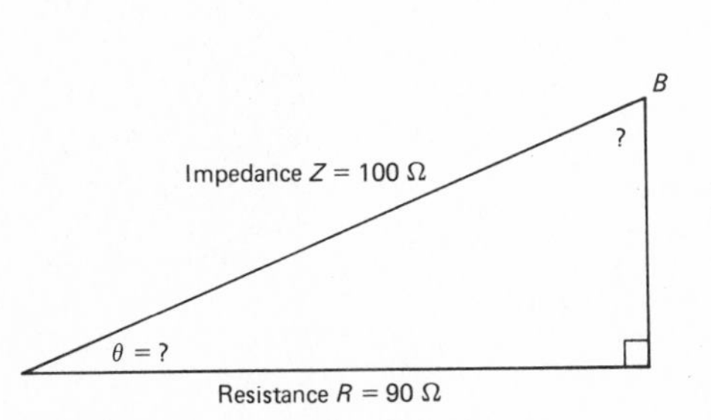

Fig. 33-2 The angle theta (θ) is the phase angle in an inductive ac circuit.

Solution

1. Name the sides which have values, using $\angle \theta$ as the reference angle.

100 Ω = hypotenuse

90 Ω = side adjacent $\angle \theta$

2. Choose the trigonometric formula which uses the *adjacent* side and the *hypotenuse.* The only formula that uses these two sides is the *cosine* formula.

$$\cos \angle \theta = \frac{a}{h} = \frac{90}{100} = 0.9000$$

3. Find the number in the cosine table closest to 0.9000.

$\angle \theta = 25°$ (nearest angle) *Ans.*

4. Since the two acute angles of a right triangle always total 90°,

$\angle B = 90° - \angle \theta$

$= 90° - 25° = 65°$ *Ans.*

Self-Test 33-3 An electrician must bend a pipe to make a 3.5-dm rise in a 5-dm horizontal distance. What is the angle at each bend?

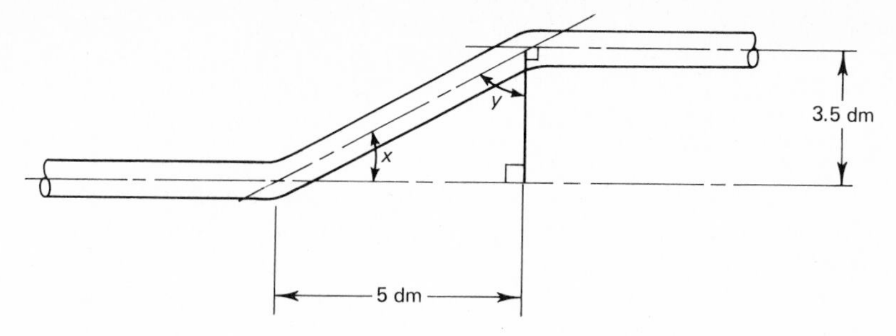

Fig. 33-3

Solution The diagram describing the conditions is shown in Fig. 33-3.

1. Name the sides which have values, using $\angle x$ as the reference angle.

   The 3.5-dm side is the side _______ $\angle x$. — opposite

   The 5-dm side is the side _______ $\angle x$. — adjacent

2. Choose the trigonometric formula which uses these sides for which we have values—the opposite side and the adjacent side. The only formula that uses these two sides is the _______ formula. — tangent

   $$\tan x = \frac{?}{?}$$ — $\frac{o}{a}$

   $$= \frac{3.5}{5} = \_\_\_\_\_\_$$ — 0.7000

3. Find the number in the _______ table that is closest to 0.7000. — tangent

   $\angle x =$ _______ *Ans.* — 35°

4. Since the two acute angles of a right triangle always total _______°, — 90

   $\angle y = 90° -$ _______ — $\angle x$

   $= 90° - 35° =$ _______ — 55°

5. The angle at the top bend $= \angle y +$ _______ — 90°

   $= 55° + 90°$

   $=$ _______° *Ans.* — 145

Problems

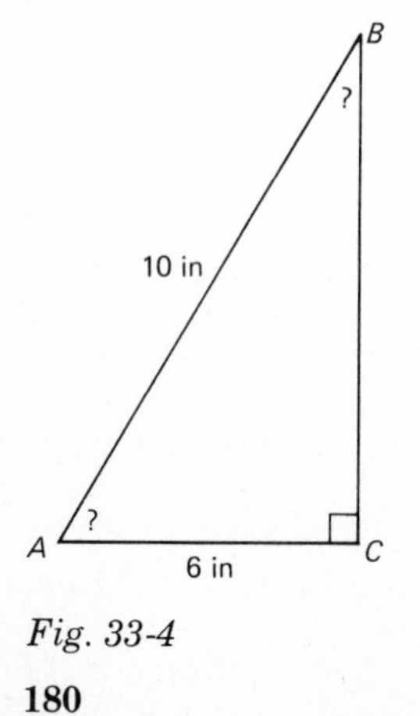

Fig. 33-4

1. Find $\angle A$ and $\angle B$ in the right triangle shown in Fig. 33-4.

Using a triangle similar to Fig. 33-4, find $\angle A$ and $\angle B$ in each of the following problems if:

2. AC = 100 ft and BC = 70 ft.
3. BC = 4 in and AB = 8 in.
4. BC = 40 Ω and AC = 25 Ω.
5. AC = 200 Ω and AB = 350 Ω.
6. BC = 300 W and AB = 1,000 W.
7. AC = 600 Ω and AB = 960 Ω.
8. BC = 7.5 V and AC = 12.5 V.
9. BC = 12.4 A and AB = 67.8 A.

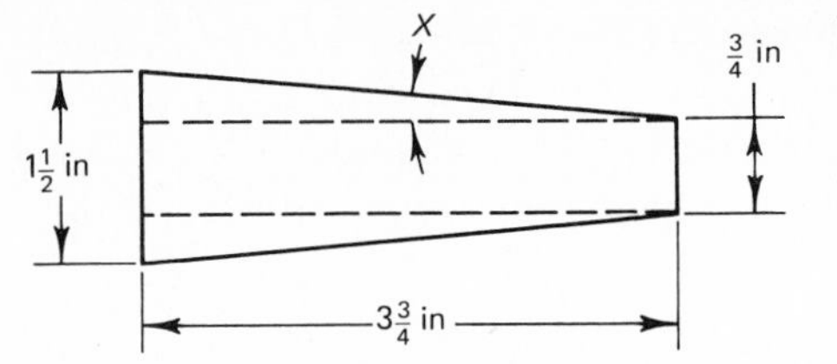

Fig. 33-5 Angle x is equal to one-half of the taper angle.

10. $AC = 62.5\ \Omega$ and $BC = 100\ \Omega$.
11. (*C, E*) At what angles must a pipe be bent in order to make a 7-dm rise in a 3-dm horizontal distance?
12. (*C, E*) At what angles must a pipe be bent in order to make a rise of 4 ft 6 in in a 2-ft 3-in horizontal distance?
13. (*C, G*) A car rises 50 ft while traveling along a road 1,000 ft long. At what angle is the road inclined to the horizontal?
14. (*M*) Find angle X in the taper shown in Fig. 33-5.
15. (*C*) A guy wire 120 ft long reaches from the top of a pole to a point 64 ft from the foot of the pole. What angle does the wire make with the ground?
16. (*C*) In Fig. 33-6, what angle does a rafter make with the horizontal if it has a rise of 7 ft in a run of 15 ft?

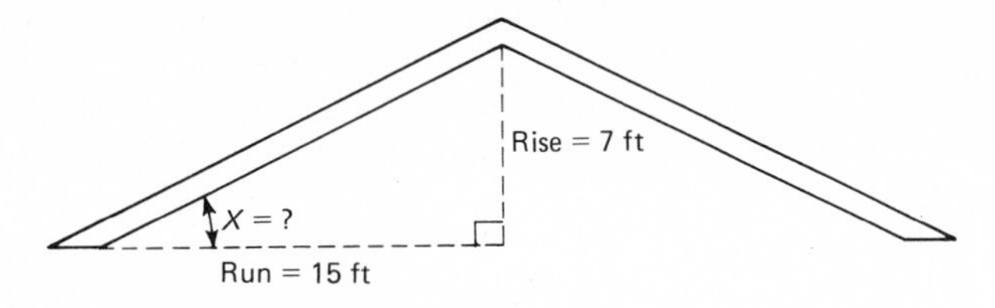

Fig. 33-6

17. (*C*) In Fig. 33-7, the stairway has a tread of 8 in and a rise of 5 in. Find the angle X that the stair stringer makes with the horizontal.

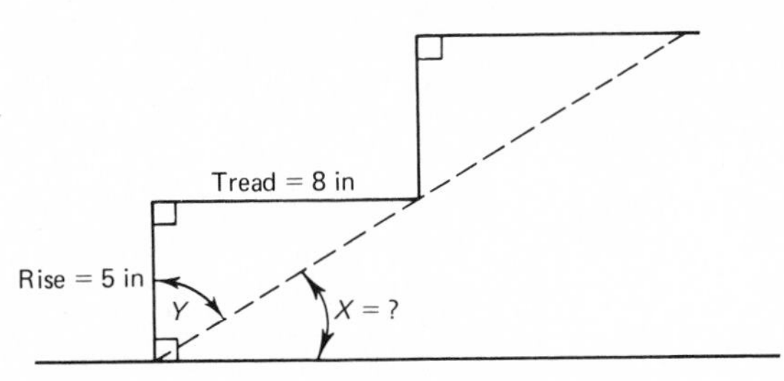

Fig. 33-7 Be careful! Angle X is not in the triangle.

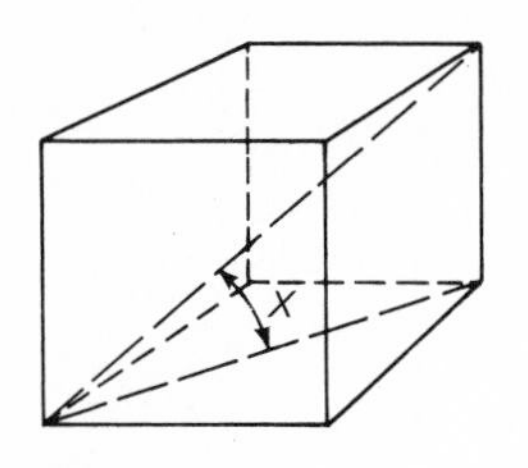

Fig. 33-8

18. (*G*) In Fig. 33-8, $\angle X$ is formed by the diagonal of a cube and a diagonal of the bottom face. Find $\angle X$.
19. (*M*) Find $\angle X$ in the lathe center shown in Fig. 33-9.
20. (*M*) Find $\angle X$ in the template shown in Fig. 33-10.

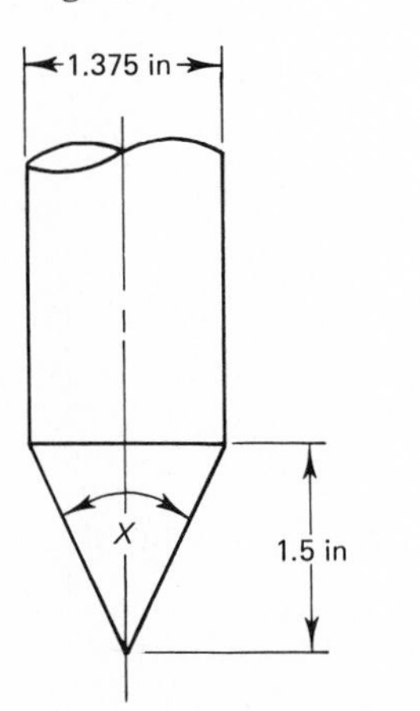

Fig. 33-9 Find the taper angle of the lathe center.

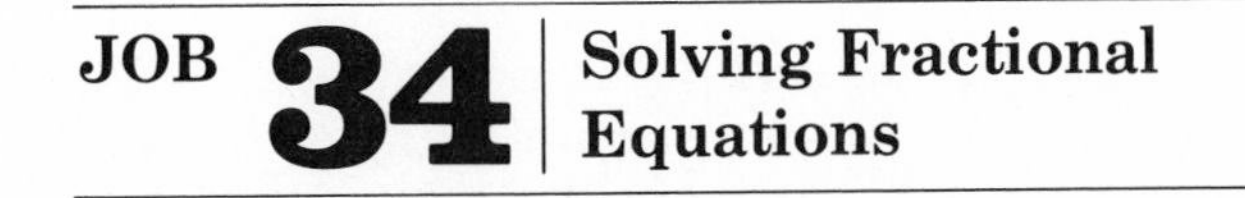

JOB 34 Solving Fractional Equations

We are now ready to find the missing sides in a right triangle. However, the equations we shall encounter require that we understand how to solve fractional equations. As soon as we have mastered this operation, we shall continue with trigonometry.

Consider the equation $\frac{E}{4} = 5$. To find the value of E, we must get E all by itself on one side of the equality sign. In other words, we must eliminate the number 4. We can do this by applying the general rule which says that we may do *anything* to one side of an equality sign provided we

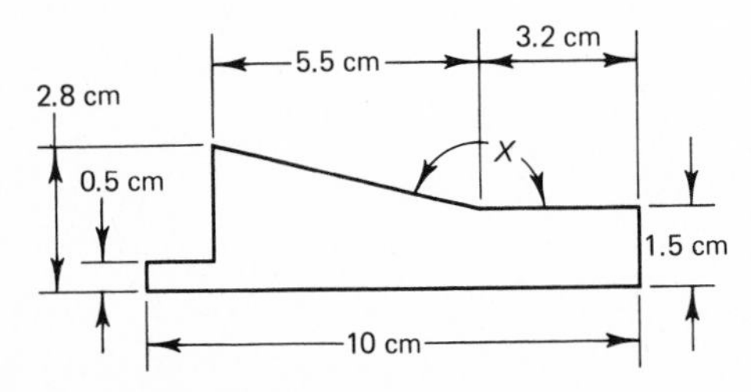

Fig. 33-10

do the *same thing* to the other side. In general, the operation to be performed on both sides of the equality sign will be the *opposite* of the operation used in the equation. Since our equation involves division, the *opposite* operation of *multiplication* should be performed on both sides of the equality sign.

RULE 1 **Both sides of an equality sign may be multiplied by the same number without destroying the equality.**

Example 34-1 Find the value of E in the equation $\frac{E}{4} = 5$.

Solution Solving an equation means getting the letter all alone on one side of the equality sign with its value on the other side. We can get E alone on the left side of the equality sign if we can get rid of the denominator 4. Since the equation says E is *divided* by 4, we can eliminate the 4 by doing the *opposite* operation, *multiplying* both sides of the equality sign by that same 4.

1. Write the equation.

$$\frac{E}{4} = 5$$

2. Multiply both sides by 4.

$$\overset{1}{\cancel{4}} \times \frac{E}{\underset{1}{\cancel{4}}} = 5 \times 4$$

3. Multiply each side separately.

$1 \times E = 20$ or $E = 20$ *Ans.*

Example 34-2 Using the formula $I = \frac{E}{R}$, find E if $I = 2$ and $R = 6$.

Solution

1. Write the formula.

$$I = \frac{E}{R}$$

2. Substitute numbers.

$$2 = \frac{E}{6}$$

Step 2 asks the question, "What number *divided* by 6 gives 2 as an answer"? We can find E if we can eliminate the denominator 6. This may easily be done by applying the *opposite* operation of *multiplying* both sides of the equality sign by that same number 6.

3. Multiply both sides by 6.

$$6 \times 2 = \frac{E}{\underset{1}{\cancel{6}}} \times \overset{1}{\cancel{6}}$$

4. Multiply each side separately.

$12 = E \times 1$ or $E = 12$ *Ans.*

Example 34-3 Using the formula $\frac{kW}{0.75}$ = hp, find the number of kilowatts that are equivalent to 4 hp.

Solution 1. Write the formula.

$$\frac{kW}{0.75} = hp$$

2. Substitute numbers.

$$\frac{kW}{0.75} = 4$$

3. Multiply both sides by 0.75.

$$\overset{1}{\cancel{0.75}} \times \frac{kW}{\underset{1}{\cancel{0.75}}} = 4 \times 0.75$$

4. Multiply each side separately.

kW = 3 *Ans.*

Notice that in each of the last three examples we eliminated the number in the denominator by *multiplying both sides* by that *same number.* If we were to multiply both sides by any other number, we would still have a number in the denominator and the letter would *not* stand alone. Therefore, to eliminate a number in the denominator of a fraction, we use the following rule:

RULE 2 **To eliminate a number divided into a letter, multiply both sides of the equality sign by that same number.**

Problems

Find the value of the unknown letter in each problem.

1. $\frac{E}{3} = 8$
2. $4 = \frac{E}{10}$
3. $\frac{P}{6} = 2$
4. $0.5 = \frac{E}{3}$
5. $\frac{M}{0.2} = 5$
6. $\frac{A}{10} = 0.35$
7. $\frac{E}{3} = \frac{2}{3}$
8. $0.4 = \frac{P}{0.8}$
9. $\frac{P}{\frac{1}{2}} = 16$
10. $\frac{E}{4} = 2\frac{1}{2}$
11. $150 = \frac{x}{45}$
12. $117 = \frac{P}{4.7}$
13. (*E*) In the formula $I = \frac{E}{R}$, find E if $I = 3$ and $R = 18$.
14. (*E*) In the formula $E = \frac{P}{I}$, find P if $E = 110$ and $I = 5$.
15. (*A, E, C*) In the formula Efficiency $= \frac{O}{I}$, find O if $I = 36$ and efficiency = 0.85.
16. (*E*) In the formula $Z = \frac{E}{I}$, find E if $Z = 2{,}000$ and $I = 0.015$.
17. (*M*) In the formula $PD = \frac{N}{P}$, find N if $PD = 4$ and $P = 12$.
18. (*M, A*) In the formula $TPI = \frac{TPF}{12}$, find TPF if $TPI = 0.052$.

19. (*C, G*) In the formula $L = \frac{A}{W}$, find A if $L = 8.2$ and $W = 3.45$.

20. (*C*) In the formula $S = \frac{P}{A}$, find P if $S = 15{,}000$ and $A = 0.125$.

USING THE LEAST COMMON DENOMINATOR TO SOLVE FRACTIONAL EQUATIONS

Example 34-4 Find E in the equation $\frac{2E}{9} = \frac{4}{3}$.

Solution When fractions appear on both sides of the equality sign, the problem will be easier to solve if we eliminate *all* denominators first. We could do this one denominator at a time by the method used in Examples 34-1 to 34-3, but it is faster to eliminate both denominators at the same time. In order to eliminate *both* the 9 and the 3 at the same time, we must multiply both sides of the equality sign by some number that will make *both* denominators cancel out. This must be a number into which *both* denominators will evenly divide—the *least common denominator.* In this problem, the LCD of 9 and 3 is 9.

1. Write the equation.

$$\frac{2E}{9} = \frac{4}{3}$$

2. Multiply both sides by the LCD (9).

$$\overset{1}{\cancel{9}} \times \frac{2E}{\underset{1}{\cancel{9}}} = \frac{4}{\underset{1}{\cancel{3}}} \times \overset{3}{\cancel{9}}$$

3. Multiply each side separately.

$$2E = 12$$

4. Solve for E.

$$E = \frac{12}{2}$$

5. Divide.

$$E = 6 \qquad \textit{Ans.}$$

Example 34-5 Find R in the equation $\frac{2R}{3} = \frac{9}{4}$

Solution The LCD for 3 and 4 is 12. That is, both 3 and 4 will divide evenly into 12.

1. Write the equation.

$$\frac{2R}{3} = \frac{9}{4}$$

2. Multiply both sides by the LCD (12).

$$\overset{4}{\cancel{12}} \times \frac{2R}{\underset{1}{\cancel{3}}} = \frac{9}{\underset{1}{\cancel{4}}} \times \overset{3}{\cancel{12}}$$

3. Multiply each side separately.

$8R = 27$

4. Solve for R.

$$R = \frac{27}{8}$$

5. Divide.

$$R = 3\frac{3}{8} \qquad \textit{Ans.}$$

Self-Test 34-6 $\frac{R}{K} = \frac{L}{A}$ is a formula that may be used to calculate the length of a particular material and cross section that will provide a desired resistance. Find the length L if $K = 60$ for pure iron wire, $A = 25$ circular mils (cmil) of area, and $R = 24\ \Omega$.

Solution

1. Write the formula.

$$\frac{R}{K} = \frac{L}{A}$$

2. Substitute numbers.

$$\frac{24}{?} = \frac{L}{?}$$ 60, 25

3. The LCD for the denominators 60 and 25 is ______. 300

4. Multiply both sides by ______. 300

$$\overset{5}{\cancel{300}} \times \frac{24}{\underset{1}{\cancel{60}}} = \frac{L}{\underset{1}{\cancel{25}}} \times \overset{12}{\cancel{300}}$$

5. Multiply each side separately.

______ $= 12L$ 120

6. Solve for L.

$$\frac{120}{?} = L$$ 12

7. Divide.

$L =$ ______ ft *Ans.* 10

Self-Test 34-7 Find L in the equation $\frac{6}{L} = \frac{2}{3}$.

Solution

1. Write the equation:

$$\frac{6}{L} = \frac{2}{3}$$

2. The LCD is the number into which both denominators will divide ______. In general, although it may not be the *least* common denominator, a workable common denominator will be the value obtained by multiplying all the denominators. Thus, in this problem, the LCD is ______. evenly 3L

3. Multiply both sides by the LCD ($3L$).

$$\overset{1}{3\cancel{L}} \times \frac{6}{\underset{1}{\cancel{L}}} = \frac{2}{\underset{1}{\cancel{3}}} \times \overset{1}{\cancel{3}L}$$

Note that L cancels L on the left side, and 3 cancels 3 on the right side.

4. Multiply each side separately.

$\_\_\_\_\_\_ = 2L$

5. Solve for L.

$$\frac{18}{?} = L$$

6. Divide.

$L =$ ______ *Ans.*

18

2

9

RULE 3 **When fractions appear on both sides of the equality sign, eliminate the denominators by multiplying both sides by the least common denominator.**

Problems

Solve each of the following equations for the value of the unknown letter:

1. $\frac{R}{3} = \frac{2}{3}$	6. $\frac{4}{5} = \frac{T}{2}$	11. $\frac{0.3}{E} = \frac{1}{4}$	16. $\frac{100}{250} = \frac{0.1}{R}$
2. $\frac{E}{8} = \frac{3}{4}$	7. $\frac{2E}{5} = \frac{3}{0.5}$	12. $\frac{3}{5} = \frac{2}{R}$	17. $\frac{N}{0.4} = \frac{3.9}{0.2}$
3. $\frac{2}{3} = \frac{P}{6}$	8. $\frac{2B}{5} = \frac{7}{10}$	13. $\frac{2}{L} = \frac{6}{30}$	18. $\frac{9}{L} = \frac{0.3}{8.9}$
4. $\frac{2E}{3} = \frac{4}{9}$	9. $\frac{N}{20} = \frac{3}{4}$	14. $\frac{10}{C} = \frac{150}{200}$	19. $\frac{21}{0.3} = \frac{2E}{9}$
5. $\frac{M}{5} = \frac{2}{3}$	10. $\frac{4}{7} = \frac{S}{2}$	15. $\frac{22}{7} = \frac{11}{R}$	20. $\frac{37}{250} = \frac{74}{R}$

21. (*A, M, C*) $\frac{D}{d} = \frac{s}{S}$ is a formula used to calculate the speeds of pulleys. In Fig. 34-1, find s if $D = 12$ in, $d = 6$ in, and $S = 100$ rpm.

22. (*A, M, C*) $\frac{N}{n} = \frac{s}{S}$ is a formula used to calculate the speeds of gears. In Fig. 34-2, find s if $N = 32$ teeth, $n = 16$ teeth, and $S = 96$ rpm.

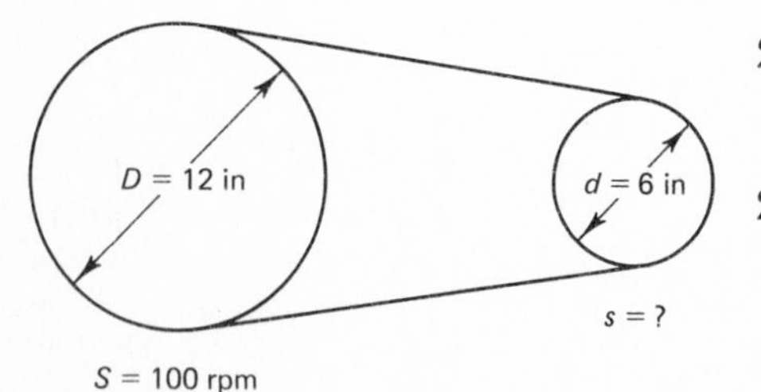

Fig. 34-1

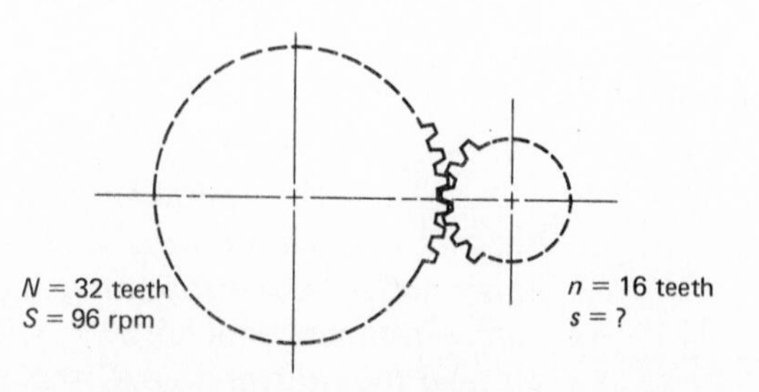

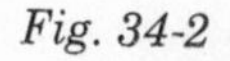

Fig. 34-2

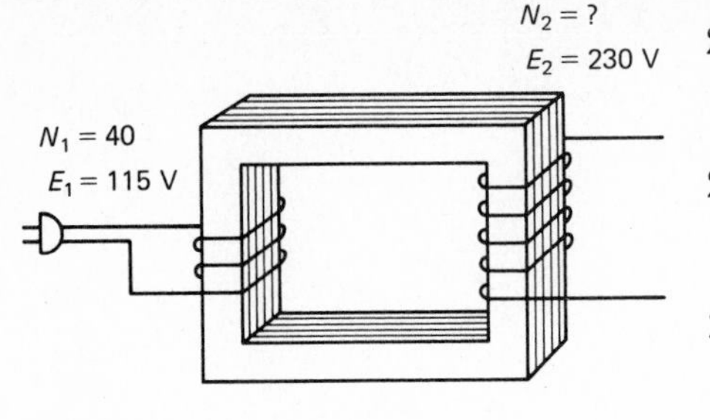

Fig. 34-3

23. (*C*) $\frac{P}{W} = \frac{d}{D}$ is a formula used to calculate the forces in levers. Find W if $P = 100$, $d = 15$, and $D = 75$.
24. (*E*) $\frac{N_2}{N_1} = \frac{E_2}{E_1}$ is a formula used in electrical transformer calculations. In Fig. 34-3, find N_2 if $N_1 = 40$ turns, $E_1 = 115$ V, and $E_2 = 230$ V.
25. (*E*) $\frac{A_2}{A_1} = \frac{R_1}{R_2}$ is a formula used to calculate the sizes of wires in electrical installations. Find A_2 if $A_1 = 100$, $R_1 = 1{,}000$, and $R_2 = 3{,}000$.

CROSS MULTIPLICATION

When a fractional equation contains *only two fractions* equal to each other, the equation may be solved by a very simple method known as *cross multiplication.* This method automatically multiplies both sides of the equality sign by the LCD and therefore eliminates one step in the solution. This method is most useful when the unknown letter is in the denominator of one of the fractions.

RULE 4 **In cross multiplication, the product of the numerator of the first fraction and the denominator of the second is set equal to the product of the numerator of the second fraction and the denominator of the first.**

It is easier to understand this rule by putting it in picture form. The multiplication of the numbers along one diagonal is equal to the multiplication of the numbers along the other diagonal.

$$\frac{2}{3} = \frac{4}{6} \qquad \frac{A}{B} = \frac{C}{D}$$

$$2 \times 6 = 4 \times 3 \qquad A \times D = C \times B$$

or   or

$$3 \times 4 = 6 \times 2 \qquad B \times C = D \times A$$

Example 34-8 Find the value of R in the equation $\frac{3}{R} = \frac{2}{5}$.

Solution

1. Write the equation.

$$\frac{3}{R} = \frac{2}{5}$$

2. Cross-multiply.

$$2 \times R = 3 \times 5$$

3. Simplify each side separately.

$$2R = 15$$

4. Solve for R.

$$R = \frac{15}{2}$$

5. Divide.

$R = 7\frac{1}{2}$ *Ans.*

Example 34-9 Using the formula $I = \frac{E}{R}$, find the voltage E needed to operate the portable sander shown in Fig. 34-4 if the current I = 4 A and the resistance R = 30 Ω.

Solution

1. Write the formula.

$$I = \frac{E}{R}$$

2. Substitute numbers for letters.

$$4 = \frac{E}{30}$$

3. Complete all fractions by placing all whole numbers over 1.

$$\frac{4}{1} = \frac{E}{30}$$

4. Cross-multiply.

$$1 \times E = 4 \times 30$$

5. Simplify each side separately.

$E = 120$ V *Ans.*

Fig. 34-4 A dustless belt sander. (Courtesy Rockwell International)

Example 34-10 $\frac{S}{s} = \frac{d}{D}$ is a formula used to calculate the speeds of pulleys. In Fig. 34-5, find the speed s of the small pulley if the speed S of the large pulley is 120 rpm, the small pulley d = 3 in, and the large pulley D = 10 in.

Fig. 34-5 A pivot-type motor mount makes belt changing easy on this drill press pulley system. (Courtesy Rockwell International)

Solution

1. Write the formula.

$$\frac{S}{s} = \frac{d}{D}$$

2. Substitute numbers for letters.

$$\frac{120}{s} = \frac{3}{10}$$

3. Cross-multiply.

$$3 \times s = 120 \times 10$$

4. Simplify each side separately.

$3s = 1{,}200$

5. Solve for s.

$$s = \frac{1{,}200}{3}$$

6. Divide.

$s = 400$ rpm *Ans.*

Problems

Solve each of the following equations for the value of the unknown letter:

1. $\frac{E}{10} = \frac{3}{5}$
2. $\frac{8}{3} = \frac{R}{6}$
3. $\frac{60}{A} = \frac{3}{4}$
4. $\frac{9}{E} = \frac{15}{40}$
5. $\frac{84}{E} = \frac{28}{17}$
6. $\frac{84}{28} = \frac{66}{R}$
7. $\frac{E}{18} = \frac{3}{2}$
8. $9 = \frac{54}{R}$
9. $\frac{50}{T} = 2$
10. $\frac{2R}{3} = 10$
11. $\frac{24}{2} = \frac{6M}{5}$
12. $\frac{3P}{0.4} = 6$
13. $\frac{3.6}{5} = \frac{A}{2}$
14. $0.88 = \frac{R}{3}$
15. $\frac{3}{12} = \frac{4}{3T}$
16. $5 = \frac{1}{0.2E}$
17. $\frac{1}{R} = 10$
18. $80 = \frac{1}{R}$
19. $\frac{1}{R} = 13$
20. $\frac{10}{R} = 50$
21. (*E*) $\frac{I_1}{I_2} = \frac{R_2}{R_1}$ is a formula used for calculating the way current divides in a parallel circuit. Find I_1 if $I_2 = 2$ A, $R_2 = 100\ \Omega$, and $R_1 = 25\ \Omega$.
22. (*A, C*) $\frac{E}{R} = \frac{A_1}{A_2}$ is a formula used to determine the effort required to raise a large weight in a hydraulic lift. Find the effort E required to raise a weight R of 200,000 lb if the area of the small piston A_1 is 4 sq in and the area of the large piston A_2 is 1,600 sq in.
23. (*E*) $\frac{R_1}{R_2} = \frac{L_2}{L_1}$ is a formula used to locate the position of a break in an underground cable. Find the length to the break in the cable L_2 if the length of the unbroken cable L_1 is 4,000 ft, R_1 is 10 Ω, and R_2 is 250 Ω.
24. (*E*) $\frac{R_s}{R_m} = \frac{I_m}{I_s}$ is the formula for finding the value of the shunt resistor needed to extend the range of an ammeter. Find R_s if $I_m = 0.001$, $R_m = 50$, and $I_s = 0.049$.
25. (*E*) $\frac{R_1}{R_2} = \frac{C_x}{C_1}$ is a formula used to measure the capacitance of an unknown capacitor. Find C_x if $R_1 = 100\ \Omega$, $R_2 = 425\ \Omega$, and $C_1 = 0.5\ \mu$F.

JOB 35 Finding the Sides of a Right Triangle

Example 35-1 Find (a) side BC and (b) $\angle B$ if the right triangle shown in Fig. 35-1.

Solution **a.** Find the side BC.

1. Select the angle to be used ($\angle A = 30°$),

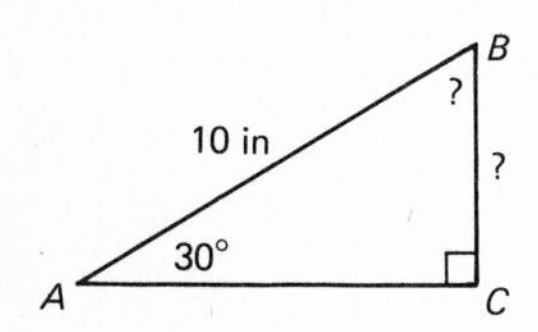

Fig. 35-1

2. Select the side to be found (BC).

3. Select one other side whose value is known ($AB = 10$ in).

4. Name these two sides.

BC = side *opposite* $\angle A$

AB = *hypotenuse*

5. Select the correct trigonometric formula. It will be the formula that uses these two sides—the *opposite* and the *hypotenuse*. Only the *sine* formula uses the *opposite* side and the *hypotenuse*.

$$\sin \angle A = \frac{o}{h} = \frac{BC}{AB}$$

$$\sin 30° = \frac{BC}{10}$$

From Table 32-1, sin 30° = 0.5000. Therefore,

$$0.5000 = \frac{BC}{10}$$

$$\frac{0.5000}{1} = \frac{BC}{10}$$

Cross-multiplying,

$BC = 0.5000 \times 10 = 5$ in *Ans.*

b. Find $\angle B$.

$\angle B = 90° - \angle A$

$\angle B = 90° - 30° = 60°$ *Ans.*

Example 35-2 Find side BC in the right triangle shown in Fig. 35-2.

Solution **1.** Select the angle to be used ($\angle A = 15°$).

2. Select the side to be found (BC).

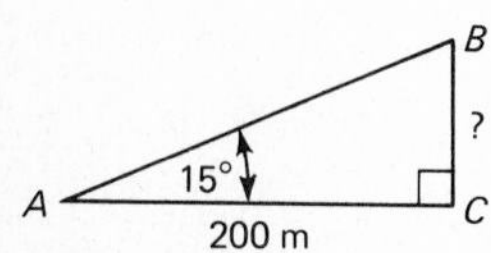

Fig. 35-2

3. Select one other side whose value is known ($AC = 200$ m).

4. Name these two sides.

BC = side *opposite* $\angle A$

AC = side *adjacent* $\angle A$

5. Select the correct trigonometric formula. It will be the formula that uses these two sides—the *opposite* and the *adjacent*. Only the tangent formula uses the *opposite* side and the *adjacent* side.

$$\tan A = \frac{o}{a} = \frac{BC}{AC}$$

$$\tan 15° = \frac{BC}{200}$$

From Table 32-1, tan 15° = 0.2679. Therefore,

$$\frac{0.2679}{1} = \frac{BC}{200}$$

Cross-multiplying,

$BC = 0.2679 \times 200 = 53.58$ m *Ans.*

Self-Test 35-3 The relationship of the impedance Z, the resistance R, and the capacitive reactance X_C of an ac circuit is shown in Fig. 35-3. Find (a) the impedance and (b) the reactance.

Solution **a.** Find the impedance. The impedance Z is represented by the side AB.

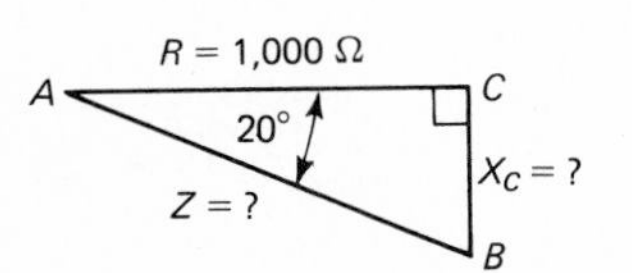

Fig. 35-3 The resistance (R), reactance, (X_c), and impedance (Z) of an ac circuit may be described in terms of the sides of a right triangle.

1. The angle to be used = ______.	20°
2. The side to be found = ______.	AB
3. Another side whose value is known is ______.	$AC = 1{,}000\ \Omega$
4. Name these two sides, using $\angle A = 20°$ as the reference angle.	
AC = the side ______ $\angle A$	adjacent
AB = the ______	hypotenuse
5. Select the correct trigonometric formula that uses these two sides. The adjacent side and the hypotenuse are connected by the ______ formula.	cosine
$\cos A = \frac{a}{h} = \frac{AC}{AB}$	
$\cos 20° = \frac{?}{AB}$	1,000
From Table 32-1, cos 20° = ______. Therefore,	0.9397
$\frac{0.9397}{1} = \frac{1{,}000}{AB}$	
Cross-multiplying,	
$AB \times$ ______ = 1,000	0.9397
Solving,	
$AB = \frac{1{,}000}{?}$	0.9397
= ______	1,064
Impedance Z = ______ *Ans.*	1,064 Ω

b. Find the reactance. The reactance X_C is represented by the side BC.

1. The angle to be used = ______.	20°
2. The side to be found = ______.	BC
3. Another side whose value is known is ______.	AC = 1,000
4. Name these two sides, using $\angle A = 20°$ as the reference angle.	
AC = the side ______ $\angle A$	adjacent
BC = the side ______ $\angle A$	opposite
5. Select the correct trigonometric formula that uses these two sides. The adjacent side and the opposite side are connected by the ______ formula.	tangent
$\tan A = \frac{o}{a} = \frac{BC}{AC}$	
$\tan 20° = \frac{BC}{?}$	1,000
From Table 32-1, tan 20° = ______. Therefore,	0.3640
$\frac{0.3640}{1} = \frac{BC}{?}$	1,000
Cross-multiplying,	
$BC = 0.3640 \times$ ______	1,000
= ______	364
Reactance X_C = ______ *Ans.*	364 Ω

Problems

Use a triangle similar to that shown in Fig. 35-1 for the following problems:

1. (*G*) Find AC and $\angle B$ if AB = 20 in and $\angle A$ = 60°.
2. (*G*) Find BC and $\angle B$ if AB = 26 cm and $\angle A$ = 40°.
3. (*G*) Find AC and $\angle A$ if BC = 75 ft and $\angle B$ = 50°.
4. (*G*) Find AC and $\angle B$ if AB = 400 W and $\angle A$ = 28°.
5. (*G*) Find AB and $\angle B$ if BC = 75.5 Ω and $\angle A$ = 30°.
6. (*G*) Find AB and $\angle B$ if AC = 30 Ω and $\angle A$ = 25°.
7. (*G*) Find BC and $\angle A$ if AC = 500 m and $\angle B$ = 28°.
8. (*G*) Find AC and $\angle A$ if AB = 36.8 in and $\angle B$ = 35°.
9. (*G*) Find BC and $\angle B$ if AB = 475 Ω and $\angle A$ = 15°.
10. (*G*) Find AC and $\angle A$ if AB = 92.8 V and $\angle B$ = 15°.

11. (*C, E*) A guy wire reaches from the top of an antenna pole to a point 30 ft from the foot of the pole and makes an angle of 70° with the ground. How long is the wire? How tall is the pole?
12. (*M*) Use Fig. 33-5. If the small diameter equals 3 in, the length equals 12 in, and angle X equals 4°, find the large diameter.
13. (*G*) Four holes are evenly spaced around a 5-in-diameter circle. Find the distance between the centers of the holes.

14. (*M*) How long is each side of the largest square bar that can be made from a piece of 6-cm-diameter round stock?
15. (*C*) A ladder 30 ft long leans against the side of a building and makes an angle of 65° with the ground. How high up the building does it reach?
16. (*G*) A sled is being pulled with a force of 50 lb. This force is exerted through a rope held at an angle of 20° with the ground, as shown in Fig. 35-4. How much of this force is useful in moving the sled horizontally? *Hint:* find F_x.

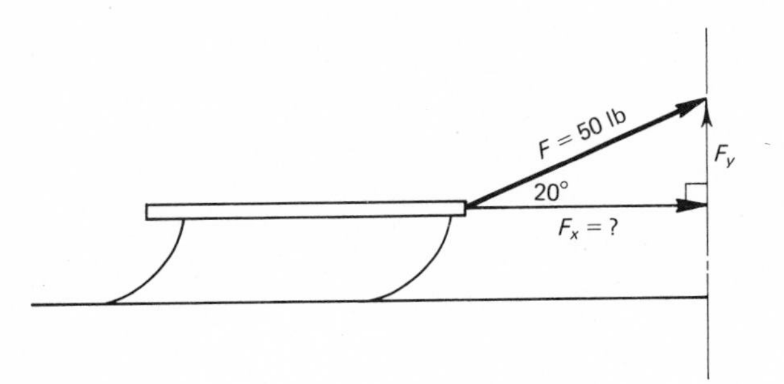

Fig. 35-4 A force may be resolved into its component parts, which are at right angles to each other.

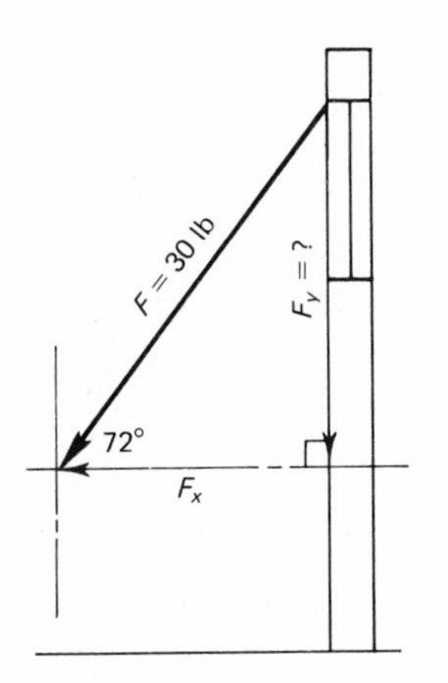

Fig. 35-5 F_y is the vertical component that is useful in opening the window.

17. (*G*) A window pole is used to pull down a window (Fig. 35-5). If the force exerted through the pole is 30 lb at an angle of 72° with the horizontal, find the useful vertical component (F_y).
18. (*G*) At a point 40 ft from the foot of a tree, the line of sight to the top of the tree makes an angle of 60° with the ground. Find the height of the tree to the nearest foot.
19. (*E*) In an impedance triangle similar to that shown in Fig. 35-3, find the reactance X_C if $R = 2{,}000\ \Omega$ and angle $A = 20°$.
20. (*E*) Find the impedance Z in the triangle used for Prob. 19.
21. (*M*) As shown in Fig. 35-6, the taper angle is 6°. Find (a) the distance x, and (b) the taper per foot. *Hint:* the taper per foot means the decrease in the diameter for each foot of length.
22. (*M*) In Fig. 35-7, find the pitch P.

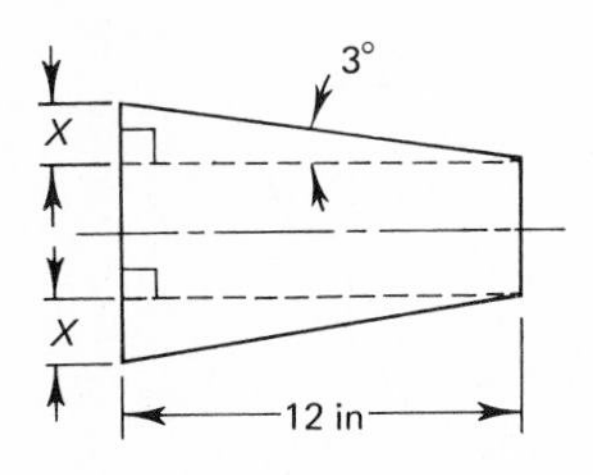

Fig. 35-6

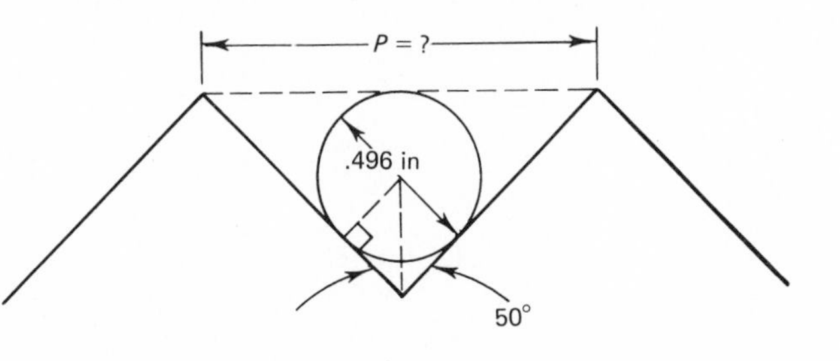

Fig. 35-7

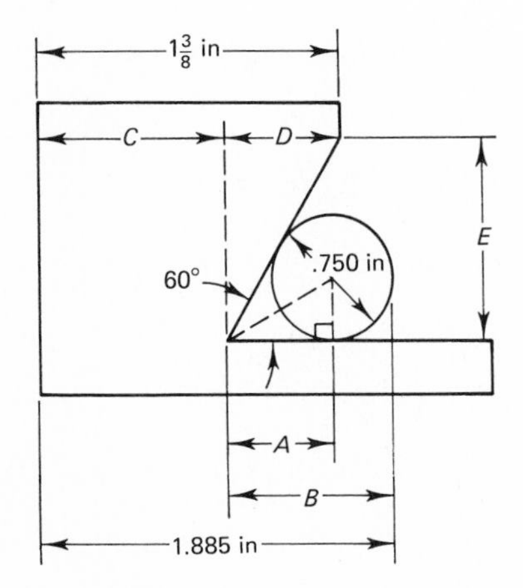

Fig. 35-8

23. (*M*) In Fig. 35-8, find dimensions A, B, C, D, and E.
24. (*C, E*) A pipe is to be offset 9 in with 45° elbows, as shown in Fig. 35-9. Find the center-to-center length AB.
25. (*C*) What is the run of each step of a stairway that has 20 risers, a total rise of 12 ft 3 in, and a slope of 37°?
26. (*C*) The grade of a highway is the ratio of the vertical distance to the horizontal distance. If the grade of a highway is 0.14, find (a) the angle that the road makes with the horizontal, and (b) the length of roadway needed to raise the road a vertical distance of 250 m.

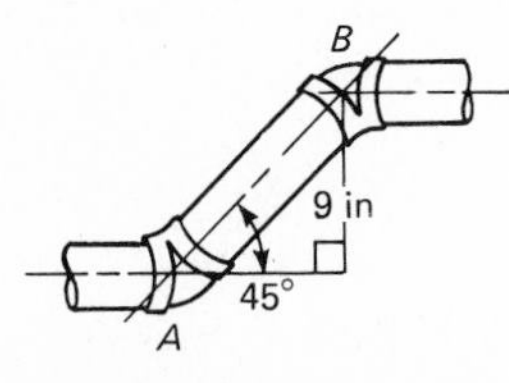

Fig. 35-9

JOB 36 Review of Trigonometry

FORMULAS

1. Tangent of an angle $= \dfrac{\text{opposite side}}{\text{adjacent side}}$.

   $\tan \angle = \dfrac{o}{a}$

2. Sine of an angle $= \dfrac{\text{opposite side}}{\text{hypotenuse}}$.

   $\sin \angle = \dfrac{o}{h}$

3. Cosine of an angle $= \dfrac{\text{adjacent side}}{\text{hypotenuse}}$.

   $\cos \angle = \dfrac{a}{h}$

4. The sum of the two acute angles of a right triangle equals 90°.

   $\angle A + \angle B = 90°$

PROCEDURE FOR FINDING ANGLES IN A RIGHT TRIANGLE

1. Name the sides which have values, using the angle to be found as the reference angle.
2. Choose the trigonometric formula which uses these sides.
3. Substitute values and divide.
4. Find the number in the table closest to this quotient. Be sure to look in the column indicated in step 2.
5. Find the angle in the angle column corresponding to the number.
6. Use Formula 4 to find the other acute angle.

PROCEDURE FOR FINDING SIDES IN A RIGHT TRIANGLE

1. Select the angle to be used.
2. Select the side to be found.
3. Select one other side whose value is known.
4. Name these two sides.
5. Select the trigonometric formula that uses these two sides.
6. Substitute values, using Table 32-1.
7. Solve the equation.

Problems

Find the value of the following functions:

1. sin 36°
2. cos 78°
3. tan 69°
4. sin 52°
5. cos 25°

Find angle A, correct to the nearest degree:

6. $\sin A = 0.5878$
7. $\cos A = 0.4226$
8. $\tan A = 5.7500$
9. $\cos A = 0.9800$
10. $\sin A = 0.7200$
11. $\tan A = 0.3113$
12. $\tan A = 0.1340$
13. $\cos A = 0.2868$
14. $\sin A = 0.7240$

Use a triangle similar to that shown in Fig. 35-1 for the following problems:

15. Find $\angle A$ if $BC = 30$ and $AC = 40$.
16. Find $\angle B$ if $AC = 50$ and $AB = 100$.
17. Find $\angle B$ if $BC = 25$ and $AB = 75$.
18. Find $\angle A$ if $BC = 16$ and $AB = 65$.
19. Find $\angle B$ if $AC = 22.5$ and $BC = 14$.
20. Find $\angle A$ if $AC = 4.5$ and $AB = 20.5$.
21. Find BC if $AB = 2{,}200$ W and $\angle A = 25°$.
22. Find AC if $AB = 600$ W and $\angle A = 8°$.
23. Find AB if $BC = 28.6$ and $\angle B = 42°$.
24. Find AB if $AC = 9.3\ \Omega$ and $\angle A = 34°$.
25. Find BC if $AC = 750\ \Omega$ and $\angle A = 48°$.
26. Find AC if $BC = 17.6$ ft and $\angle B = 60°$.
27. Find AC if $AB = 4{,}000$ W and $\angle B = 45°$.
28. Find BC if $AB = 2{,}500$ W and $\angle A = 45°$.
29. Find AB if $BC = 142$ V and $\angle A = 37°$.
30. Find AC if $AB = 85.8\ \Omega$ and $\angle A = 83°$.
31. (*C, E*) A pipe must be bent to provide a 6-m rise in a 1-m horizontal distance. Find the angle at each bend.
32. (*C, G*) The foot of a ladder 30 ft long rests on the ground 12 ft from the side of a building. What angle does the ladder make with the ground?
33. (*M*) In a taper similar to that shown in Fig. 33-5, the large diameter equals 0.75 in and the small diameter equals 0.47 in. Find angle X if the length equals 2.8 in.
34. (*C, G*) How long is a ladder that reaches 20 ft up the side of a building if the angle between the ladder and the ground is 65°?
35. (*M*) In a taper similar to that shown in Fig. 33-5, the small diameter equals 1.25 in, the length equals 2.25 in, and angle X equals 6°. Find the large diameter.
36. (*M*) In the large bevel gear shown in Fig. 36-1, the diameter of the 6-hole bolt circle is 12 in. Find the distance X between two adjacent holes.
37. (*G*) A rectangle measures 3 in by 7 in. Find the angle that the diagonal makes with the longer side.
38. (*G*) Find the length of the diagonal in Prob. 37.
39. (*E*) In an impedance triangle similar to that shown in Fig. 35-3, find the impedance Z if $R = 150\ \Omega$ and angle $A = 50°$.
40. (*M*) Find dimensions A and B used to locate the toolmaker's buttons shown in Fig. 36-2.

Fig. 36-1 A set of bevel gears. (Courtesy Arrow Gear Company)

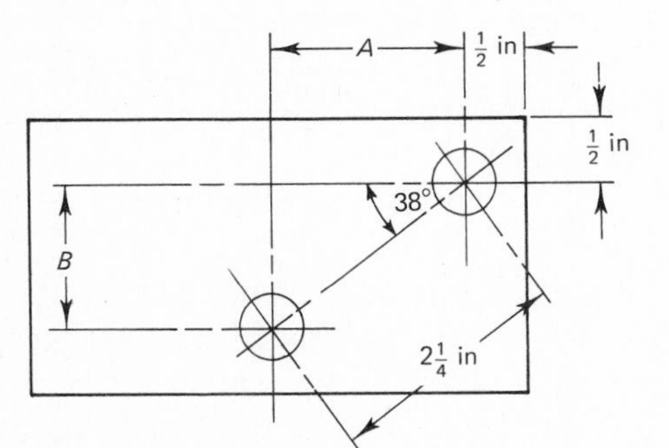

Fig. 36-2

41. (*C*) A hole is counterbored in a log to provide $\frac{1}{4}$-in clearance to get at the hex-head bolt shown in Fig. 36-3. Find the diameter of the counterbored hole.

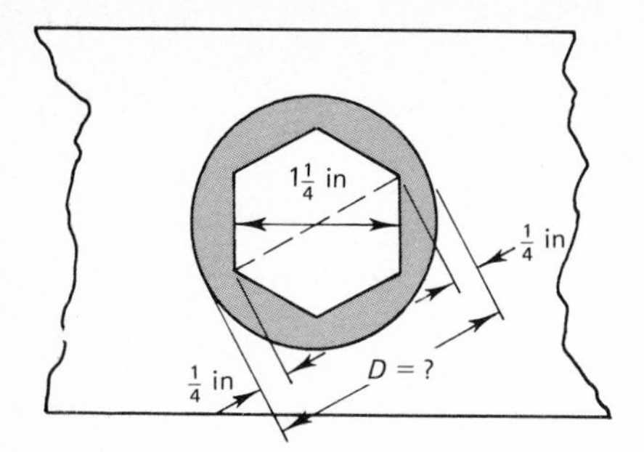

Fig. 36-3

42. (*M*) Find dimensions A and B in Fig. 36-4.

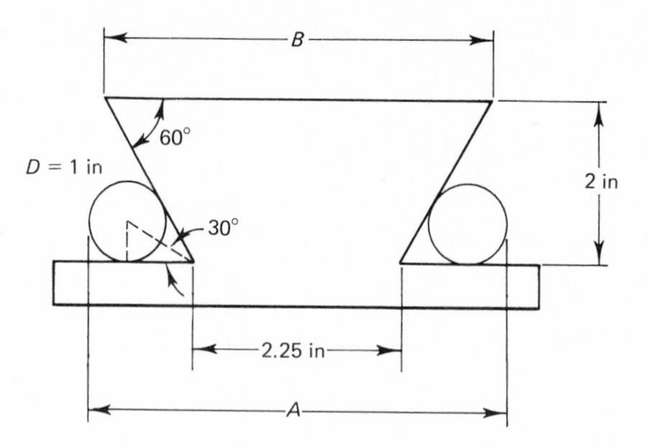

Fig. 36-4

43. (*M*) Find dimension D in Fig. 36-5.

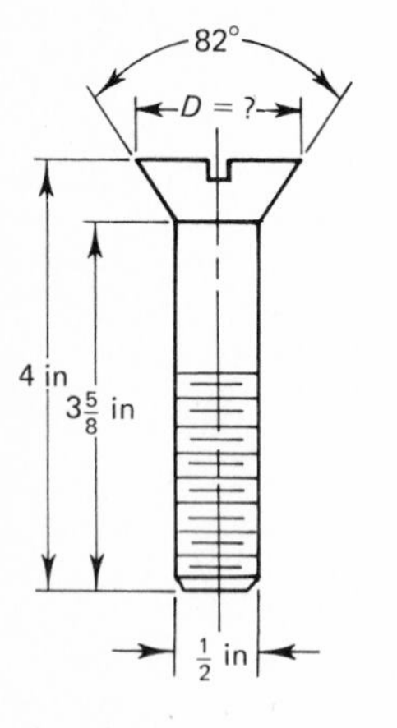

Fig. 36-5

44. (*M*) Using Fig. 36-6, find (a) $\angle 1$, (b) distance OA, (c) $\angle 2$, (d) $\angle 3$, and (e) $\angle 4$.

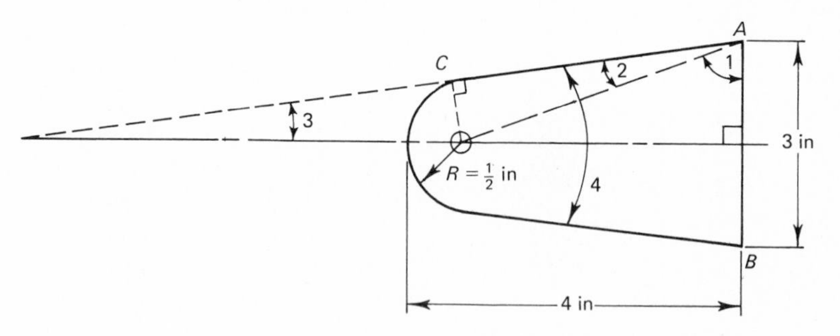

Fig. 36-6

(See Answer Key for Test 8—Trigonometry)

A tapered pin and the tapered hole into which it fits is shown in Fig. 37-1. A round piece of work which gradually decreases in diameter is said to be *tapered*. The shanks of drills, reamers, and lathe centers are all tapered to fit into correspondingly tapered holes to provide positive and secure fits. The advantage of the taper over straight shanks is that the taper may be removed from its mating part without damaging either part.

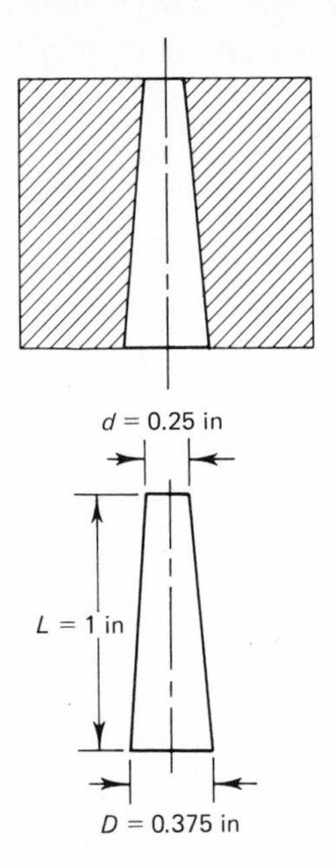

Fig. 37-1 The total taper is the difference between the diameters at the ends of the piece.

TAPER

The *taper* of a piece is the difference in the diameters at the ends of the piece. Thus, in Fig. 37-1,

$$\text{Taper} = D - d$$

$$= 0.375 - 0.25 = 0.125 \text{ in} \qquad \textit{Ans.}$$

TAPER PER INCH

A taper of $\frac{1}{4}$ in/in means that diameters 1 in apart differ by $\frac{1}{4}$ in, as shown in Fig. 37-2.

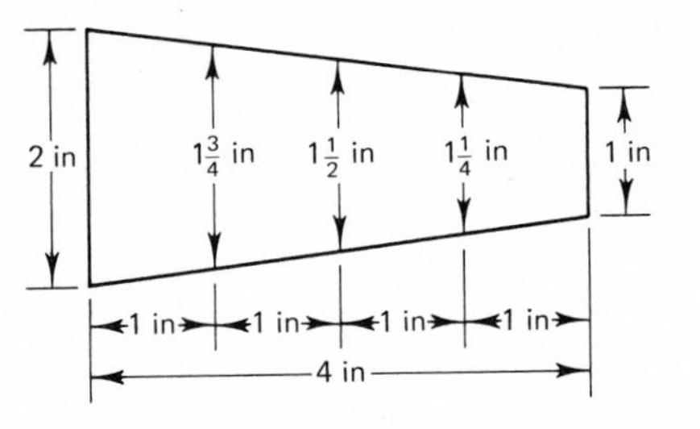

Fig. 37-2 A taper of ¼ in/in means that the diameter decreases ¼ in for every inch of length.

The taper per inch (TPI) $= \dfrac{\text{taper}}{\text{length}}$ or

FORMULA $$\text{TPI} = \frac{D - d}{L} \qquad (37\text{-}1)$$

where TPI = the taper per inch of length
D = large diameter, in
d = small diameter, in
L = length of the taper, in

Example 37-1 Find the TPI of the piece shown in Fig. 37-1.

Solution $$\text{TPI} = \frac{D - d}{L}$$

$$= \frac{0.375 - 0.25}{1}$$

$$= \frac{0.125}{1} = 0.125 \text{ in/in} \qquad \textit{Ans.}$$

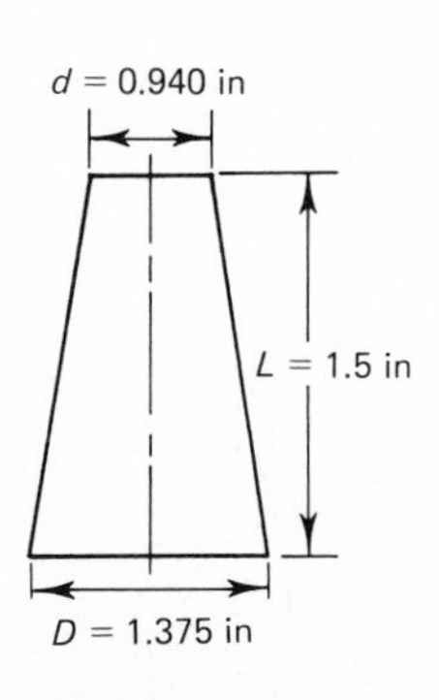

Fig. 37-3

Example 37-2 Find the taper per inch for the taper plug shown in Fig. 37-3.

Solution $$\text{TPI} = \frac{D - d}{L}$$

$$= \frac{1.375 - 0.940}{1.5}$$

$$= \frac{0.435}{1.5} = 0.29 \text{ in/in} \qquad \textit{Ans.}$$

Example 37-3 The taper per inch of the Morse standard taper No. 7 is 0.052 in/in. Find the total taper in a length of 6 in.

Solution

FORMULA **Total taper = TPI × length in inches** (37-2)

Total taper $= 0.052 \times 6 = 0.312$ in *Ans.*

TAPER PER FOOT

The taper per foot (TPF) means the amount of taper in 1 *foot* of length of the taper. The taper in 1 ft will be equal to 12 times the TPI, since there are 12 in in 1 ft.

FORMULA **TPF = 12 × TPI** (37-3)

Example 37-4 If the taper per inch is 0.0625 in, what is the taper per foot? What is the taper in 9 in?

Solution TPF = 12 × TPI

$= 12 \times 0.0625 = 0.75$ in/ft *Ans.*

Total taper in 9 in = TPI × L

$= 0.0625 \times 9 = 0.5625$ in *Ans.*

CHANGING TAPER PER FOOT TO TAPER PER INCH

We can find the taper in only 1 in of length by dividing the taper in a foot by the 12 in that are contained in 1 ft.

FORMULA $$\mathbf{TPI} = \frac{\mathbf{TPF}}{\mathbf{12}} \qquad (37\text{-}4)$$

Example 37-5 The Jarno standard taper has a taper of 0.6 in/ft. Find the taper per inch.

Solution $$\text{TPI} = \frac{\text{TPF}}{12}$$

$$= \frac{0.6}{12} = 0.05 \text{ in/in} \qquad \textit{Ans.}$$

Problems

Find the taper per inch and the taper per foot in Prob. 1–10.

	LARGE DIAMETER D, in	SMALL DIAMETER d, in	LENGTH L, in
1.	5	3	10
2.	6	$4\frac{1}{2}$	10
3.	$1\frac{1}{4}$	1	4
4.	$1\frac{1}{2}$	$1\frac{1}{4}$	6
5.	1.500	1.125	2
6.	0.625	0.510	0.75
7.	1.345	1.123	1.50
8.	$2\frac{1}{4}$	1.85	14
9.	4.085	3.875	6
10.	$1\frac{1}{4}$	1.05	$4\frac{1}{2}$

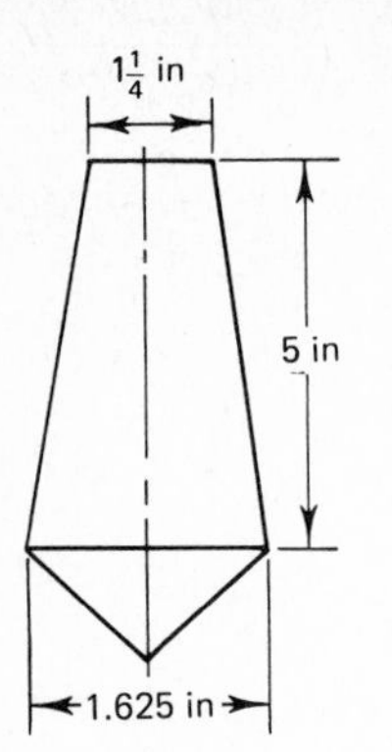

Fig. 37-4 A lathe center.

11. Find the taper per inch and the taper per foot for the lathe center shown in Fig. 37-4.
12. A tapered reamer is 0.825 in in diameter at the large end and 0.700 in at the small end. If the tapered portion is 6 in long, find the taper per inch and the taper per foot.
13. If the taper per inch is 0.42 in, find the total taper for a piece 8 in long.
14. Find the taper per foot of a piece 9 in long with a taper per inch of 0.248 in/in.
15. Find the taper per foot of a piece 14 in long which tapers from a diameter of $2\frac{1}{4}$ in to a diameter of 1.950 in.
16. Find the taper per foot of the tapered portion of the piece shown in Fig. 37-5.
17. Find the taper per inch and the taper per foot for the piece shown in Fig. 37-6.
18. Find the taper per inch and the taper per foot on the shank of the drill shown in Fig. 37-7.

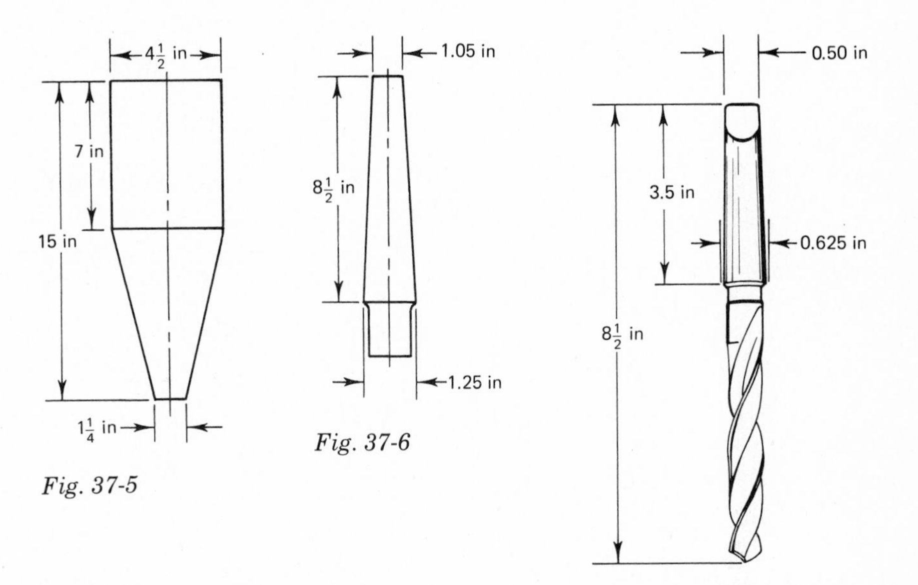

Fig. 37-5

Fig. 37-6

Fig. 37-7 The shank of a drill is tapered.

19. Find the taper per inch and the taper per foot of the tapered portion of the piece shown in Fig. 37-8.
20. Find the taper per foot for the tapered portion of the piece shown in Fig. 37-9.

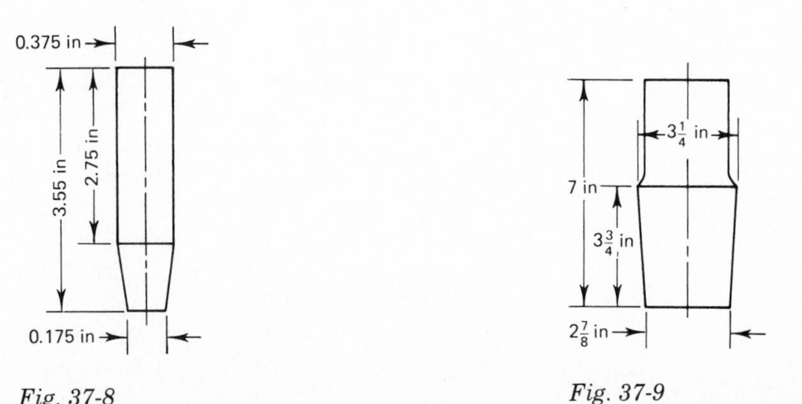

Fig. 37-8

Fig. 37-9

FINDING THE LARGE OR SMALL DIAMETER

Example 37-6 What is the small diameter of a round piece of work 9 in long if the large diameter is 1.125 in and the taper per foot is 0.6 in/ft? See Fig. 37-10.

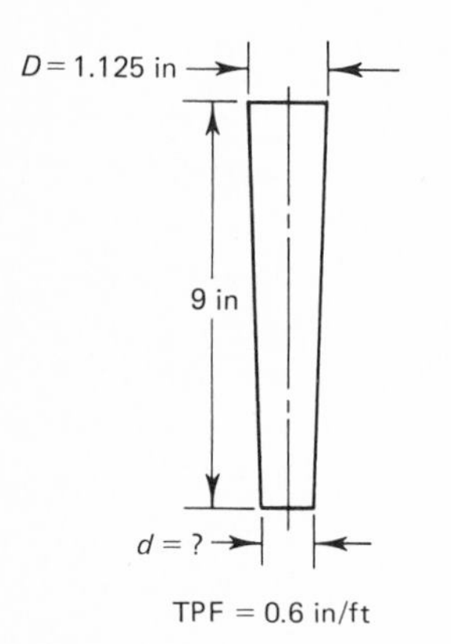

Fig. 37-10

Solution

1. Find the taper per inch.

$$\text{TPI} = \frac{\text{TPF}}{12}$$

$$= \frac{0.6}{12} = 0.05 \text{ in/in}$$

2. Find the total taper over the length of 9 inches.

$$\text{Total taper} = \text{TPI} \times \text{length}$$

$$= 0.05 \times 9 = 0.45 \text{ in}$$

3. Find the small diameter. The small diameter will be less than the large diameter by an amount equal to the total taper.

$$\text{Small diameter} = \text{large diameter} - \text{total taper}$$

$$= 1.125 - 0.45 = 0.675 \text{ in} \qquad \textit{Ans.}$$

Example 37-7 The taper pin reamer shown in Fig. 37-11 has a taper of $\frac{1}{4}$ in/ft. Find the diameter at the large end of the flutes.

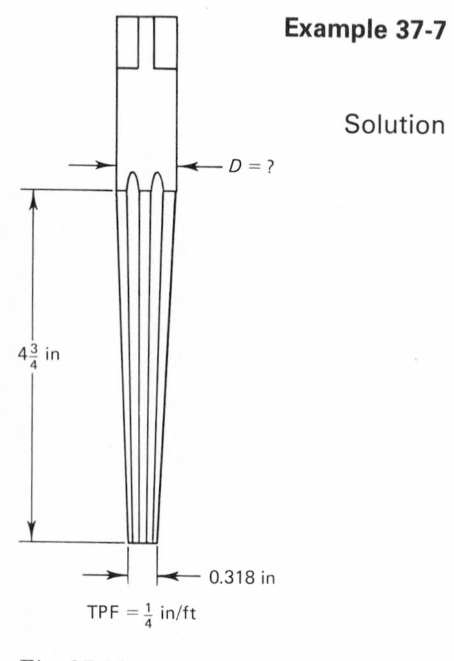

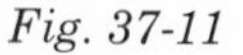

Fig. 37-11

Solution

1. Find the taper per inch.

$$\text{TPI} = \frac{\text{TPF}}{12} = \frac{0.25}{12} = 0.21 \text{ in/in}$$

2. Find the total taper over the length of 4.75 in.

$$\text{Total taper} = \text{TPI} \times \text{length}$$

$$= 0.021 \times 4.75 = 0.100 \text{ in}$$

3. Find the large diameter.

$$\text{Large diameter} = \text{small diameter} + \text{total taper}$$

$$= 0.318 + 0.100 = 0.418 \text{ in} \qquad \textit{Ans.}$$

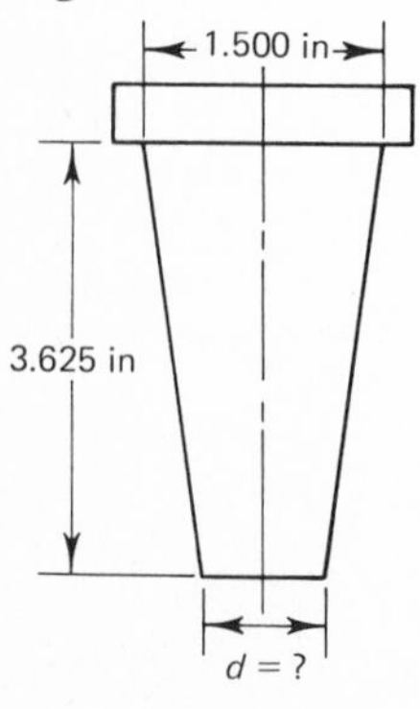

Fig. 37-12 A tapered bushing.

Problems

1. A taper reamer has a taper of $\frac{3}{4}$ in/ft, and the flutes are 5 in long. If the large diameter is 0.875 in, find the diameter at the small end.
2. A taper reamer has a taper of $\frac{5}{8}$ in/ft. If the diameter at the large end is $1\frac{1}{4}$ in and the flutes are $4\frac{1}{2}$ in long, find the diameter at the small end.
3. Find the large diameter of a tapered piece 6 in long if the small diameter is 0.784 in and the taper is $\frac{3}{4}$ in/ft.
4. Find the small diameter of the tapered bushing shown in Fig. 37-12 if the taper is 0.602 in/ft.
5. A tapered piece is 10 in long. The diameter at the small end is $1\frac{1}{4}$ in and the taper is $\frac{3}{4}$ in/ft. Find the diameter at the large end.

6. Find the large diameter D in the taper shown in Fig. 37-13.

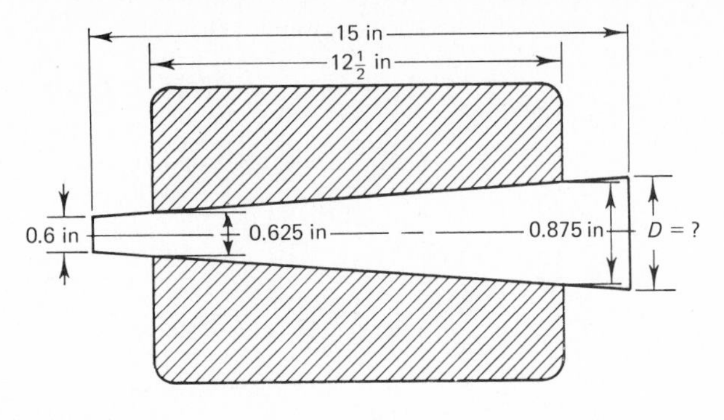

Fig. 37-13

7. Find the distance d in the taper shown in Fig. 37-14.

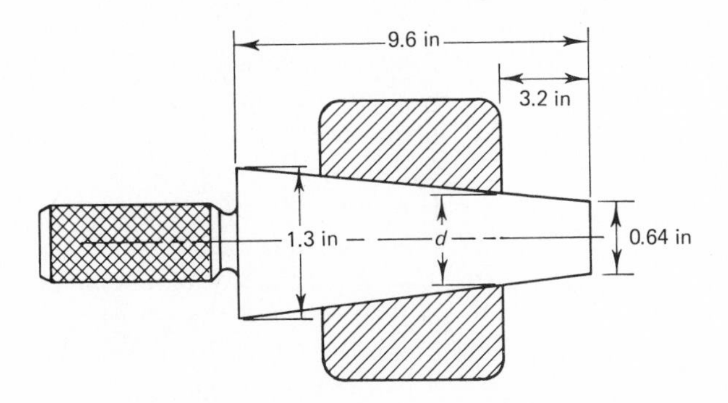

Fig. 37-14

FINDING THE LENGTH OF A TAPER

Formula 37-2 was used to find the total taper when we knew the TPI and the length of the taper.

$$\text{Total taper} = \text{TPI} \times L \qquad (37\text{-}2)$$

We can solve this equation to find the length by dividing both sides by the TPI, as we learned in Job 14.

$$\frac{\text{Total taper}}{\text{TPI}} = \frac{\overset{1}{\cancel{\text{TPI}}} \times L}{\underset{1}{\cancel{\text{TPI}}}}$$

The TPI will cancel on the right side and give us a new formula for the length of the taper.

FORMULA

$$\frac{\textbf{Total taper}}{\textbf{TPI}} = L \qquad (37\text{-}5)$$

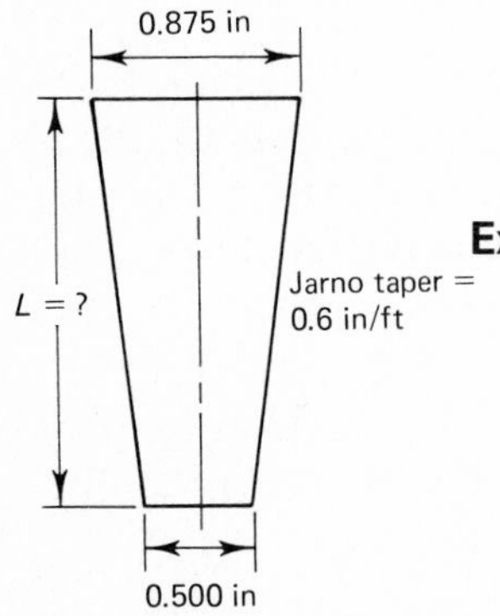

Example 37-8 A Jarno taper of 0.6 in/ft is turned on the piece shown in Fig. 37-15. Find its length.

0.875 in

$L = ?$

Jarno taper = 0.6 in/ft

0.500 in

Fig. 37-15

Solution 1. Find the taper per inch.

$$\text{TPI} = \frac{\text{TPF}}{12}$$

$$= \frac{0.6}{12} = 0.05 \text{ in/in}$$

2. Find the total taper.

$$\text{Total taper} = D - d$$

$$= 0.875 - 0.500 = 0.375 \text{ in}$$

3. Find the length of the taper.

$$L = \frac{\text{Total taper}}{\text{TPI}}$$

$$= \frac{0.375}{0.05} = 7.5 \text{ in} \qquad \textit{Ans.}$$

Problems

1. A piece of work tapers from a diameter of $2\frac{1}{2}$ in to a diameter of 2 in at the rate of $\frac{1}{4}$ in/in. Find the length of the piece.
2. Find the length of a piece that tapers 0.125 in/in from 2.625 in to 2.250 in.
3. How deep (L = ?) should the hole be bored in Fig. 37-16 before cutting the taper indicated?
4. The taper per foot of a lathe center is 0.602 in/ft. If the large diameter is 1.250 in and the small diameter is 1.097 in, find the length of the tapered part.
5. A milling machine arbor has a taper of $\frac{5}{8}$ in/ft. Find the length of the arbor if the large diameter is 1.125 in and the small diameter is 0.825 in.
6. Find the length L of the taper shown in Fig. 37-17.

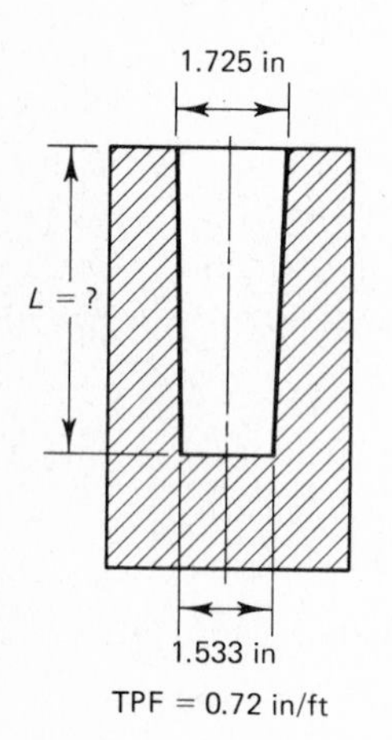

Fig. 37-16

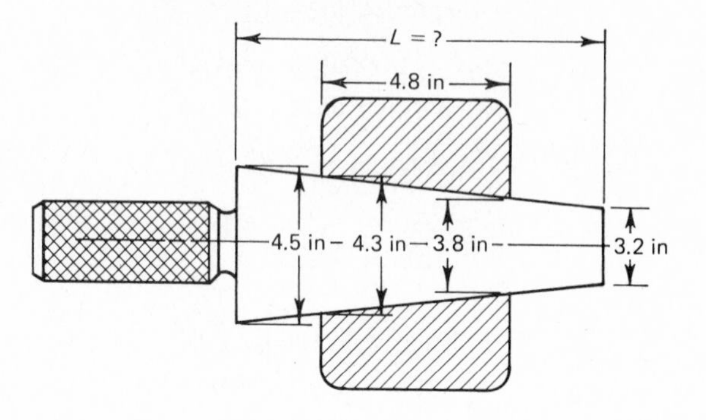

Fig. 37-17

STANDARD COMMERCIAL TAPERS

The shanks of drills, reamers, lathe centers, etc., are tapered to fit securely into sockets of similar taper. In order to make these parts interchangeable, it is necessary to standardize the tapers. The tapers most commonly used are:

Morse Standard

Brown & Sharpe Standard

Taper-pin Standard

Jarno Standard

1. The Morse Standard tapers are:

No. 0 = 0.6246 in/ft	No. 4 = 0.6233 in/ft
No. 1 = 0.5986 in/ft	No. 5 = 0.6315 in/ft
No. 2 = 0.5994 in/ft	No. 6 = 0.6257 in/ft
No. 3 = 0.6024 in/ft	No. 7 = 0.6240 in/ft

2. The Brown & Sharpe taper is $\frac{1}{2}$ in/ft.

3. The Taper-pin Standard is $\frac{1}{4}$ in/ft.

4. The Jarno Standard has a taper of 0.6 in/ft. In the Jarno series, numbered from 1 to 20, the various parts may be found without any mathematical calculations by applying the following formulas.

$$\text{Length of taper} = \frac{\text{no. of taper}}{2}$$

$$\text{Diameter of large end} = \frac{\text{no. of taper}}{8}$$

$$\text{Diameter of small end} = \frac{\text{no. of taper}}{10}$$

Example 37-9 Find the diameters and length of a No. 4 Jarno taper.

Solution

$$\text{Length of taper} = \frac{\text{no. of taper}}{2} = \frac{4}{2} = 2.0 \text{ in} \qquad \textit{Ans.}$$

$$\text{Diameter of large end} = \frac{\text{no. of taper}}{8} = \frac{4}{8} = 0.5 \text{ in} \qquad \textit{Ans.}$$

$$\text{Diameter of small end} = \frac{\text{no. of taper}}{10} = \frac{4}{10} = 0.4 \text{ in} \qquad \textit{Ans.}$$

Example 37-10 A Morse No. 7 taper has a large diameter of 1.875 in and a small diameter of 1.625 in. Find the length of the taper.

Solution

1. Find the total taper.

$$\text{Total taper} = D - d$$
$$= 1.875 - 1.625 = 0.250 \text{ in}$$

2. Find the TPI. A Morse No. 7 taper has a taper of 0.624 in/ft.

$$\text{TPI} = \frac{\text{TPF}}{12} = \frac{0.624}{12} = 0.052 \text{ in/in}$$

3. Find the length.

$$L = \frac{\text{Total taper}}{\text{TPI}} = \frac{0.250}{0.052} = 4.8 \text{ in} \qquad \textit{Ans.}$$

Problems

1. What is the taper per inch of a Morse No. 3 taper?
2. What is the taper per inch of a Brown & Sharpe taper?
3. What is the total taper for a Brown & Sharpe taper 9 in long?
4. Find the diameter at the large end of a No. 12 Jarno taper.

5. What is the length of a No. 5 Jarno taper?
6. What is the total taper for a Morse No. 2 taper which is 10 in long?
7. Find the diameter at the small end of a No. 6 Jarno taper.
8. A Brown & Sharpe taper $2\frac{1}{2}$ in long has a large diameter of 1.125 in. Find the small-end diameter.
9. A standard taper pin $1\frac{1}{2}$ in long has a small-end diameter of 0.125 in. Find the diameter at the large end.
10. A tapered piece has a large-end diameter of 1.375 in and a small-end diameter of 1.124 in. If the length is 5 in, find the taper per foot. Which Morse Standard taper is closest to this taper?
11. A Morse No. 2 taper has a large-end diameter of 1.618 in. If the piece is $4\frac{1}{4}$ in long, find the small-end diameter.
12. A piece $5\frac{1}{2}$ in long with a small-end diameter of 2.475 in is to be turned to a Brown & Sharpe taper. Find the large-end diameter.
13. Figure 37-18 has been turned to a standard Jarno taper. Find the number of the taper and the small diameter.

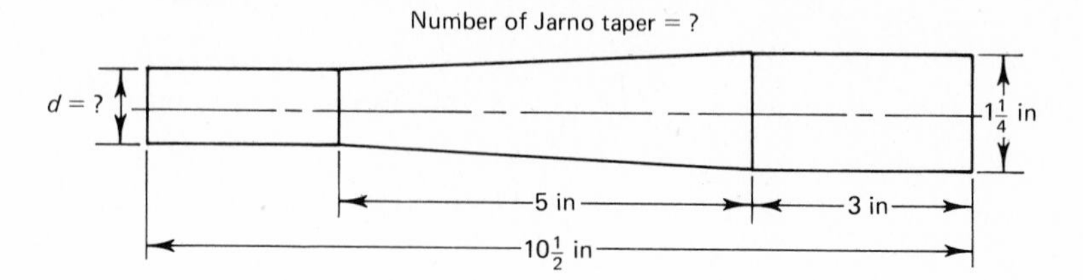

Fig. 37-18

14. A machinist is making a taper reamer for standard taper pins. The small-end diameter must be 0.580 in and the large-end diameter must be 0.750 in. How long should the fluted part be made?
15. A tapered hole 5 in deep must have a large-end diameter of 2.825 in and a small-end diameter of 2.562 in. What standard reamer should be used?

CUTTING TAPERS BY OFFSETTING THE TAILSTOCK

When a cylinder is to be turned in a lathe, the center line of the work lies on the center line of the lathe between the headstock and the tailstock, as shown in Fig. 37-19*a*. The tool moves parallel to the center line of the lathe and cuts a perfect cylinder. Now suppose that the tailstock is moved *off* the center line of the lathe into the position shown in Fig. 37-19*b*. The

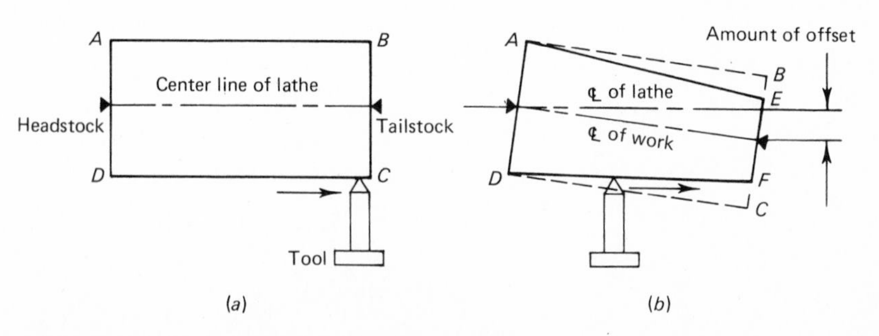

Fig. 37-19 (a) *A cylinder is turned when the center line of the work lies on the center line of the lathe.* (b) *A taper is turned when the center line of the work is offset from the center line of the lathe.*

tailstock is said to be *offset* by that amount. The work will now be in the position shown by the dotted lines *ABCD*. As the tool moves to the right, it will still move parallel to the center line of the lathe, cutting off more and more material as it moves from D to F. The radius of the work will gradually be reduced by an amount equal to the offset, but the diameter of the work will be reduced by *twice* this amount because the metal is removed on *both* sides of the work. The total decrease in diameter, or the taper, must then be equal to 2 times the offset.

$2 \times \text{offset} = \text{taper}$

and, by solving for the offset, we get the following formula:

FORMULA $$\textbf{Offset} = \frac{\textbf{taper}}{\textbf{2}} \qquad (37\text{-}6)$$

Example 37-11 Find the tailstock offset required to turn the taper shown in Fig. 37-20.

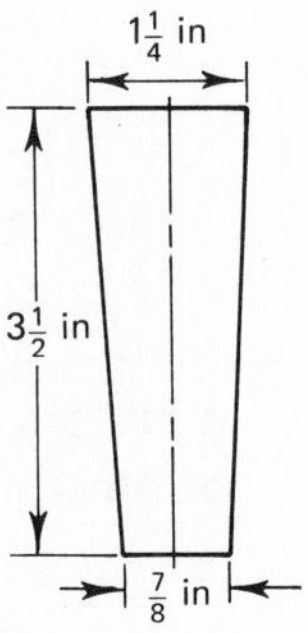

Solution **1.** Find the taper.

$$\text{Taper} = D - d$$

$$= 1\frac{1}{4} - \frac{7}{8} = \frac{3}{8}\text{ in}$$

2. Find the offset.

$$\text{Offset} = \frac{\text{taper}}{2}$$

$$= \frac{\frac{3}{8}}{2} = \frac{3}{16}\text{ in} \qquad \textit{Ans.}$$

Fig. 37-20

Example 37-12 Find the tailstock offset required to turn a Morse No. 3 taper over a piece 10 in long.

Solution Given: Morse No. 3 = 0.6024 in/ft Find: Tailstock offset
Length = 10 in

1. Find the taper per inch.

$$\text{TPI} = \frac{\text{TPF}}{12}$$

$$= \frac{0.6024}{12} = 0.0502\text{ in/in}$$

2. Find the total taper over the 10-in length.

$$\text{Taper} = \text{TPI} \times L$$

$$= 0.0502 \times 10 = 0.502\text{ in}$$

3. Find the offset.

$$\text{Offset} = \frac{\text{taper}}{2} = \frac{0.502}{2} = 0.251\text{ in} \qquad \textit{Ans.}$$

With the same amount of offset, pieces of different lengths will be machined with different tapers, as shown in Fig. 37-21. Therefore, when only a part of a piece is to be tapered, since the *entire* piece is set over, and not just the part that is being tapered, we must calculate the total taper as though the *entire* piece were being tapered.

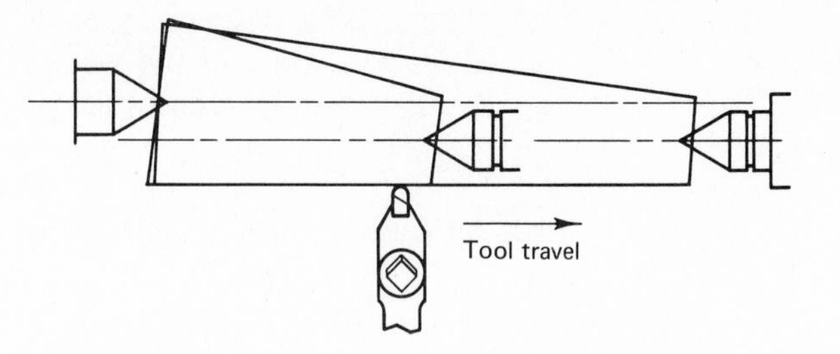

Fig. 37-21 With the same amount of offset, pieces of different lengths are turned with different tapers.

Example 37-13 Find the tailstock offset for turning a Brown & Sharpe taper for a length of 5 in on a piece of work which is 12 in long (Fig. 37-22).

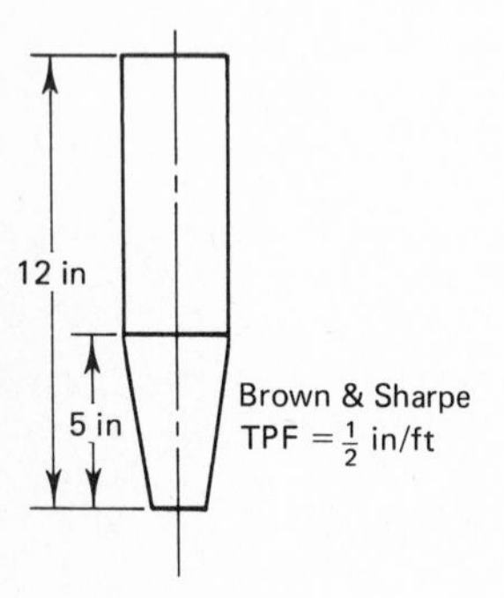

Fig. 37-22

Solution

1. Find the TPI. (The Brown & Sharpe TPF = $\frac{1}{2}$ in/ft.)

$$\text{TPI} = \frac{\text{TPF}}{12}$$

$$= \frac{\frac{1}{2}}{12} = \frac{1}{2} \times \frac{1}{12} = \frac{1}{24} \text{ in/in.}$$

2. Find the total taper. Since the *entire* piece is set over, and not just the part that is being tapered, we must calculate the total taper on the basis of the *entire* length.

$$\text{Total taper} = \text{TPI} \times \text{entire length}$$

$$= \frac{1}{24} \times 12 = \frac{1}{2} \text{ in}$$

3. Find the offset.

$$\text{Offset} = \frac{\text{taper}}{2} = \frac{\frac{1}{2}}{2} = \frac{1}{4} \text{ in} \qquad \textit{Ans.}$$

Example 37-14 Find the tailstock offset to turn the piece shown in Fig. 37-23.

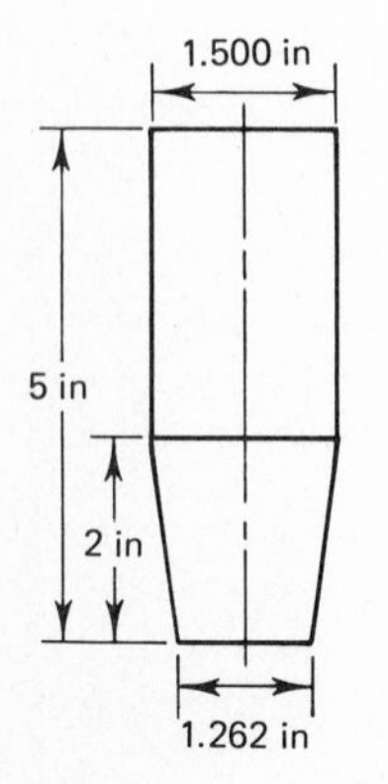

Fig. 37-23

Solution

1. Find the taper in the tapered portion.

$$\text{Taper} = D - d$$

$$= 1.500 - 1.262 = 0.238 \text{ in}$$

2. Find the TPI in the tapered portion.

$$\text{TPI} = \frac{\text{taper}}{\text{length}}$$

$$= \frac{0.238}{2} = 0.119 \text{ in}$$

3. Find the total taper over the *entire* length.

$$\text{Total taper} = \text{TPI} \times \text{entire length}$$

$$= 0.119 \times 5 = 0.595 \text{ in}$$

4. Find the offset.

$$\text{Offset} = \frac{\text{taper}}{2} = \frac{0.595}{2} = 0.298 \text{ in} \qquad \textit{Ans.}$$

Problems

1. The diameter of a tapered piece decreases from $1\frac{1}{4}$ in to $\frac{7}{8}$ in over a length of 10 in. Find the tailstock offset needed to cut the taper.
2. Find the offset required to turn a taper of $\frac{3}{4}$ in/ft over a length of 12 in.
3. Find (a) the length of the taper, (b) the large diameter, (c) the small diameter, (d) the total taper, and (e) the offset required to turn the following Jarno tapers: No. 7, No. 8, No. 10, and No. 15.
4. What is the offset required to turn the following tapers: (a) Morse No. 1, 12 in long, (b) Morse No. 2, 8 in long, (c) Morse No. 5, 10 in long, and (d) Morse No. 7, 9 in long?
5. Find the offset needed to turn a taper of $\frac{1}{4}$ in/ft over a $6\frac{1}{2}$-in length.
6. Find the offset needed to turn a Morse No. 5 taper for a length of 4 in on a piece of work 12 in long.
7. Find the offset needed to turn the piece of work shown in Fig. 37-24.
8. Find the offset needed to turn the piece of work shown in Fig. 37-25.
9. Find the offset needed to turn the piece of work shown in Fig. 37-26.
10. Find the tailstock offsets needed to turn the tapers A and B in the part shown in Fig. 37-27.

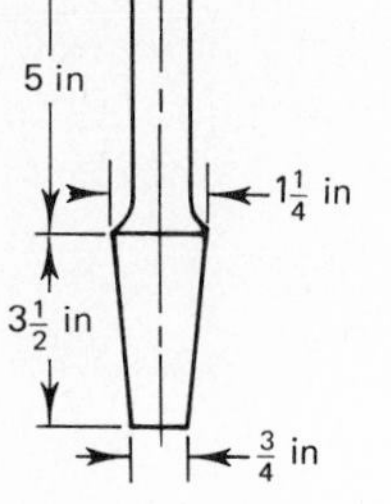

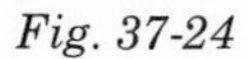

Fig. 37-24

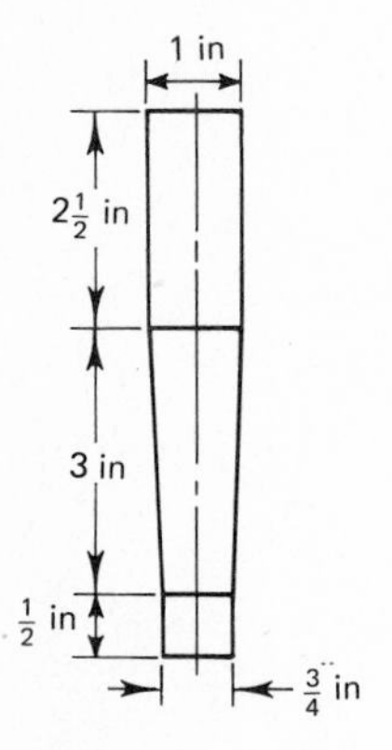

Fig. 37-25

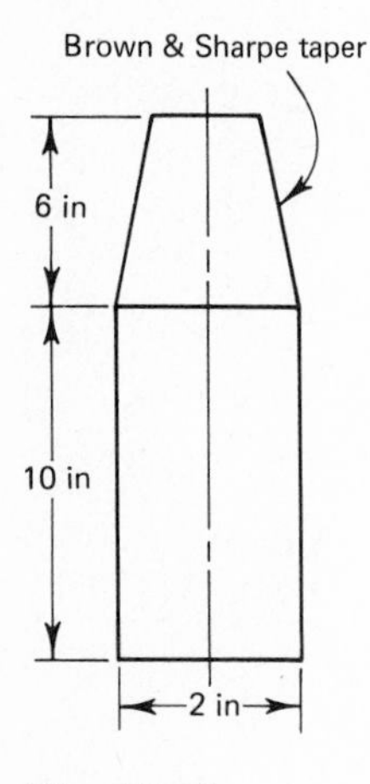

Fig. 37-26

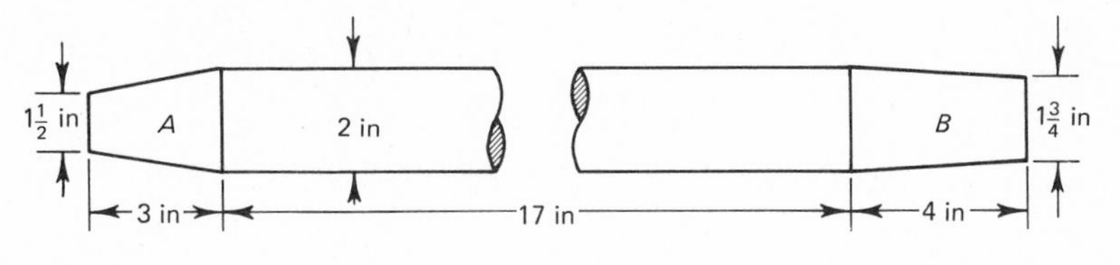

Fig. 37-27

SUMMARY OF TAPERS

1. A round piece of work which gradually decreases in ______ is said to be tapered. — diameter
2. The taper of a piece is the ______ in the diameters at the ______ of the piece. — difference; ends
3. The taper per inch (TPI) is the decrease in diameter for each inch of ______ of the taper. — length
4. $\text{TPI} = \dfrac{D - ?}{L}$ — d
5. Total taper = TPI × ______ — length
6. TPF = TPI × ______ — 12
7. $\text{TPI} = \dfrac{\text{TPF}}{?}$ — 12
8. To find the small diameter:

   Small diameter = large diameter − ______ — total taper

9. To find the large diameter:

   Large diameter = small diameter ______ total taper

10. To find the length of a taper:

   $L = \dfrac{\text{total taper}}{?}$

+

TPI

STANDARD COMMERCIAL TAPERS

1. The Morse standard tapers are:

No. 0 = 0.6246 in/ft	No. 4 = 0.6233 in/ft
No. 1 = 0.5986 in/ft	No. 5 = 0.6315 in/ft
No. 2 = 0.5994 in/ft	No. 6 = 0.6257 in/ft
No. 3 = 0.6024 in/ft	No. 7 = 0.6240 in/ft

2. The Brown & Sharpe taper is ______ in/ft.

3. The Taper-pin standard is ______ in/ft.

4. The Jarno Standard has a taper of ______ in/ft. In this series,

   $\text{Length of taper} = \dfrac{\text{no. of taper}}{?}$

   $\text{Diameter of large end} = \dfrac{\text{no. of taper}}{?}$

   $\text{Diameter of small end} = \dfrac{\text{no. of taper}}{?}$

5. Taper turning by offsetting the tailstock can be determined by the formula

   $\text{Offset} = \dfrac{?}{2}$

$\frac{1}{2}$

$\frac{1}{4}$

0.6

2

8

10

taper

(See Answer Key for Test 9—Tapers)

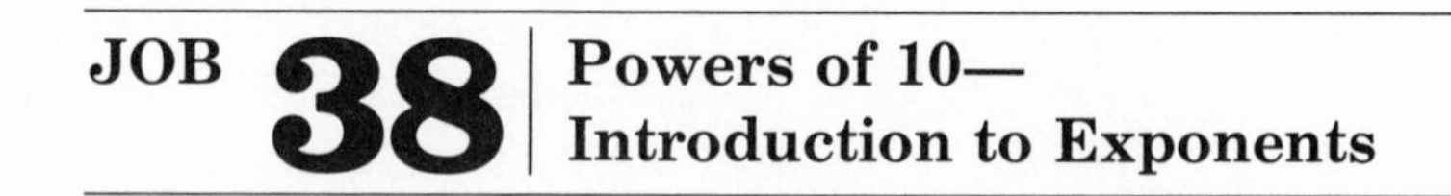

JOB 38 Powers of 10—Introduction to Exponents

MEANING OF AN EXPONENT

Exponents are a convenient shorthand method for writing and expressing many mathematical operations. For example, $2 \times 2 \times 2$ may be written as 2^3. The number 3 above and to the right of the 2 is called an *exponent*. The exponent says, "Write the number beneath it as many times as the exponent indicates, and then multiply." An exponent may be used with letters as well as numbers. Thus, R^3 means $R \times R \times R$. The exponent 2 is read as the word "squared." The exponent 3 is read as the word "cubed." When the exponent is any other number, it is read as "to the fourth power," "to the seventh power," etc. For example:

3^2 (3 squared) $= 3 \times 3 = 9$

5^3 (5 cubed) $= 5 \times 5 \times 5 = 125$

2^4 (2 to the fourth power) $= 2 \times 2 \times 2 \times 2 = 16$

10^1 (10 to the first power) $= 10$

10^2 (10 squared) $= 10 \times 10 = 100$

10^3 (10 cubed) $= 10 \times 10 \times 10 = 1{,}000$

10^6 (10 to the sixth power) $= 10 \times 10 \times 10 \times 10 \times 10 \times 10 = 1{,}000{,}000$

In many problems in electronics the usual units of amperes, volts, and ohms are either too large or too small. It has been found more convenient to use new units of measurement. These new units are formed by placing a special word or prefix in front of the unit. Each of these prefixes has a definite meaning.

Milli unit (m unit) means one one-thousandth of the unit

Kilo unit (k unit) means one thousand of these units

Micro unit (μ unit) means one one-millionth of the unit

Pico unit (p unit) means one-millionth of one-millionth of the unit

Mega unit (M unit) means one million of these units.

As we learned above, these numbers may be written as powers of ten.

$10 = 10^1 =$ 10 to the *first* power

$100 = 10^2 =$ 10 to the *second* power

$1{,}000 = 10^3 =$ 10 to the *third* power

$10{,}000 = 10^4 =$ 10 to the *fourth* power

$100{,}000 = 10^5 =$ 10 to the *fifth* power

$1{,}000{,}000 = 10^6 =$ 10 to the *sixth* power

The use of these units will clearly involve multiplying and dividing by numbers of this size and notation. Our work will be considerably simplified if we learn to apply the following rules.

MULTIPLYING BY POWERS OF 10

Please review Examples 3-1 and 3-2 on page 19.

RULE 1 **To multiply values by numbers expressed as 10 raised to some power, move the decimal point to the right as many places as the exponent indicates.**

Example 38-1

$0.345 \times 10^2 = \odot 34.5$ or 34.5 (move point two places right)

$0.345 \times 10^3 = \odot 345.$ or 345 (move point three places right)

$0.345 \times 10^6 = \odot 345000.$ or 345,000 (move point six places right)

$0.0065 \times 10^3 = \odot 006.5$ or 6.5 (move point three places right)

Problems

1. $0.0072 \times 1{,}000$
2. 45.76×100
3. $3.09 \times 1{,}000$
4. 0.0045×100
5. 37×100
6. 0.08×10^2
7. 0.0006×10^3
8. 0.00056×10^6
9. 27×10^2
10. 15×10^3
11. 0.00078×10^2
12. 7.8×10^6
13. 15.4×10^3
14. $3 \times 15 \times 10^2$
15. $0.005 \times 2 \times 10^2$
16. 6×10^4
17. 0.008×10^4
18. 0.009×10^5
19. 2.34×10^5
20. 0.0000005×10^8

DIVIDING BY POWERS OF 10

Please review Examples 3-3 and 3-4 on page 19.

RULE 2 **To divide values by numbers expressed as 10 raised to some power, move the decimal point to the left as many places as the exponent indicates.**

Example 38-2

$$467.9 \div 10^2 = 4.67\odot 9 = 4.679$$ (move point two places left)

$$59 \div 10^2 = .59\odot = 0.59$$ (move point two places left)

$$8.5 \div 10^3 = .008\odot 5 = 0.0085$$ (move point three places left)

$$5{,}500{,}000 \div 10^8 = .05500000\odot = 0.055$$ (move point eight places left)

Problems

1. $6{,}500 \div 100$
2. $7{,}500 \div 10^3$
3. $880{,}000 \div 1{,}000$
4. $32 \div 10^2$
5. $6 \div 10^2$
6. $17.8 \div 10$
7. $835 \div 10^3$
8. $550 \div 10^6$
9. $653.8 \div 10^3$
10. $100{,}000{,}000 \div 10^6$
11. $0.45 \div 10^2$
12. $0.08 \div 10$
13. $8.5 \div 10^3$
14. $7 \div 10^6$
15. $\dfrac{28.6}{10^3}$
16. $\dfrac{2 \times 1{,}000}{10^3}$
17. $0.02 \div 10^3$
18. $\dfrac{180}{2 \times 10^2}$
19. $\dfrac{5{,}000}{10^6}$
20. $\dfrac{50 \times 10^6}{10^8}$

NEGATIVE POWERS OF TEN

By definition:

$$10^{-n} = \frac{1}{10^n} \quad \text{and} \quad \frac{1}{10^n} = 10^{-n}$$

Thus,

$50 \times \dfrac{1}{10^2}$ is the same as 50×10^{-2}

or,

$\dfrac{50}{10^2}$ is the same as 50×10^{-2}

Therefore, if dividing by 10^2 means to move the decimal point to the left for two places, then the − sign in the exponent (−2) must mean to do the same thing.

MULTIPLYING BY NEGATIVE POWERS OF TEN

RULE 3 **To multiply values by numbers expressed as 10 raised to some negative power, move the decimal point to the *left* as many places as the exponent indicates.**

Example 38-3

$50 \times 10^{-2} = .50\odot$ or 0.5 (move point two places left)

$64.9 \times 10^{-2} = .64\odot 9$ or 0.649 (move point two places left)

$50{,}000 \times 10^{-6} = .050000\odot$ or 0.05 (move point six places left)

$0.2 \times 10^{-3} = .000\odot 2$ or 0.0002 (move point three places left)

Problems

1. $25{,}000 \times 10^{-3}$
2. 250×10^{-2}
3. 1.5×10^{-1}
4. 0.5×10^{-2}
5. $6{,}250 \times 10^{-3}$
6. $100{,}000 \times 10^{-5}$
7. $250{,}000 \times 10^{-8}$
8. 6×10^{-3}
9. 75.4×10^{-4}
10. 16.5×10^{-2}

DIVIDING BY NEGATIVE POWERS OF TEN

Since by definition:

$$\frac{1}{10^n} = 1 \times 10^{-n}$$

and

$$1 \times 10^{-n} = \frac{1}{10^n}$$

we can transfer any power of 10 from numerator to denominator, or vice versa, by simply changing the sign of the exponent.

Example 38-4

$$\frac{15}{\mathbf{10^{-2}}} = 15 \times \mathbf{10^2} = 1{,}500$$

$$\frac{0.05}{\mathbf{10^{-3}}} = 0.05 \times \mathbf{10^3} = 50$$

$$\frac{15{,}000}{\mathbf{10^3}} = 15{,}000 \times \mathbf{10^{-3}} = 15$$

$$\frac{5 \times 0.2}{\mathbf{10^{-4}}} = 1.0 \times \mathbf{10^4} = 10{,}000$$

Problems

1. $\dfrac{19.2}{10^{-2}}$
2. $\dfrac{25.6}{10^{2}}$
3. $\dfrac{0.85}{10^{-3}}$
4. $\dfrac{85}{10^{3}}$
5. $\dfrac{0.0072}{10^{-6}}$
6. $\dfrac{0.9}{10^{-4}}$
7. $\dfrac{0.0045}{10^{2}}$
8. $\dfrac{0.96}{10^{-3}}$
9. $\dfrac{880{,}000}{10^{-2}}$
10. $\dfrac{0.005}{10^{-2} \times 10^{-3}}$
11. $\dfrac{6 \times 5}{10^{-2}}$
12. $\dfrac{0.5 \times 0.04}{10^{-3}}$
13. $\dfrac{30 \times 10^{3}}{10^{-2}}$
14. $\dfrac{50 \times 10^{2}}{10^{-3}}$
15. $\dfrac{60 \times 10^{-2}}{10^{3}}$
16. $\dfrac{64.9 \times 10^{3}}{10^{5}}$

EXPRESSING NUMBERS AS POWERS OF 10

As you have seen, multiplying and dividing numbers by powers of 10 is simple to do mentally. It certainly will be to our advantage, then, if we can express the numbers in any problem as powers of 10 before we multiply or divide.

EXPRESSING NUMBERS LARGER THAN 1 AS A SMALL NUMBER TIMES A POWER OF 10

RULE 4 **To express a large number as a smaller number times a power of 10, move the decimal point to the *left* as many places as desired. Then multiply the number obtained by 10 to a power which is equal to the number of places moved.**

Example 38-5

a. $3{,}000 = 3{,}000\odot$ (Moved three places left)

$3{,}000 = 3 \times 10^{3}$

b. $4{,}500 = 45{,}00\odot$ (Moved two places left)

$4{,}500 = 45 \times 10^{2}$

c. $4{,}500 = 4{,}500\odot$ (Moved three places left)

$4{,}500 = 4.5 \times 10^{3}$

d. $770{,}000 = 77{,}0000\odot$ (Moved four places left)

$770{,}000 = 77 \times 10^{4}$

e. $800{,}000 = 8{,}00000\odot$ (Moved five places left)

$800{,}000 = 8 \times 10^{5}$

f. $5{,}005.2 = 5{,}005\odot 2$ (Moved three places left)

$5{,}005.2 = 5.0052 \times 10^{3}$

g. Express the following number to three significant figures and express it as a number between 1 and 10 times the proper power of 10. (Review rounding off numbers on page 14.)

7,831 = 7,830 (Since only three significant figures are wanted)

7,830 = 7.830⊙ (Moved three places left)

7,830 = 7.83×10^3

h. Express 62,495 using three significant figures as a number between 1 and 10 times the proper power of 10.

62,495 = 62,500 (Written as three significant figures)

62,500 = 6.2500⊙ (Moved four places left)

62,500 = 6.25×10^4

Problems

Express the following numbers to three significant figures and write them as numbers between 1 and 10 times the proper power of 10.

1. 6,000
2. 5,700
3. 150,000
4. 500,000
5. 235,000
6. 7,350,000
7. 4,960
8. 62,500
9. 980
10. 175
11. 12.5
12. 7,303
13. 48.2
14. 12,600
15. 880,000,000
16. 54,009
17. 38,270
18. 8,019.7
19. 1,754,300
20. 2,395,000
21. 482,715

EXPRESSING NUMBERS LESS THAN 1 AS A WHOLE NUMBER TIMES A POWER OF 10

RULE 5 **To express a decimal as a whole number times a power of 10, move the decimal point to the right as many places as desired. Then multiply the number obtained by 10 to a *negative* power which is equal to the number of places moved.**

Example 38-6 **a.** 0.005 = 0⊙005. (Moved three places right)

0.005 = 5×10^{-3}

b. 0.00672 = 0⊙006.72 (Moved three places right)

0.00672 = 6.72×10^{-3}

c. 0.758 = 0⊙75.8 (Moved two places right)

0.758 = 75.8×10^{-2}

d. 0.0758 = 0⊙07.58 (Moved two places right)

0.0758 = 7.58×10^{-2}

e. 0.0000089 = 0⊙000008.9 (Moved six places right)

0.0000089 = 8.9×10^{-6}

f. Express the number 0.0003578 to three significant figures and then write it as a number between 1 and 10 times the proper power of 10.

$0.0003578 = 0.000358$ (Written as three significant figures)

$0.000358 = 0 \odot 0003.58$ (Moved four places right)

$0.000358 = 3.58 \times 10^{-4}$

Problems

Express the following numbers to three significant figures and write them as numbers between 1 and 10 times the proper power of 10.

1. 0.006
2. 0.0075
3. 0.0035
4. 0.08
5. 0.456
6. 0.0357
7. 785×10^{-2}
8. 0.00000012
9. 0.0965
10. 0.00482
11. 0.5
12. 0.0003743
13. 0.008147
14. 0.000007949
15. 0.000725×10^{5}
16. 0.01333
17. 0.0006×10^{3}
18. $3{,}200 \times 10^{-5}$
19. 360×10^{-4}
20. 0.000008×10^{4}
21. *(F, G)* The calcium salts in a sample of hard water measured 19 parts per million. Express this concentration as a number between 1 and 10 multiplied by the proper power of 10.
22. *(C, G)* The concentration of chlorine in a swimming pool is 0.87 parts per million. Express this concentration as a number between 1 and 10 multiplied by the proper power of 10.

MULTIPLYING WITH POWERS OF TEN

If x^3 means $x \cdot x \cdot x$, and x^2 means $x \cdot x$, then $x^3 \cdot x^2$ means $x \cdot x \cdot x \cdot x \cdot x = x^5$ or, $x^3 \cdot x^2 = x^{(3+2)} = x^5$, which gives us the following rule.

RULE 6 **The multiplication of two or more powers using the *same* base is equal to that base raised to the sum of the powers.**

Example 38-7

a. $a^4 \times a^5 = a^{(4+5)} = a^9$

b. $10^2 \times 10^3 = 10^5$

c. Multiply $10{,}000 \times 1{,}000$.

If $10{,}000 = 10^4$ and $1{,}000 = 10^3$, then

$10{,}000 \times 1{,}000 = 10^4 \times 10^3 = 10^{(4+3)} = 10^7$ *Ans.*

d. Multiply $25{,}000 \times 4{,}000$.

If $25{,}000 = 25 \times 10^3$ and $4{,}000 = 4 \times 10^3$, then

$$25{,}000 \times 4{,}000 = 25 \times 10^3 \times 4 \times 10^3$$
$$= 25 \times 4 \times 10^3 \times 10^3$$
$$= 100 \times 10^6$$
$$= 10^2 \times 10^6 = 10^8 \quad \textit{Ans.}$$

e. Multiply 0.00005×0.003.

If $0.00005 = 5 \times 10^{-5}$ and $0.003 = 3 \times 10^{-3}$, then

$$\begin{aligned} 0.00005 \times 0.003 &= 5 \times 10^{-5} \times 3 \times 10^{-3} \\ &= 5 \times 3 \times 10^{-5} \times 10^{-3} \\ &= 15 \times 10^{[-5+(-3)]} \\ &= 15 \times 10^{-8} \quad \textit{Ans.} \end{aligned}$$

Note: to add two minus numbers, add the numbers and prefix the sum with a minus sign.

f. Multiply $7{,}000 \times 0.00091$.

If $7{,}000 = 7 \times 10^3$ and $0.00091 = 9.1 \times 10^{-4}$, then

$$\begin{aligned} 7{,}000 \times 0.00091 &= 7 \times 10^3 \times 9.1 \times 10^{-4} \\ &= 7 \times 9.1 \times 10^3 \times 10^{-4} \\ &= 63.7 \times 10^{[3+(-4)]} \\ &= 63.7 \times 10^{-1} \\ &= 6.37 \quad \textit{Ans.} \end{aligned}$$

Note: to add a plus number and a minus number, *subtract* the numbers and prefix the difference with the sign of the larger number.

g. Multiply $0.00005 \times 20{,}000 \times 1{,}500$.

If $0.00005 = 5 \times 10^{-5}$, and $20{,}000 = 2 \times 10^4$, and $1{,}500 = 1.5 \times 10^3$, then

$$\begin{aligned} 0.00005 \times 20{,}000 \times 1{,}500 &= 5 \times 10^{-5} \times 2 \times 10^4 \times 1.5 \times 10^3 \\ &= 5 \times 2 \times 1.5 \times 10^{-5} \times 10^4 \times 10^3 \\ &= 15 \times 10^2 \\ &= 1{,}500 \quad \textit{Ans.} \end{aligned}$$

Problems

Multiply the following numbers.

1. $5{,}000 \times 0.001$
2. $850 \times 2{,}000$
3. $16 \times 10^2 \times 4 \times 10^3$
4. $0.0004 \times 5 \times 10^2$
5. $250 \times 4{,}000 \times 3 \times 10^{-2}$
6. $1{,}000 \times 10^{-4} \times 0.02$
7. $3 \times 10^{-5} \times 4 \times 10^6$
8. $15 \times 10^{-4} \times 20{,}000 \times 0.04$
9. $200{,}000 \times 0.000005 \times 3 \times 10^{-2}$
10. $0.004 \times 0.0005 \times 5{,}000$
11. $0.005 \times 5 \times 10^{-3} \times 0.02$
12. $6{,}000{,}000 \times 0.00025 \times 0.3 \times 10^{-2}$
13. $0.3 \times 10^{-2} \times 800{,}000 \times 400 \times 10^{-3}$
14. $(500)^2 \times 0.0002 \times 4{,}000$
15. $500{,}000{,}000 \times 0.000004 \times 3.14 \times 10^2$

16. (*E*) A coil of wire offers a special resistance to the flow of an alternating electric current. This resistance is called the inductive reactance X_L and is measured in ohms of resistance. The reactance is given by the formula

$$X_L = 6.28\,fL$$

where f = frequency, hertz (Hz)
L = inductance of the circuit, henries (H)
X_L = reactance, Ω

Find the inductive reactance when:
a. $f = 60$ Hz and $L = 0.025$ H
b. $f = 1{,}000{,}000$ Hz and $L = 0.25$ H
c. $f = 10{,}000$ Hz and $L = 0.000025$ H

17. (*G*) If light travels 186,000 mi/sec, find the number of miles that light will travel in 1 hr. Express the answer as a number between 1 and 10 multiplied by the proper power of 10.

DIVIDING WITH POWERS OF 10

As noted in the section on Dividing by Negative Powers of 10, we can transfer any power of 10 from numerator to denominator, or vice versa, by simply changing the sign of the exponent. This will permit us to change *all* division problems into multiplications, which are generally easier to do.

Example 38-8

a. $10^6 \div 10^2 = \frac{10^6}{10^2} = 10^6 \times 10^{-2} = 10^4$ *Ans.*

b. $\frac{4{,}000}{10^2} = 4 \times 10^3 \times 10^{-2} = 4 \times 10^1 = 40$ *Ans.*

c. $\frac{35{,}000}{0.005} = \frac{35 \times 10^3}{5 \times 10^{-3}} = 7 \times 10^3 \times 10^3$

$= 7 \times 10^6$ *Ans.*

d. $\frac{144{,}000}{12 \times 10^3} = \frac{144 \times \cancelto{1}{10^3}}{12 \times \cancelto{1}{10^3}} = 12$ *Ans.*

Note that *any* factor divided by itself cancels out to 1, and that it is not necessary to transfer any powers. That is, $10^3/10^3 = 10^{(3-3)} = 10^0 = 1$.

e. $\frac{0.00075}{500} = \frac{75 \times 10^{-5}}{5 \times 10^2} = 15 \times 10^{-5} \times 10^{-2}$

$= 15 \times 10^{-7}$ *Ans.*

f. $\frac{60}{0.0003 \times 40{,}000} = \frac{60}{3 \times 10^{-4} \times 4 \times 10^4} = \frac{5}{10^0}$

Now, since $10^0 = 1$, $\frac{5}{10^0} = \frac{5}{1} = 5$ *Ans.*

Problems

Divide the following numbers.

1. $\frac{10^8}{10^3}$
2. $\frac{10^3}{10^5}$
3. $\frac{60{,}000}{5 \times 10^2}$
4. $\frac{50{,}000}{0.05}$
5. $\frac{10}{50{,}000}$
6. $\frac{20}{0.0005}$
7. $\frac{0.0001}{50}$
8. $\frac{20}{4{,}000 \times 0.005}$
9. $\frac{1{,}000 \times 0.008}{0.002 \times 500}$
10. $\frac{150{,}000}{3 \times 10^5}$
11. $\frac{0.00015}{3 \times 10^{-2}}$
12. $\frac{1}{4 \times 100{,}000 \times 0.00005}$

13. A capacitor offers a special resistance to the flow of an alternating electric current. This resistance is called the capacitive reactance X_C and is measured in ohms of resistance. The reactance is given by the formula

$$X_C = \frac{1}{2\pi fC}$$

where f = frequency, Hz
C = capacitance, farads (F)
π = 3.14
X_C = reactance, Ω

Find the capacitive reactance when:

a. f = 60 Hz and C = 0.00005 F
b. f = 1,000 Hz and C = 0.0000025 F
c. f = 1,000,000 Hz and C = 0.00000005 F

(See Answer Key for Test 10—Powers of 10)

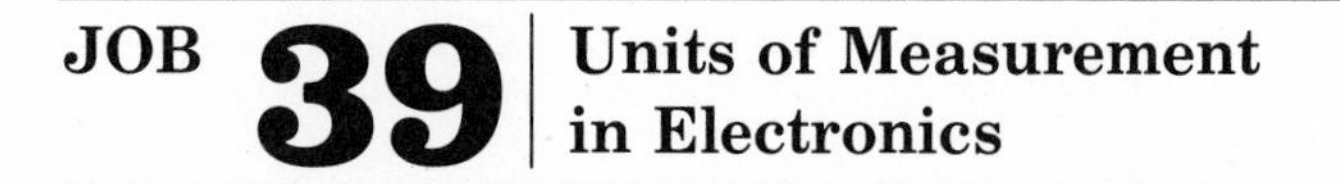

JOB 39 Units of Measurement in Electronics

Ohm's law and other electrical formulas use the simple electrical units of volts, amperes, and ohms. However, if the measurements given or obtained in a problem were stated in kilovolts, milliamperes, or megohms, it would be necessary to change these units of measurement into the units required by the formula.

CHANGING UNITS OF MEASUREMENT

There are two factors to be considered when describing any measurement: (1) how many of the measurements and (2) what *kind* of measurement. For example, $1 may be described as 2 half-dollars, 4 quarters, 10 dimes, 20 nickels, or 100 pennies. When the $1 was changed into each of the new measurements, *both* the unit of measurement *and* the number of them were changed. For example:

Since 3 ft = 1 yd, 2 yd = 2 × 3 = 6 ft

Since 2,000 lb = 1 ton, 3 tons = 3 × 2,000 = 6,000 lb

Since 100¢ = 1 dollar, 4 dollars = 4 × 100 = 400¢

4 = small number; dollars = large unit; 400 = large number; ¢ = small unit

Notice that a *small number* of *large units* is always changed into a *large number* of *small units* by *multiplying* by the number showing the relationship between the units.

RULE 1 **To change from a large unit into a small unit, multiply by the number showing the relationship between the units.**

This rule is illustrated for various units in Table 39-1.

Table 39-1. Changing Large Units into Small units

TO CHANGE	INTO	MULTIPLY BY
Mega units	Units	10^6
Mega units	Kilo units	10^3
Kilo units	Units	10^3
Units	Milli units	10^3
Units	Micro units	10^6
Units	Pico units	10^{12}
Milli units	Micro units	10^3
Milli units	Pico units	10^9
Micro units	Pico units	10^6

Example 39-1

1. Change 2.4 V to millivolts.

   $2.4 \times 10^3 = 2{,}400$ mV

2. Change 0.56 A to milliamperes.

   $0.56 \times 10^3 = 560$ mA

3. Change 0.5 W to milliwatts.

   $0.5 \times 10^3 = 500$ mW

4. Change 0.3 kV to volts.

   $0.3 \times 10^3 = 300$ V

5. Change 0.15 kW to watts.

   $0.15 \times 10^3 = 150$ W

6. Change 880 kHz to hertz.

   $880 \times 10^3 = 880{,}000$ Hz

7. Change 0.0004 A to microamperes.

   $0.0004 \times 10^6 = 400\ \mu$A

8. Change 0.00005 F to microfarads.

$0.00005 \times 10^6 = 50\ \mu F$

9. Change 0.25 MΩ to ohms.

$0.25 \times 10^6 = 250{,}000\ \Omega$

10. Change 3.2 MHz to hertz.

$3.2 \times 10^6 = 3{,}200{,}000$ Hz

11. Change 0.00035 μF to picofarads.

$0.00035 \times 10^6 = 350$ pF

12. Change 0.000000005 F to picofarads.

$0.000000005 \times 10^{12} = 5{,}000$ pF

Now let us reverse the process and change small units into large units. For example:

Since 3 ft = 1 yd, 6 ft = 6 ÷ 3 = 2 yd

Since 2,000 lb = 1 ton, 6,000 lb = 6,000 ÷ 2,000 = 3 tons

Since 100¢ = 1 dollar, 400¢ = 400 ÷ 100 = 4 dollars

small unit ⟶ large unit

large number ⟶ small number

Notice that a *large number* of *small units* is always changed into a *small number* of *large units* by *dividing* by the number showing the relationship between the units.

RULE 2 **To change from a small unit into a large unit, divide by the number showing the relationship between the units.**

This rule is illustrated for various units in Table 39-2.

Table 39-2. Changing Small Units into Large Units

TO CHANGE	INTO	DIVIDE BY	OR	MULTIPLY BY
Units	Mega units	10^6		10^{-6}
Kilo units	Mega units	10^3		10^{-3}
Units	Kilo units	10^3		10^{-3}
Milli units	Units	10^3		10^{-3}
Micro units	Units	10^6		10^{-6}
Pico units	Units	10^{12}		10^{-12}
Micro units	Milli units	10^3		10^{-3}
Pico units	Milli units	10^9		10^{-9}
Pico units	Micro units	10^6		10^{-6}

Example 39-2

1. Change 500,000 Ω to megohms.

$500{,}000 \div 10^6 = 0.5\ M\Omega$ or $500{,}000 \times 10^{-6} = 0.5\ M\Omega$

2. Change 660 kHz to megahertz.

$660 \times 10^{-3} = 0.66$ MHz

3. Change 600 V to kilovolts.

$600 \times 10^{-3} = 0.6$ kV

4. Change 14.5 mA to amperes.

$14.5 \times 10^{-3} = 0.0145$ A

5. Change 2.5 μF to farads.

$2.5 \times 10^{-6} = 0.0000025$ F

6. Change 30,000,000 pF to farads.

$30{,}000{,}000 \times 10^{-12} = 0.00003$ F

7. Change 400 μV to millivolts.

$400 \times 10^{-3} = 0.4$ mV

8. Change 350 pF to microfarads.

$350 \times 10^{-6} = 0.00035$ μF

9. Change 4,000 W to kilowatts.

$4{,}000 \times 10^{-3} = 4$ kW

10. Change 1,010,000 Hz to kilohertz.

$1{,}010{,}000 \times 10^{-3} = 1{,}010$ kHz

11. Change 356 mV to volts.

$356 \times 10^{-3} = 0.356$ V

12. Change 15,000 μV to volts.

$15{,}000 \times 10^{-6} = 0.015$ V

Table 39-1 and Table 39-2 are shown in picture form in Fig. 39-1.

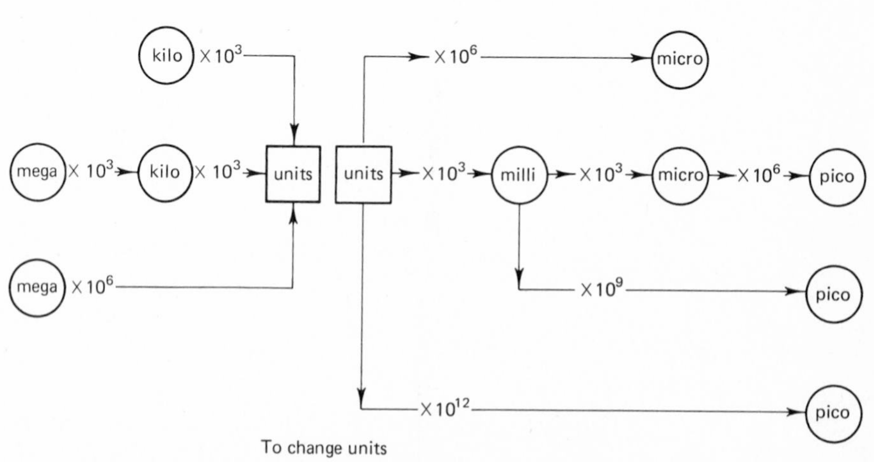

Fig. 39-1

Problems

Change the following units of measurement.

1. 225 mA to amperes
2. 0.076 V to millivolts
3. 3.5 MΩ to ohms
4. 5 kW to watts
5. 550 kHz to hertz
6. 700,000 Hz to kilohertz
7. 70,000 Ω to megohms
8. 0.00008 F to microfarads
9. 0.065 A to milliamperes
10. 6,500 W to kilowatts
11. 75 mV to volts
12. 2.3 MHz to hertz
13. 6,000 μA to amperes
14. 0.007 F to microfarads
15. 3.9 mA to amperes
16. 75,000 W to kilowatts
17. 0.005 μF to picofarads
18. $\frac{1}{4}$ A to milliamperes
19. 1,000 kHz to hertz
20. 0.5 MΩ to ohms
21. 0.008 V to millivolts
22. 0.0045 W to milliwatts
23. 0.00006 μF to picofarads
24. 0.15 A to milliamperes
25. 0.15 μF to farads
26. 125 mV to volts
27. 8,000 W to kilowatts
28. 4.16 kW to watts
29. 0.000004 A to microamperes
30. 0.6 MHz to hertz

JOB 40 Using Electronic Units of Measurement in Simple Circuits

All the formulas for Ohm's law, series circuits, parallel circuits, and power demand that the measurements be given in the units of amperes, volts, and ohms only. If a certain problem gives the measurements in units other than these, we must change all the measurements into amperes, volts, and ohms before we can use any of these formulas.

Example 40-1 Find the voltage that will force 28.6 μA of current through a 70-kΩ resistor in the base circuit of a transistor.

Solution Given: $I = 28.6\ \mu\text{A}$ Find: $E = ?$
$R = 70\ \text{k}\Omega$

1. Change μA to A.

$$28.6\ \mu\text{A} = 28.6 \times 10^{-6}\ \text{A}$$

2. Change 70 kΩ to Ω.

$$70\ \text{k}\Omega = 70 \times 10^{3}\ \Omega$$

3. Find the voltage E.

$$E = I \times R$$

$$= 28.6 \times 10^{-6} \times 70 \times 10^{3}$$

$$= 28.6 \times 7 \times 10^{-?}$$

$$= 200.2 \times 10^{-?}$$

$$= 2\ \text{V} \qquad \textit{Ans.}$$

A bias voltage of 300 mV is developed across a 2-MΩ resistor. Find the current flowing.

Solution Given: $E = 300$ mV Find: $I = ?$
$R =$ ______ 2 MΩ

1. Change 300 mV to V.

   300 mV = 300 × ______ V 10^{-3}

2. Change 2 MΩ to Ω.

   2 MΩ = 2 × ______ Ω 10^{6}

3. Find the current I.

$$E = I \times R$$

$$300 \times 10^{-3} = I \times 2 \times 10^{6}$$

$$I = \frac{300 \times 10^{-3}}{?}$$ 2×10^{6}

$$= 150 \times 10^{-?}$$ 9

$$= 0.15 \times 10^{-?} \text{ A}$$ 6

$= 0.15$ ______ A *Ans.* μ

Problems

1. How many milliamperes of current will flow through a 100-Ω resistor if the voltage across its ends is 20 mV?
2. If 2 μA of current flow in an antenna whose resistance is 50 Ω, find the voltage drop in the antenna.
3. An emf of 200 μV sends 10 mA of current through the primary of a transformer. Find the total resisting effect of the coil.
4. Find the number of microamperes flowing through a 2-MΩ grid leak if the voltage drop across it is 1,000 mV.
5. Find the voltage drop across the 5-kΩ load of a transistor if the collector current is 2 mA.
6. The three branches of a parallel circuit carry the following currents respectively: 40 mA, 6,000 μA, and 0.013 A. Find the total current by adding the branch currents.
7. A 0.2-MΩ, a 5-kΩ, and a 10,000-Ω resistor are connected in series. Find the total resistance by adding the resistances.
8. Using the formula for capacitances in parallel, $C_T = C_1 + C_2$, find the total capacitance of a 0.0025-μF and a 125-pF capacitor in parallel.
9. The time constant of an RC circuit is equal to the product of the resistance (ohms) and the capacitance (farads). Find the time constant of a circuit if $R = 10$ kΩ and $C = 0.004$ μF.
10. How many ohms of resistance are used in the volume control of a complementary-coupled audio amplifier if it passes 0.0002 A at 1.5 V?
11. The Lafayette LR-1500T radio uses an output stage similar to that shown in Fig. 40-1. If the dc resistance of the primary of the transformer is 0.2 kΩ and it passes 5 mA of current, find (a) the voltage lost in the primary, and (b) the voltage E_C which is available at the collector ($E_C = E_T - E_R$).
12. The radiation resistance of a shortwave antenna is 100 Ω. If the transmitter delivers 900 mA to the antenna, find the number of watts radiated.

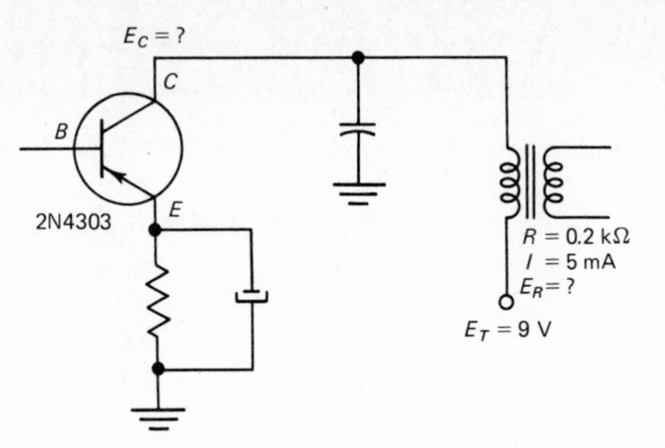

Fig. 40-1 The output stage of a transistor radio.

13. A 470-kΩ resistor in the base circuit of a 2N2924 phase-shift oscillator circuit carries a current of 30 μA. Find the voltage drop across the resistor.
14. A photoelectric cell circuit contains a resistance of 0.12 MΩ and carries a current of 50 μA. Find the voltage drop in the resistor.

(See Answer Key for Test 11—Electronic Units of Measurement)

ANSWER SECTION

These are the answers to the odd-numbered problems.

Job 1: Page 3

1. 0.7
3. 0.114
5. 0.006
7. 0.018
9. 0.011
11. 0.013
13. 0.045
15. 0.0545
17. 0.0062
19. 0.0005
21. 0.4; 0.16; 0.007
23. 0.8; 0.496; 0.02
25. 0.5; 0.18; 0.051
27. 0.4; 0.236; 0.1228
29. 0.19; 0.08; 0.004
31. 0.13; 0.02; 0.004
33. No; 0.0031 is smaller than 0.0032

Job 1: Page 4

1. 2.3
3. 3.144
5. 2.025
7. 2.020
9. 1.002
11. 9.145

Job 2: Page 8

1. 0.25
3. 0.625
5. 0.4
7. 0.15
9. 0.875
11. 0.444
13. 0.094
15. 0.625
17. 0.02
19. 0.192
21. 0.4
23. 7.5
25. 0.4
27. 0.67
29. 1.33
31. 0.0625
33. 21.5 ft
35. 0.12
37. 4.25
39. 0.18
41. 0.35 hp/cu in
43. 6.67
45. 15.4 ft
47. 0.302
49. 7.2; yes

Job 2: Page 11

1. 0.625
3. 0.5625
5. 2.25
7. 1.140625
9. $\frac{3}{8}$
11. $\frac{29}{64}$
13. $\frac{1}{32}$
15. $\frac{13}{32}$
17. $\frac{3}{4}$
19. $\frac{9}{64}$
21. $2\frac{13}{16}$
23. $1\frac{27}{32}$
25. $4\frac{21}{64}$ in
27. Yes
29. $A = 0.297$; $B = 0.333$; B has the better average
31. No
33. $1\frac{1}{2} = 1.500$; $\frac{19}{32} = 0.59375$; $\frac{13}{32} = 0.40625$; $1\frac{3}{16} = 1.1875$; $1\frac{7}{8} = 1.875$; $3\frac{1}{4} = 3.25$; $2\frac{5}{8} = 2.625$; $\frac{11}{32} = 0.34375$; $\frac{7}{16} = 0.4375$; $3\frac{11}{16} = 3.6875$

Job 2: Page 16

1. 0.785
3. 45.7
5. 18.0
7. 4.064
9. 26.74
11. 0.312
13. 0.38
15. 0.66
17. 0.88
19. 0.422
21. 0.444
23. 0.67
25. 0.571
27. 0.0139 in

Job 3: Page 19 (Middle)

1. 7.2
3. 3,090
5. 3,700
7. 0.6
9. 2,700
11. 0.078
13. 15,400
15. 0.5
17. 80
19. 2,340

Job 3: Page 19 (Bottom)

1. 6.5

3. 880
5. 0.06
7. 0.835
9. 0.6538
11. 0.00045
13. 0.0085
15. 0.0286
17. 0.0398
19. 0.002

Job 3: Page 22

1.

	mm	cm
AB	20	2
AC	25	2.5
AD	36	3.6
AE	52	5.2
AF	67	6.7
AG	80	8.0
AH	91	9.1
AI	98	9.8
AJ	110	11
AK	121	12.1
AL	129	12.9
AM	145	14.5
BC	5	0.5
CD	11	1.1
GH	11	1.1
JK	11	1.1
KL	8	0.8
MN	11	1.1

3. 4.8
5. 8
7. 0.3
9. 3.25
11. 6.5
13. 68
15. 3.6
17. 0.8
19. 1
21. 3.498
23. 20 cm; 180 mm; 0.15 m
25. 345.6 mm
27. A = 31 mm = 3.1 cm; B = 31 mm = 3.1 cm; C = 61 mm = 6.1 cm
29. A = 30 mm = 3 cm; B = 9 mm = 0.9 cm; C = 27 mm = 2.7 cm
31. A = 28 mm = 2.8 cm; B = 11 mm = 1.1 cm; C = 19 mm = 1.9 cm; D = 13 mm = 1.3 cm; E = 9 mm = 0.9 cm; F = 95 mm = 9.5 cm

Job 4: Page 25

1. 13.492
3. 22.832
5. 11.967
7. 5.59
9. 5.29
11. 58.5
13. 25.3
15. 25.8
17. 2,830.0 lb
19. $14.57
21. 0.1019 in
23. 2,104.3 ft
25. (a) 3.125 in; (b) 4.740 in; (c) 9.830 in
27. $3,796.34
29. 2.558 Ω
31. 0.100 in; 0.008 in; 0.0015 in
33. 6.63 mm
35. 1.501 in
37. 0.243 in
39. $242.68
41. 0.0315 A
43. (a) 0.8125 in; (b) 0.9375 in
45. 3.385 in
47. $44.90
49. 0.1875 in
51. 0.750 in
53. 3.895 in
55. 79 mm

Job 5: Page 31

1. 0.23
3. 0.34
5. 7.56
7. 0.317
9. 2.92
11. 5.9
13. 2.39
15. 0.262
17. 2.62
19. 0.515
21. (a) 0.304; (b) 1.106; (c) 0.39; (d) 5.75; (e) 1.98
23. 0.914
25. 6.41
27. 0.2215 in
29. 0.005 in
31. $0.92/hr
33. 25.9 acres
35. 0.375 in
37. 0.075 in
39. 8.29 Ω
41. 0.109 in
43. 0.113
45. 725.9 gal
47. 0.455 MHz
49. 0.00465 μF
51. (a) 0.027 in; (b) 0.083 in; (c) 0 in; (d) 0.053 in; (e) 0.016 in
53. 134.2 mm
55. 1967 car: top—OK, bottom—OK; 1968 car: bottom—OK, top—discard; it is 0.0003 in too large.

Job 6: Page 37

1. 3.128; 3.123
3. 0.0006 in
5. A = 0.749, 0.747; B = 0.624, 0.623; C = 0.752, 0.750; D = 0.626, 0.625; E = 0.750, 0.748; F = 1.000, 0.998; G = 0.752, 0.750; H = 1.002, 0.998
7. 0.022 in
9. (a) shaft = 0.874 in, bearing = 0.877 in; (b) shaft = 0.875 in, bearing = 0.876 in
11. (a) yes; (b) yes; (c) yes; (d) yes; (e) no
13. 0.0022 in
15. 0.0005 in

Job 7: Page 41

1. 0.1002 + 0.125 + 0.750
3. 0.1009 + 0.147 + 0.100 + 2.000
5. 0.1003 + 0.104 + 0.850
7. 0.1003 + 0.134 + 0.650
9. 0.1001 + 0.128 + 0.450 + 2.000
11. 0.1004 + 0.121 + 0.800 + 1.000
13. 0.106 + 0.150 + 1.000
15. 0.1004 + 0.115 + 0.550 + 2.000
17. 0.1004 + 0.104 + 0.800 + 1.000

Job 8: Page 44

1. (a) 0.017; (b) 0.245; (c) 0.75; (d) 0.9; (e) 0.003
3. (a) 0.62; (b) 0.47; (c) 0.188; (d) 0.333; (e) 0.078; (f) 0.281; (g) 0.556; (h) 0.54
5. 17.7 mm
7. 0.0013 in
9. 7,000 bushels
11. (a) 865; (b) 92,600,000; (c) 3.5; (d) 0.05
13. (a) 0.1001 + 0.121 + 0.350; (b) 0.1002 + 0.146 + 0.550 + 2.000
15. $19.23
17. 300,000 lb
19. 1.125 in
21. 330,000,000 ft-lb
23. 78.02 percent
25. A = 0.425 in; B = 0.473 in
27. 0.0025 in
29. 0.3662 in
31. 3.091 in
33. A = 0.188 in; B = 1.375 in
35. X = 2.375 in; Y = 0.750 in
37. 12 cm = 120 mm

Job 9: Page 50

1. 0.41
3. 0.4633
5. 30.31
7. 0.221
9. 0.0186
11. 4.77
13. 4.55 A
15. $202.50
17. $124.80
19. 0.275 in
21. $4,060
23. (a) 250 mA; (b) 25 mA; (c) 2,500 mA
25. 335 tons
27. 8,070 lb
29. 1,680 sq ft
31. 1.68 cm
33. 377.2 W
35. $29.75
37. 0.0288 in
39. $11.25
41. 0.075 gal/mi
43. 62.63 acres
45. $16.50
47. $28.13
49. $120.57
51. 41.850 in
53. (a) $94.46; (b) $87.02; (c) $7.44
55. (a) $432.00; (b) $6.43; (c) $31.00; (d) $337.03; (e) $9.24; (f) $67.23
57. $736.25

Job 10: Page 56

1. 13
3. 161
5. 4.7
7. 786
9. 2.32
11. 0.1032 Ω/ft
13. 8.45 gal
15. 33.2 cu in
17. 8 ft
19. 250 rev
21. 35 strips
23. 0.0526 in/in
25. 148 strokes
27. 2.25 min
29. 14.69 ft
31. 18 rolls
33. 0.075 in
35. 2.786 in
37. 11.5 ft
39. 0.942 in
41. 75 copies
43. 104 rev; 1.3 min

Job 11: Page 61

1. A = 2.8475 in; B = 3.090 in
3. (a) 0.0271 in; (b) 0.0406 in; (c) 0.0650 in; (d) 0.0812 in; (e) 0.1083 in
5. A = 0.625 in; B = 0.300 in; H = 0.225 in; T = 0.1875 in
7. (a) 9.92 in; (b) 17.36 in
9. 0.082
11. 144 bolts

13. 56.25 rev
15. 3.359 in
17. 186.68 cu ft
19. $6.22
21. $50,750
23. 0.6975 lb
25. 1.588 in
27. 0.57
29. 12 A
31. $10.00
33. 68.63 lb
35. 9 rev
37. 20,250 lb
39. (a) 0.6495 in; (b) 1.299 cm; (c) 0.54 in; (d) 14.98 mm
41. $7.22/100 lb
43. 12.6 in
45. $5\frac{5}{8}$ min
47. $63.05
49. 10 gage

Job 12: Page 78

1. 110 V
3. 120 V
5. 55.8 V
7. 117 V
9. 111.3 V
11. 28 V
13. 6.25 V
15. 9.75 V
17. 352 V
19. No; it uses 0.27 V

Job 13: Page 81

1. 480 W
3. 22 W
5. 264 W
7. 110 W
9. 0.06 W
11. 40 W
13. 807.3 W
15. 1 W
17. 2,508 W
19. 15 W
21. $0.08
23. Yes, since it develops only 0.135 W.
25. 2,246 W
27. $0.81/yr

Job 14: Page 87

1. 5
3. 6
5. 9
7. 13.67
9. 5.85
11. 40
13. 400
15. 200
17. 800
19. 390
21. $\frac{1}{16}$

Job 14: Page 88

1. 2 Ω
3. 13 Ω
5. 1.5 A
7. 20 A
9. 0.133 Ω
11. 1,200 Ω
13. 2.25 A
15. 3.67 A
17. 0.0015 Ω
19. 1.76 Ω
21. 0.2 A
23. 5 A
25. 28.9 Ω

Job 14: Page 89

1. 5 A
3. 50 A
5. 0.5 A
7. 5.45 A
9. 20 A

Job 14: Page 91

1. 720 W
3. 6.25 W
5. 0.006 W
7. 3,361 W
9. 661 W
11. 0.055 W

Job 15: Page 94

1. 110 V
3. Yes
5. 18 V
7. 702 W
9. 720 W
11. (a) 750 W; (b) 0.75 kW; (c) $0.39
13. 24 V
15. 0.07 Ω
17. 10,000 Ω
19. 0.625 W
21. 0.81 W
23. 0.03 A
25. 86.5 Ω
27. 75 V
29. 4.35 V
31. 1.2 V
33. 300,000,000 Ω
35. 6.4 V

Job 16: Page 100

1. 0.175 in

3. 0.207 in
5. 0.382 in
7. 0.152 in
9. 0.318 in
11. 0.498 in
13. 0.278 in
15. 0.051 in
17. 0.102 in
19. 0.002 in
21. 0.198 in
23. 0.243 in

Job 17: Page 103

a. 0.1867 in
c. 0.1235 in
e. 0.2997 in
g. 0.2583 in
i. 0.1726 in
k. 0.4058 in
m. 0.0983 in
o. 0.3023 in
q. 0.3347 in
s. 0.2282 in

Job 18: Page 109

a. 2.025 in
c. 0.665 in
e. 2.869 in
g. 1.458 in
i. 3.617 in
k. 3.958 in

Job 18: Page 113

a. inside = 2.414 in; outside = 2.114 in
c. inside = 2.285 in; outside = 1.985 in

Job 19: Page 114

1. A = 1.2 cm = 12 mm; B = 2.6 cm = 26 mm; C = 4.5 cm = 45 mm; D = 6.7 cm = 67 mm; E = 8.3 cm = 83 mm; F = 9.9 cm = 99 mm; G = 10.7 cm = 107 mm; H = 12.1 cm = 121 mm
3. A = 2.9 cm = 29 mm; B = 3.8 cm = 38 mm; C = 6.6 cm = 66 mm
5. (a) 0.2784 in; (b) 0.0625 in; (c) 0.3125 in; (d) 0.1114 in
7. (a) 0.775 in; (b) 0.531 in
9. (a) inside = 2.734 in; outside = 2.434 in; (b) inside = 2.069 in; outside = 1.769 in

Job 20: Page 120

1. 10.16 cm = 101.6 mm
3. 8.89 cm
5. 805 m
7. 1.38 in
9. 0.016 in
11. $4\frac{3}{8}$ in = 11.1 cm; $8\frac{1}{2}$ in = 21.59 cm
13. 5 ft = 1.525 m; 5.9 ft = 1.8 m
15. (a) 15 in = 38.1 cm; (b) 16 in = 40.64 cm; (c) $16\frac{1}{2}$ in = 41.91 cm; (d) $17\frac{1}{2}$ in = 44.45 cm
17. (a) $\frac{23}{64}$ in; (b) $\frac{7}{16}$ in; (c) $\frac{33}{64}$ in; (d) $\frac{19}{32}$ in; (e) $\frac{3}{4}$ in
19. 3.66 m
21. 28 ft
23. A = 1.5 cm = 15 mm = 0.591 in; B = 3 cm = 30 mm = 1.182 in; C = 4.4 cm = 44 mm = 1.734 in; D = 6.2 cm = 62 mm = 2.443 in; E = 7.8 cm = 78 mm = 3.073 in; F = 10.5 cm = 105 mm = 4.137 in; G = 11.3 cm = 113 mm = 4.452 in
25. (a) $2\frac{5}{8}$ in; (b) 6.67 cm = 66.7 mm

Job 20: Page 125

1. 5,635 km
3. 2,496 km
5. 6,923 km
7. 3,770 mi
9. 3,968 mi
11. 40,250 km
13. 3.1:1

Job 20: Page 127

1. 76 l
3. 66 gal
5. 26.5 mpg
7. 10.9 kg
9. 57 l
11. 91.2 l
13. $64.20
15. 140 cu in
17. 20 kg
19. 5.58 lb/gal
21. 84.9 cu m
23. 15 mpg
25. 62.3 lb/cu ft

Job 20: Page 129

1. 9.04 sq ft
3. 1,000 A/sq in
5. (a) 45.15 sq cm; (b) 406.35 kg

Job 21: Page 131 (Top)

1. 3.86 mm
3. 5.12 mm
5. 7.04 mm
7. 0.14 mm
9. 9.73 mm

Job 21: Page 131 (Bottom)

1. 23.80 mm
3. 44.78 mm
5. 73.38 mm

Job 21: Page 133

1. outside = 1.173 in = 29.80 mm; inside = 1.473 in = 37.42 mm
3. outside = 3.098 in = 78.70 mm; inside = 3.398 in = 86.32 mm
5. outside = 1.908 in = 48.46 mm; inside = 2.208 in = 56.08 mm

Job 22: Page 139

1. $A = 13$; $B = 38$; $C = 69$; $D = 101$; $E = 136$
3. $A = 110$; $B = 280$; $C = 470$; $D = 630$; $E = 820$; $F = 970$
5. $A = 0.4$; $B = 1.2$; $C = 1.75$; $D = 2.30$; $E = 2.95$
7.

Point	Scale 1	Scale 2	Scale 3
A	13	1.3	0.26
B	37	3.7	0.74
C	93	9.3	1.86
D	118	11.8	2.36
E	146	14.6	2.92

9.

F°	C°
$A = 70$	$A = 22$
$B = 340$	$B = 170$
$C = 570$	$C = 300$
$D = 850$	$D = 450$
$E = 1{,}150$	$E = 620$
$F = 1{,}440$	$F = 780$

Job 23: Page 142

1. 100,000 Ω
3. 82,000 Ω
5. 75 Ω
7. 4.7 Ω
9. 1.2 Ω
11. 1,800,000 Ω
13. 620 Ω
15. 9,100 Ω
17. 0.47 Ω
19. 82 Ω
21. Red, yellow, yellow
23. Green, brown, black
25. Brown, black, blue
27. Orange, white, red
29. Brown, red, black
31. Blue, gray, silver
33. Red, red, gold
35. Brown, gray, gold
37. Orange, blue, brown
39. Brown, green, gold

Job 24: Page 145

15. $A = 0.7$ cm = 7 mm; $B = 2$ cm = 20 mm; $C = 3.5$ cm = 35 mm; $D = 5.6$ cm = 56 mm; $E = 7.2$ cm = 72 mm; $F = 8.8$ cm = 88 mm; $G = 10$ cm = 100 mm
17. 29.92 in
19. $\frac{3}{4}$ in = 19.05 mm; $\frac{7}{16}$ in = 11.11 mm; $\frac{3}{8}$ in = 9.53 mm; $1\frac{1}{4}$ in = 31.75 mm
21. 8 turns
23. $874\frac{2}{3}$ yd
25. 1 dm 5 cm = 5.91 in; 7.5 cm = 2.96 in; 2 dm 1 cm = 8.27 in; 2.8 cm = 1.10 in; 2 dm 2 cm = 8.67 in
27. $DD = 9.1$ mm; minor diameter = 1 in
29. 88.6 km/hr
31. 739 mi/hr
33. 1.27 m
35. 17.1 cents/l
37. 4.75 l
39. 305 cu in
41. (a) 150 l; (b) 39.6 gal
43. (a) 38.76 mm; (b) 16.90 mm
45. $A = 1.6$; $B = 2.35$; $C = 3.05$; $D = 3.8$; $E = 4.45$; $F = 4.95$
47.

Point	Scale 1	Scale 2	Scale 3	Scale 4	Scale 5	Scale 6
A	60	12	2.4	7.5	3	1.5
B	180	36	7.2	19	7.6	3.8
C	270	54	10.8	26.6	10.7	5.36

49. (a) 25,000 Ω; (b) 15 Ω; (c) 300,000 Ω; (d) 4,000,000 Ω; (e) 72 Ω; (f) 4.8 Ω
51. $231
53. 1.06 kg/sq cm
55. 9,935 lb/sq in

Job 25: Page 151

1. 0.38
3. 0.06
5. 0.04
7. 0.036
9. 1.25
11. 0.167
13. 0.625
15. 0.125
17. 0.0225
19. 0.0425

Job 25: Page 152

1. 50%
3. 20%
5. 145%
7. 100%
9. 62.5%
11. 22.2%
13. 70%
15. 5.5%

Job 25: Page 153

1. 25%
3. 40%
5. 62.5%
7. 30%
9. 50%
11. 65%

13. 66.7%
15. 45.4%
17. $33\frac{1}{3}$%
19. 22.2%

Job 26: Page 156

1. 33
3. $7.50
5. Copper = 4.81 lb; zinc = 2.59 lb
7. 9.555 cm
9. 315 lb
11. 72 sq cm
13. 17.25 A
15. 34 kg
17. 1,806 bd ft
19. Wheat (7.6 lb)
21. Lead = 155 lb; tin = 13 lb; antimony = 32 lb
23. 225 lb
25. 49 hp
27. 57 cents/basket
29. 50 defective; 893 finished
31. (a) 11 V; (b) 209 V
33. 656.25 sq ft
35. $450
37. (a) 1,250 sq ft; (b) 1,420 sq ft
39. Max = 176 lb; min = 144 lb
41. No
43. 0.7672 in
45. 271.9 bd ft
47. 248.9 bd ft
49. 438 blocks
51. 3.3 tons
53. 38,200 lb
55. 0.07 mm

Job 27: Page 161

1. (a) $1.26; (b) $22.26
3. $352.50
5. $2,845.80
7. (a) 1,500 lb; (b) $480.00; (c) $38.40; (d) $441.60
9. $17.06
11. $283.00
13. $138.01
15. $263.16
17. (a) Spark plugs = $14; gaskets = $23.50; rings = $35.50; valves = $38.40; springs = $22.40; (b) $133.80; (c) $53.52; (d) $80.28
19. $18,431.25

Job 28: Page 163

1. 25%
3. 20%
5. 35%
7. $12\frac{1}{2}$%
9. 40%
11. 66.7%
13. 10%
15. 5%
17. Copper = 62%; zinc = 37%; tin = 1%
19. 7.56%
21. $33\frac{1}{3}$%
23. 56.3%
25. 5%
27. 40%
29. (a) $11.26; (b) 40%

Job 29: Page 167

1. 20
3. 1,000
5. 150 V
7. 25 lb
9. 200 tiles
11. 20 l
13. 20,000 lb
15. $400
17. $210
19. 119.4 V

Job 30: Page 168

1. (a) 0.62; (b) 0.03; (c) 0.056; (d) 0.008; (e) 1.16; (f) 0.045; (g) 0.0625
3. (a) 75%; (b) 60%; (c) 42.9%; (d) 30%; (e) 23.1%; (f) $16\frac{2}{3}$%; (g) 46.1%
5. 5.4
7. 60
9. $13.20; $178.20
11. 1.5%
13. $766.02
15. Copper = 6 lb; tin = 132 lb; antimony = 12 lb
17. 2.4 V; 117.6 V
19. $197
21. 729 bd ft
23. 683.7 bd ft
25. 225 hp
27. 85%
29. $12\frac{1}{2}$%
31. 80%
33. $33\frac{1}{3}$%
35. 25%

Job 31: Page 175

1. For ∠ C:
   H = 40 cm
   O = 24 cm
   A = 32 cm
   For ∠ D:
   H = 40 cm
   O = 32 cm
   A = 24 cm
   For ∠ M:
   H = 8.5 in
   O = 7.5 in

A = 4 in
For ∠N:
H = 8.5 in
O = 4 in
A = 7.5 in

3. Sin B = 0.2800
   Cos B = 0.9600
   Tan B = 0.2917
5. Sin D = 0.9756
   Cos D = 0.2195
   Tan D = 4.4444

Job 32: Page 178 (Top)

1. 18°
3. 80°
5. 30°
7. 30°
9. 60°

Job 32: Page 178 (Bottom)

1. 15°
3. 58°
5. 65°
7. 16°
9. 46°

Job 33: Page 180

1. ∠A = 53°; ∠B = 37°
3. A = 30°; B = 60°
5. A = 55°; B = 35°
7. A = 51°; B = 39°
9. A = 11°; B = 79°
11. 67°; 113°
13. 3°
15. 58°
17. 32°
19. 49°

Job 34: Page 183

1. 24
3. 12
5. 1
7. 2
9. 8
11. 6,750
13. 54
15. 30.6
17. 48
19. 28.29

Job 34: Page 186

1. 2
3. 4
5. $3\frac{1}{3}$
7. 15
9. 15
11. 1.2
13. 10
15. $3\frac{1}{2}$
17. 7.8
19. 315
21. 200 rpm
23. 500
25. 33.3

Job 34: Page 189

1. 6
3. 80
5. 51
7. 27
9. 25
11. 10
13. 1.44
15. 5.33
17. 0.1
19. 0.077
21. 8 A
23. 160 ft
25. 0.1176 μF

Job 35: Page 192

1. AC = 10 in; ∠B = 30°
3. AC = 89.4 ft; ∠A = 40°
5. AB = 151 Ω; ∠B = 60°
7. BC = 940 m; ∠A = 62°
9. BC = 123 Ω; ∠B = 75°
11. 87.7 ft; 82.4 ft
13. 3.54 in
15. 27.2 ft
17. 28.53 lb
19. 728 Ω
21. (a) 0.629 in; (b) 1.258 in/ft
23. A = 0.650 in; B = 1.025 in; C = 0.860 in; D = 0.515 in; E = 0.892 in
25. 0.813 ft = $9\frac{3}{4}$ in

Job 36: Page 195

1. 0.5878
3. 2.6051
5. 0.9063
7. 65°
9. 11°
11. 17°
13. 73°
15. 37°
17. 71°
19. 58°
21. BC = 930 W
23. AB = 38.5
25. BC = 833 Ω
27. AC = 2,828 W
29. AB = 236 V
31. 81°; 99°
33. 3°

35. 1.722 in
37. 23°
39. $Z = 233\ \Omega$
41. 1.943 in = $1\frac{15}{16}$ in
43. $D = 1.152$ in

Job 37: Page 198
1. 0.2 in/in; 2.4 in/ft
3. $\frac{1}{16}$ in/in; $\frac{3}{4}$ in/ft
5. $\frac{3}{16}$ in/in; $2\frac{1}{4}$ in/ft
7. 0.148 in/in; 1.776 in/ft
9. 0.035 in/in; 0.42 in/ft
11. 0.075 in/in; 0.9 in/ft
13. 3.36 in
15. 0.257 in/ft
17. 0.0235 in/in; 0.282 in/ft
19. $\frac{1}{4}$ in/in; 3 in/ft

Job 37: Page 200
1. 0.5625 in
3. 1.159 in
5. 1.875 in
7. $d = 0.860$ in

Job 37: Page 202
1. 2 in
3. 3.2 in
5. 5.76 in

Job 37: Page 203
1. 0.0502 in/in
3. 0.375 in
5. 2.5 in
7. 0.6 in
9. 0.156 in
11. 1.406 in
13. Jarno No. 10; d = 1 in
15. Morse No. 5

Job 37: Page 207
1. 0.1875 in
3. No. 7: $L = 3.5$ in, $D = 0.875$ in, $d = 0.700$ in, $T = 0.175$ in, offset = 0.0875 in; No. 8: $L = 4$ in, $D = 1.0$ in, $d = 0.8$ in, $T = 0.2$ in, offset = 0.1; No. 10: $L = 5$ in, $D = 1.25$ in, $d = 1.00$ in, $T = 0.25$ in, offset = 0.125 in; No. 15: $L = 7.5$ in, $D = 1.875$ in, $d = 1.500$ in, $T = 0.375$ in, offset = 0.1875 in
5. 0.0676 in
7. 0.6071 in
9. 0.333 in

Job 38: Page 210 (Top)
1. 7.2
3. 3,090
5. 3,700
7. 0.6
9. 2,700
11. 0.078
13. 15,400
15. 1
17. 80
19. 234,000

Job 38: Page 210 (Middle)
1. 65
3. 880
5. 0.06
7. 0.835
9. 0.6538
11. 0.0045
13. 0.0085
15. 0.0286
17. 0.00002
19. 0.005

Job 38: Page 211
1. 25
3. 0.15
5. 6.25
7. 0.0025
9. 0.00754

Job 38: Page 212
1. 1,920
3. 850
5. 7,200
7. 0.000045
9. 88,000,000
11. 3,000
13. 3,000,000
15. 0.0006

Job 38: Page 213
1. 6×10^3
3. 1.5×10^5
5. 2.35×10^5
7. 4.96×10^3
9. 9.8×10^2
11. 1.25×10
13. 4.82×10
15. 8.8×10^8
17. 3.83×10^4
19. 1.75×10^6
21. 4.83×10^5

Job 38: Page 214
1. 6×10^{-3}
3. 3.5×10^{-3}
5. 4.56×10^{-1}
7. 7.85
9. 9.65×10^{-2}
11. 5×10^{-1}
13. 8.15×10^{-3}

15. 7.25×10
17. 6×10^{-1}
19. 3.6×10^{-2}
21. 1.9×10^{-5}

Job 38: Page 215

1. 5
3. 6.4×10^{6}
5. 3×10^{4}
7. 120
9. 0.03
11. 5×10^{-7}
13. 960
15. 6.28×10^{5}
17. 6.7×10^{8} mi

Job 38: Page 217

1. 10^{5}
3. 120
5. 0.0002
7. 2×10^{-7}
9. 8
11. 5×10^{-3}
13. (a) 53 Ω; (b) 63 Ω; (c) 3.18 Ω

Job 39: Page 221

1. 0.225 A
3. 3,500,000 Ω
5. 550,000 Hz
7. 0.07 MΩ
9. 65 mA
11. 0.075 V
13. 0.006 A
15. 0.0039 A
17. 5,000 pF
19. 1,000,000 Hz
21. 8 mV
23. 60 pF
25. 0.00000015 F
27. 8 kW
29. 4 μ A

Job 40: Page 222

1. 0.2 mA
3. 0.02 Ω
5. 10 V
7. 215 kΩ
9. 4×10^{-5}
11. (a) 1V; (b) $E_C = 8$V
13. 14.1 V